# Optimum Structural Design

# Optimum Structural Design
## Theory and Applications

*Edited by*

**R. H. Gallagher**

*Department of Structural Engineering*
*Cornell University, Ithaca, N.Y., U.S.A.*

*and*

**O. C. Zienkiewicz**

*Department of Civil Engineering*
*University College of Swansea, Swansea, Wales*

**JOHN WILEY & SONS**
Chichester · New York · Brisbane · Toronto

Library of Congress Catalog card No. 72-8600

ISBN 0 471 29050 5


Reprinted October 1977

Printed in Great Britain by
J. W. Arrowsmith Ltd.,
Winterstoke Road, Bristol BS3 2NT

# Contributing Authors

DR. L. BERKE — Technical Manager, Synthesis Group, Solid Mechanics Branch, Wright-Patterson Air Force Base, Ohio, U.S.A.

PROFESSOR D. BOND — Department of Civil Engineering, Queen's University, Belfast, Northern Ireland.

DR. G. A. BUTLIN — Department of Engineering, The University, Leicester, England

J. S. CAMPBELL — Department of Civil Engineering, University College of Swansea, Swansea, Glamorgan, Wales.

DR. A. CELLA — Research Specialist, Consiglio Nazionale della Ricerche, . Instituto di Elaborazione della Informazione, Pisa, Italy.

DR. R. FLETCHER — Mathematics Branch, Atomic Energy Research Establishment, Harwell, Didcot, Berkshire, England.

PROFESSOR R. H. GALLAGHER — Department of Structural Engineering, Cornell University, Ithaca, N.Y., U.S.A.

PROFESSOR C. GAVARINI — Instituto di Costruzioni, Università degli Studi dell'Aquila degli Abruzzi, Italy.

DR. R. A. GELLATLY — Structural Systems Department, Bell Aerosystems Company, Buffalo, N.Y., U.S.A.

PROFESSOR M. R. HORNE — Department of Civil Engineering, University of Manchester, Manchester, England.

PROFESSOR F. A. LECKIE — Department of Engineering, The University, Leicester, England.

DR. R. K. LIVESLEY — Cambridge University, Cambridge, England.

PROFESSOR G. MAIER — Politecnico di Milano, Milan, Italy.

PROFESSOR J. MOE — Department of Ship Structures, Norges Tekniske Høgskole, Trondheim, Norway.

DR. L. J. MORRIS — Department of Civil Engineering, University of Manchester, Manchester, England.

PROFESSOR F. MOSES — School of Engineering, Case Western Reserve University, Cleveland, Ohio, U.S.A.

DR. A. C. PALMER — Cambridge University, Cambridge, England.

M. J. PLATTS — Churchill College, Cambridge University, Cambridge, England.

DR. B. M. E. DE SILVA — *Department of Mathematics, University of Technology, Loughborough, Leicestershire, England.*

DR. K. SOOSAAR — *Staff Structures Specialist, M.I.T. C.S. Draper Laboratory, Cambridge, Mass., U.S.A.*

J. WRIGHT — *Senior Lecturer, Engineering Science, University of Warwick, Coventry, Warwickshire, England.*

PROFESSOR O. C. ZIENKIEWICZ — *Department of Civil Engineering, University College of Swansea, Swansea, Wales.*

# Preface

In the last two decades outstanding progress has been achieved in the field of structural analysis. With the aid of the computer nearly all structural problems can be solved within the limits of our knowledge of the materials. While these achievements are, *per se*, of the greatest importance in allowing the behaviour of a particular design to be assessed, their full benefits for our society will not be materialized until they are reflected in the improved design of structures.

The aim of devising better design solutions which, while satisfying safety and performance 'constraints', do it at least cost, is clearly not a new one. From time immemorial a self-respecting engineer has investigated several alternatives and chosen the 'best' one of these. Unfortunately, cost and time usually limit very severely the number of alternatives that can be investigated. With the 'computerization' of the analysis process, it is natural that a development of more effective and rapid techniques for the search of the 'absolute best', or 'optimum' solution is required. Many possibilities obviously exist, with two clearly defined extremes. One extreme is to utilize the computer capability to the fullest and *automate* this search; the other extreme is to utilize human intuition in an *interactive manner* to guide the computer in its calculations.

Much work has been done in recent years on both approaches by various researchers, and a stage has now been reached at which an appraisal of the developments and of their practical possibilities should be made and presented to the engineering profession. This appraisal is the subject of this book. To achieve a representative picture of the 'state of the art', the editors invited contributions from some leading exponents of both the theoretical and practical aspects of structural optimization, assigning to each a certain coverage of the subject and specifying in some detail the objectives and scope of presentation so that a coherent volume could be obtained.

The major part of this volume, Chapters 1–15, is concerned with approaches in the first of the above classes—'automated' approaches. Chapter 1 outlines the background of automated optimum structural design, broadly classifies the relevant mathematical procedures, and refers to the principal sources of already published information. Chapter 2 sets the stage for most of the theoretical work that follows by defining terminology and presenting certain basic definitions and theorems.

Chapter 3 explores what has been, historically, the most appealing approach to design analysis, the *fully stressed design* philosophy and procedure. These ideas are expanded in Chapter 4 to form an approach, related also to so-called *optimality criteria*, that does not suffer the limitations of fully stressed design, but nevertheless retains its computational economies *vis-à-vis* the more sophisticated procedures.

The largest share of modern activity in optimum structural design revolves about the utilization of mathematical-programming procedures. Chapter 5 outlines these procedures, irrespective of their role in the structural context, and gives a 'road map' for the pursuit of appropriate alternative techniques for the sundry mathematical classification of problems. In the process, this chapter presents a résumé of many mathematical papers and books and gives the structural engineer a comprehensive insight into available techniques.

Chapters 6–13 follow up this review by presenting, from the viewpoint of researchers in structural engineering, the background, fundamental theory and considerations, and some applications experience in various major alternative techniques in mathematical programming. Linear programming (Chapter 6), iterative linear programming (Chapter 7), feasible-direction methods (Chapter 8), penalty-function procedures (Chapter 9), dynamic programming (Chapter 10) and discrete-variable methods (Chapters 11 and 12) are dealt with. Each of these is based on or infers a deterministic design philosophy. There is a considerable trend towards probability-based design philosophy, however, and the implications of this for optimum structural design are treated in Chapter 13. Chapters 14–16 will be of direct interest to the practising civil engineer. Here, present-day application and results for structural steel and concrete show how much has already been achieved in practical engineering and that real benefits are already being achieved. Other practical structural-engineering applications and similar achievements in such fields as aerospace, mechanical engineering and naval architecture are discussed in the prior chapters.

As noted earlier, another approach to optimization is via the *interactive* mode. This approach is specifically dealt with in Chapters 16 and 17, and also enters to some extent in other contributions.

Our thanks go out to the respective authors for entering into this venture with an enthusiasm and a spirit of collaboration which often necessitated considerable patience and rewriting. The editors have endeavoured with their blue pencil to keep the notation consistent, if not uniform, and to avoid excessive repetition. A unified notation has been sought in those chapters that deal almost exclusively with mathematical programming in structural optimization; this was not possible, however, for those chapters with heavy emphasis on both analysis and design technologies, i.e. Chapters 6, 12 and 13. To what extent the objectives of coherence and logical sequence are fully achieved will be for the reader to judge. It is seldom possible for perfection to be approached with a multiauthored text.

R. H. GALLAGHER
O. C. ZIENKIEWICZ

*ACKNOWLEDGEMENTS*

The separate chapters of this book were presented at an international symposium on 'Optimization of Structural Design' held at the University of Wales, Swansea, in January 1972. The symposium and the ensuing discussions allowed many alterations to be made to the manuscript. Our thanks go to Dr. K. G. Stagg, the organizing officer of this symposium, and to his staff for their most valuable help, and to Mrs. Helen Wheeler for her support in the typing of the major part of the final manuscript.

# Contents

# Chapter 1

# *Introduction*

*Richard H. Gallagher*

In contrast to analysis technology, optimum-structural-design technology has not yet enjoyed intensive study, and computational aspects of practical design today largely depend on iterative analysis. Modern developments in optimum structural design, represented by attempts to introduce the (then) novel accomplishments of mathematical programming into structure technology, first appeared over ten years ago. Shortly thereafter, the earliest systems for interactive, or computer-aided, design, achieved operational status. Neither of these aspects of the total design technology has enjoyed the utilization predicted of it, although certain features have proved successful.

It is difficult to ascertain the full range of considerations responsible for the slow rate of acceptance of the available design technology. These include such factors as unfamiliarity of the practitioner with mathematical-programming concepts, the bewildering array of alternative paths in mathematical programming and the costs of optimum-design analysis beyond the costs of simple analysis. Nevertheless, it appears that these problems are now being overcome. An extremely large backlog of optimum-structural-design literature has accumulated meanwhile, and, if the practitioner is to make use of it, some evaluation of the alternatives must first be made, followed by detailed study of the applicable procedures. This chapter helps to evaluate the various approaches; subsequent chapters examine techniques in detail.

Four previously independent major areas (and, to some extent, chronological phases) can be identified in the development of optimum-structural-design technology. We shall term these the (a) theory of layout, (b) simultaneous mode of failure, (c) optimality-criteria-based and (d) mathematical-programming formulations.

The first of these was the *theory of layout*, which seeks the arrangement of uniaxial structural members that produces a minimum-volume structure for specified loads and materials. The basic theorems of this approach were established by Maxwell as early as 1854[1], but these ideas were amplified and given their first significant application by Michell[2] in 1904. Such theorems, since they are applied without meaningful constraints on the geometric form of the structure, yield impractical solutions. The theory of layout has been reconsidered in the work of Cox[3] and Hemp[4], however, and many researchers are now developing further the related concepts.

1

The *simultaneous mode of failure approach* presumes that optimality is achieved when each component element of the complete structure is at its limit of strength as failure of the complete structure impends. The term 'simultaneous' implies a single load condition, and this restriction governs nearly all of the work which flourished during the 1940s and 1950s and is recorded in the books by Shanley[5], Gerard[6] and Cox[3]. These efforts, pre-dating the electronic digital computer, deal with simple structural forms and depend on classical ideas of function minimization. The fact that there are only a small number of simple situations, together with their limited applic-ability to practical design, has resulted in very little new work in this area during the past decade.

If the concept of the simultaneous mode of failure approach is broadened to admit more than one load condition, and at the same time restricted to strength limitations applying only to stress, the *fully stressed design approach* is produced. This approach generally consists of the iterative application of analysis, leading to a design in which each member is subjected to its limiting stress under at least one of the specified load conditions. Although the result is the designer's traditional view of an optimal structure, the concept has not been subjected to a broad, rational study. Chapter 3 of this book correlates the relevant published information.

The concept of a *criterion of optimality* as the basis of selection of a mini-mum-volume structure emerged in the early 1960s. This approach derives from the extremum principles of structural mechanics, and for the most part has been limited to simple structural forms and loading conditions. Prager[7] and Taylor[8] have been instrumental in the development of much of this work, and the procedures of Venkayya[9] and Gellatly and Berke[10] are described in this book. Chapter 4, by Gellatly and Berke, elaborates on their procedures.

Finally, we come to procedures characterized as *mathematical-program-ming formulations*. The basic ideas of mathematical programming are out-lined in Chapter 2. To define these concisely, it may be said that they seek the minimum or maximum of a function of many variables subject to limita-tions (*constraints*) that are expressed as equalities or inequalities. The representation of inequality constraints is of critical importance, since this permits the design to be identified as one in which not all members are subject to limiting conditions under specified loads, avoiding a restriction inherent in certain of the aforementioned approaches. Also, the orientation of mathe-matical-programming formulations towards many-variable problems fits quite well with the trend in analysis towards finite-element representations which require large-order systems. It should be emphasized, however, that procedures described in subsequent chapters are not, in general, tied to a particular method of analysis.

Mathematical programming was first applied to structural optimization in the late 1950s. Early contributions included Livesley's[11] and Pearson's[12] treatments of limit design as a linear-programming problem, and Schmit's casting[13] of elastic design as the more general non-linear-programming problem. We will not attempt to delineate here the detailed historical development of this avenue of activity in structural optimization, because recent surveys have been published and are cited below for reference, and also because many of the later chapters summarize detailed facets of the total approach.

There are other modern approaches to structural optimization which do not fit in the classifications that have already been mentioned. Certain of these fall in the category of control theory[14], while others are of a special character and have defied classification in a particular mathematical discipline (e.g. the work of Melosh and Luik[15]). These approaches may indeed be highly promising, but they have not yet received widespread attention.

To conclude this chapter, the following comments are presented to direct the reader to literature which amplifies the historical categorization of optimum structural design, fills in additional details of procedures described in subsequent chapters and enables the investigation of approaches not covered by this book.

The first comprehensive survey of related literature to appear in a widely circulated journal was written by Wasiutynski and Brandt[16] in 1963. Subsequently, in the same journal, Prager and Sheu[17] reviewed developments to 1968. More restricted surveys, with differing vantage points, have been prepared by Barnett[18], McNunn and Jorgenson[19] and Gerard[20]. The most complete and authoritative surveys of optimum structural design in the context of mathematical-programming procedures have been written by Schmit[21-23]. These are especially valuable because of their development, in elementary terms and via simple examples, of the basic concepts of these procedures.

Literature which develops the concepts and procedures of mathematical programming from first principles has not taken account, to any significant extent, of the structural design problem. A notable exception is the book by Fox[24], whose interests in applications have been principally associated with structural optimization. Also, the book by Whittle[25] describes the application of linear programming to the theory of layout.

Developments from the opposite direction, in which structural theorists and designers have compiled studies of structural optimization, have appeared. Pope and Schmit[26] edited one such document, published under the aegis of the Advisory Group for Aeronautical Research and Development (AGARD), NATO, and another, edited by Moe and Gisvold[27] emerged from a short course held at the University of Trondheim, Norway. In the same vein, the proceedings of a 1969 AGARD symposium on structural optimization have been published[28], as have a series of seminar lectures at the University of Waterloo, Canada[29]. The book by Spunt[30] covers some

# 4     *Optimum Structural Design*

ideas relating to mathematical programming, but is principally devoted to the more classical schemes cited earlier.

I believe that the existing literature of structural optimization, although growing rapidly, may be regarded as 'wieldy' and capable of assimilation by the interested individual. In certain respects, the expansion of published literature assists one in gaining acquaintance with the subject, since a large part of new work is expository, rather than investigative, in nature. The references associated with the respective chapters of this book, supplemented by the contents of previously cited surveys, comprise an almost complete bibliography of the topic to date.

## References

1. J. C. Maxwell, *Scientific Papers*, Vol. 2, 1869, p. 175.
2. A. G. M. Michell, 'The limits of economy of material in framed structures', *Phil. Mag. (Series 6)*, **8**, 589–597 (1904).
3. H. Cox, *The Design of Structures of Least Weight*, Pergamon, Oxford, 1965.
4. W. Hemp, 'Theory of structural design', Report 214, AGARD, October 1958.
5. F. Shanley, *Weight–Strength Analysis of Aircraft Structures*, Dover, New York, 1960.
6. G. Gerard, *Minimum Weight Analysis of Compression Structures*, New York University Press, 1956.
7. W. Prager and P. Marcal, 'Optimality criteria in structural design', AFFDL-TR-70-166, May 1971.
8. J. Taylor, 'Optimal design of structural systems: an energy formulation', *AIAA J.*, **7**, 1404–1406 (1969).
9. V. Venkayya, 'Design of optimum structures', *Computers and Structures*, **1**, No. 1/2, 265–309 (1971).
10. R. A. Gellatly and L. Berke, 'Optimal structural design', AFFDL-TR-70-165, February 1971.
11. R. K. Livesley, 'The automatic design of structural frames', *Quart. J. Mech. Appl. Math.*, **9**, Part 3 (1956).
12. C. Pearson, 'Structural design by high-speed computing machines', *Proc. of ASCE Conf. on Electronic Computation, Kansas City, Missouri, 1958*.
13. L. Schmit, 'Structural design by systematic synthesis', *Proc. of ASCE 2nd Conf. on Electronic Computation, Pittsburgh, Pa., 1960*.
14. E. Haug and P. Kirmser, 'Minimum weight design of beams with inequality constraints on stress and deflection', *Trans. ASME, J. Appl. Mech.*, **89**, 999–1004 (1967).
15. R. Melosh and R. Luik, 'Approximate multiple configuration analysis and allocation for least weight structural design', AFFDL-TR-67-59, April 1969.
16. E. Wasiutynski and A. Brandt, 'The present state of knowledge in the field of optimum design of structures', *Applied Mechanics Reviews*, May (1963).
17. W. Prager and C. Sheu, 'Recent developments in optimal structural design', *Applied Mechanics Reviews*, October (1968).
18. R. Barnett, 'Survey of optimum structural design', *Experimental Mechanics*, **6**, No. 12, December (1966).

19. J. McNunn and G. Jorgenson, 'A review of the literature on optimization techniques and minimax structural response problems', TR 66-5, Dept. of Aero. and Engrg. Mech., University of Minnesota, 1966.
20. G. Gerard, 'Optimum structural design concepts for aerospace vehicles: bibliography and assessment', Allied Research Associates Technical Report 272-2, March 1965.
21. L. Schmit, 'Structural synthesis: 1959–1969: a decade of progress', in *Recent Advances in Matrix Methods of Structural Analysis and Design* (Ed. R. H. Gallagher *et al.*), University of Alabama Press, 1971.
22. L. Schmit, 'Structural engineering applications of mathematical programming techniques', in *Symposium on Structural Optimization, AGARD Conf. Proc. No. 36* (Ed. R. Gellatly), Advisory Group for Aero. Res. and Devel., NATO, October 1970.
23. L. Schmit, 'Automated design', *Int. Science and Technology*, June (1966).
24. R. L. Fox, *Optimization Methods for Engineering Design*, Addison–Wesley, Reading, Mass., 1971.
25. P. Whittle, *Optimization Under Constraints*, Wiley–Interscience, London, 1971.
26. G. G. Pope and L. A. Schmit (Eds.), *Structural Design Applications of Mathematical Programming Techniques*, AGARDograph 149, 2nd ed., Advisory Group for Aero. Res. and Devel., NATO, February 1972.
27. J. Moe and K. M. Gisvold (Eds.), *Optimization and Automated Design of Structures*, Division of Ship Structures, Norges Tekniske Høgskole, University of Trondheim, Meddelelse SK/M 21, December 1971.
28. R. A. Gellatly (Ed.), *Symposium on Structural Optimization, AGARD Conf. Proc. No. 36*, Advisory Group for Aero. Res. and Devel., NATO, October 1970.
29. M. Z. Cohn (Ed.), *An Introduction to Structural Optimization*, S. M. Study No. 1, University of Waterloo Press, 1969.
30. L. Spunt, *Optimum Structural Design*, Prentice–Hall, Englewood Cliffs, N.J., 1971.

# Chapter 2

# Terminology and Basic Concepts

*Richard H. Gallagher*

## 2.1 Introduction

This chapter outlines the features of the mathematical characterization of the optimum-structural-design problem. These features are common to the approaches described in the independently written subsequent chapters, i.e. processes classified as mathematical-programming methods. Thus the basic concepts and terminology of mathematical programming are defined and reviewed. The information presented is largely drawn from standard sources (e.g. References 1–3) in which more detailed and rigorous treatments may be found, although a special effort is made to establish the relevance to the *structural* optimization problem.

Structural optimization seeks the selection of *design variables* to achieve, within the limits (*constraints*) placed on the structural behaviour, geometry, or other factors, its goal of optimality defined by the *objective function* for specified loading or environmental conditions. The three basic features—design variables, objective function and constraints—contrive to form the design problem in the geometry of the *design space*. Each of these aspects of the problem is now discussed in turn.

## 2.2 Design variables

The design variables of an optimum-structural-design problem may consist of the member sizes, parameters that describe the structural configuration, and the mechanical or physical properties of the material, as well as other quantifiable aspects of the design. The topology (i.e. the pattern of connexion of members in the mathematical model) of the complete structure is difficult to take into account, although it is treated to a limited extent when the algorithm employed permits members to reach zero size. Procedures do not permit a transition from one form of behaviour mechanism to another [e.g. from truss (axial) to rigid-frame (flexural) behaviour] within a continuous design process.

A natural hierarchy, or order of complexity, exists among the different classes of design variables. The simplest design variable is the 'size' of a member, representing the cross-sectional area of a truss member, the moment of inertia of a flexural member, or the thickness of a plate. The majority of

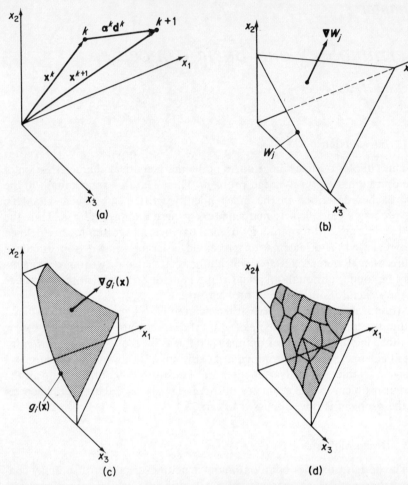

Reproduced by permission from R. A. Gellatly and R. H. Gallagher, 'A procedure for automated minimum weight
structural design. Part I: Theoretical basis', *Aeronautical Quarterly*, **XVII**, August (1966)

**Figure 2.1**　Three-variable design space: (a) design space, (b) weight surface, (c) typical side and
behaviour constraints and (d) composite constraint surface

papers published on optimum structural design deal (as do many examples
in this book) exclusively with the selection of member sizes because of the
relative simplicity of the problem, and because many practical structures
have fixed geometry and material properties. Configuration variables, often
represented by the coordinates of element joints, are next in order of difficulty, followed by material properties.

Material selection presents a special problem with conventional materials,
as they have discrete properties. This problem is also encountered in member

selection, but very often a considerable range of member properties is available. Methods for the selection of discrete variables in optimum design, now in the early stages of development, are described in Chapter 11. Dynamic programming (Chapter 10) also applies to such cases. Nevertheless, most algorithms for design assume the availability of a continuous range of design variables.

The $i$th design variable is designated herein as $x_i$, and the full set of variables for a given structure is listed in the vector $\mathbf{x}$. Perhaps the most important concept associated with this designation is the *design space*, described by axes representing the respective design variables. Figure 2.1(a) shows a three-variable, and consequently three-dimensional, design space. The three-bar truss (Figure 2.2) is representative of a design problem which

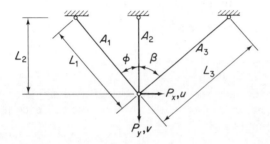

**Figure 2.2**   Three-bar truss

is described in this space. The number of design variables $n$ is generally very much greater than three, and therefore defies illustration; the $n$-dimensional space is termed a *hyperspace*. A point $k$ in design space is a design with variables $\mathbf{x}^k$.

Many of the design algorithms to be discussed in subsequent chapters employ the strategy of a *direct search*, in which a series of directed design changes (moves) are made between successive points in design space. A typical move is between the $k$th and $(k + 1)$th points, given by the equation:

$$\mathbf{x}^{k+1} = \mathbf{x}^k + \alpha^k\mathbf{d}^k. \tag{2.1}$$

The vector $\mathbf{d}^k$ [see Figure 2.1(a)] defines the direction of the move, and $\alpha^k$ gives its amplitude.

## 2.3   Objective function

The *objective function*, also termed the *cost function* or *merit function*, is the function whose least (or greatest) value is sought in an optimization procedure, and constitutes a basis for the selection of one of several alternative acceptable designs. The objective function is a scalar function of the design variables, and is designated in this book by $W(\mathbf{x})$. It represents the most important single property of a design, such as cost or weight, but it is also possible to represent the objective function as a weighted sum of a number of desirable properties.

Weight is most frequently employed as the objective function—hence the symbol $W$. However, the procedures that will be described are not dependent on a specific type of objective function. Weight is the most easily quantified measure of merit, and, although cost is of wider practical importance, it is often difficult to obtain sufficient data for the construction of a cost objective function.

The three-bar truss (Figure 2.2) gives a simple illustration of the weight function. With $\rho_i$ as the density per unit volume of the member $i$, we have

$$W = \rho_1 L_1 A_1 + \rho_2 L_2 A_2 + \rho_3 L_3 A_3 \tag{2.2}$$

and, for the truss with $m$ members,

$$W = \sum_{i=1}^{m} \rho_i L_i A_i. \tag{2.3}$$

Often only the member sizes are treated as design variables, and the density and member lengths are the given constants. Hence, it is convenient to write equation (2.3) as

$$W = \rho \mathbf{L}^T \mathbf{A}. \tag{2.4}$$

Furthermore, the generalization to structures with design variables other than cross-sectional areas is often of the form

$$W = \mathbf{c}^T \mathbf{x} \tag{2.5}$$

where $\mathbf{c}^T = [c_1 \dots c_n]$ comprises the geometric and material property constants. Clearly, equations (2.4) and (2.5) are linear functions of the design variables.

It is useful to illustrate the linear objective function in design space [Figure 2.1(b)]. A linear function in three-dimensional space is a plane, representing here the locus of all design points with a single value $W_j$. In $n$-dimensional space, the surface so defined is a *hyperplane*. When the objective function has non-linear design variables, a *hypersurface* is described in design space.

An important concept, employed continually in the chapters that follow, is that of the *gradient* of the objective function, $\nabla W$. The gradient is a vector composed of the derivatives of $W$ with respect to each of the design variables, i.e.

$$\nabla W = \left[ \frac{\partial W}{\partial x_1} \dots \frac{\partial W}{\partial x_n} \right]^T. \tag{2.6}$$

Thus, for the linear objective function given in equation (2.5), we have

$$\nabla W = \mathbf{c} \tag{2.7}$$

while, for a non-linear objective function represented as $W = [\hat{\mathbf{c}}(\mathbf{x})]^{\mathrm{T}}\mathbf{x}$, we have

$$\nabla W = \mathbf{c}(\mathbf{x}). \tag{2.8}$$

The gradient represents a vector normal to the function in question at the point at which it is taken. The linear function [equation (2.4)] describes a plane so that all gradients to this function are identical (constant). The gradient vector derives its utility from the fact that it defines the direction of design change, or *travel*, in which the objective function is increased most rapidly for a given amplitude of change. Our interest is principally in the reduction of the objective-function value, representing the negative of the gradient vector, so that, by substitution of equation (2.8) (with $\mathbf{d}^k = -[\mathbf{c}(\mathbf{x})]^k$) into equation (2.1), we get the greatest change

$$\mathbf{x}^{k+1} = \mathbf{x}^k - \alpha^k[\mathbf{c}(\mathbf{x})]^k. \tag{2.9}$$

This search algorithm is the method of *steepest descent*.

## 2.4  Constraints

A *constraint*, in any class of problem, is a restriction to be satisfied in order for the design to be acceptable. It may take the form of a limitation imposed directly on a variable or group of variables (*explicit constraint*), or may represent a limitation on quantities whose dependence on the design variables cannot be stated directly (*implicit constraint*).

An *equality constraint*, which may be either explicit or implicit, is designated herein as

$$g_i(\mathbf{x}) = 0, \qquad i = 1, \ldots, E \tag{2.10}$$

for $E$ such constraints. In theory, each equality constraint is an opportunity to remove a design variable from the optimization process and thereby reduce the number of dimensions of the problem. However, as the elimination procedure may be awkward and algebraically complicated, this approach is not always adopted.

An inequality constraint is of the form

$$g_i(\mathbf{x}) \leqslant 0, \qquad i = 1, \ldots, I \tag{2.11}$$

where a total of $I$ constraint conditions prevail. The idea of an inequality constraint is of major importance in optimum structural design. If equality constraints only were stipulated in a design limited by stresses alone, all procedures would lead to *fully stressed* designs. As will be demonstrated in later chapters, fully stressed designs are not necessarily minimum-weight designs, and, for optimality, it is essential to permit designs in which not all stress constraints are satisfied identically, i.e. inequality constraints.

Another important division of constraints is into *side* and *behaviour* constraints. A side constraint is a specified limitation (minimum or maximum) on a design variable or a relationship which fixes the relative value of a group of design variables. Side constraints are therefore explicit in form. The shortcomings of many classical methods of structural optimization can be traced to their inability to stipulate minimum member dimensions.

The behaviour constraints in structural design are usually limitations on stresses or displacements, but they may also take the form of restrictions on such factors as vibrational frequency or buckling strength. Explicit and implicit behaviour constraints are both encountered in practice. Typically, explicit behaviour constraints are given by formulae presented in design specifications. Behaviour constraints are generally implicit, however, as illustrated in the subsequent discussion of their relationship to structural analysis.

Each constraint condition appears in design space as a surface, representing the locus of design points which cause the constraint to be satisfied as an equality constraint. For continuous design variables, the surface is, in general, continuous, and is curved for the elastic structural design of an indeterminate structure. Figure 2.1(c) shows a typical behaviour constraint in design space.

The property of *convexity* is of value in ascertaining the uniqueness of a solution for an optimum design point. A function $f(x)$ is said to be convex if the line segment drawn between two points on the graph of a function never lie below the graph. For this case, the tangent planes at a point, which may be introduced to linearize a problem, lie entirely below the surface being approximated. The curved line of Figure 2.3 is convex. In algebraic terms, a function $f(x)$ is convex if, for $0 < \xi < 1$,

$$f[\xi x^1 + (1 - \xi)x^2] \leqslant \xi f(x^1) + (1 - \xi)f(x^2). \tag{2.12}$$

In the general context of constraints we define 'above' as the region permitted by inequality (2.11).

The *gradient*, or normal vector, of a constraint surface is employed in many procedures in later chapters. Thus, adopting the symbolism previously employed for the objective-function gradient,

$$\mathbf{\nabla} g_i(\mathbf{x}) = \left[ \frac{\partial g_i(\mathbf{x})}{\partial x_1}, \cdots, \frac{\partial g_i(\mathbf{x})}{\partial x_n} \right]^{\mathrm{T}}. \tag{2.13}$$

A representative gradient vector is shown in Figure 2.1(c). Since the constraint surfaces in structural design are most often non-linear functions of the design variables, as described below, it must be anticipated that the

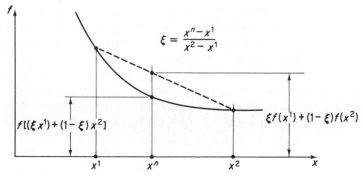

**Figure 2.3**   Convex function

constraint gradients will be a function of the point at which they are taken, rather than constants.

Constraint equations are non-linear functions of the design variables when the behaviour of the structure is elastic and statically indeterminate, even though the analysis technology is linear. To illustrate this point, as well as the implicit nature of the common form of constraint equations, we employ the symbolism of matrix finite-element analysis by the stiffness (or displacement, or stationary-potential-energy) approach. The stiffness method is the prevalent analysis tool in current computational practice in optimum structural design, but it must be emphasized that the methods of optimum structural design are not restricted to any specific method of analysis.

In the stiffness method[4-6], the analysis begins with the construction of element stiffness equations of the form

$$\mathbf{F} = \mathbf{K}^e\mathbf{\Delta} \tag{2.14}$$

where $\mathbf{F}$ represents the element nodal forces, $\mathbf{\Delta}$ are the corresponding displacements, and $\mathbf{K}^e$ is the set of *element stiffness coefficients*. The latter are linear functions of the parameters which are conventionally the design variables, i.e. the element dimensions and material properties. By application of the conventional process of assembly to form the equations of the complete system, we have

$$\mathbf{P} = \mathbf{K}\mathbf{\Delta} \tag{2.15}$$

and, by solution for the displacements,

$$\mathbf{\Delta} = \mathbf{K}^{-1}\mathbf{P} \tag{2.16}$$

where the inverse of $\mathbf{K}$ is symbolic, it generally being more efficient to employ equation-solving techniques.

Considering, for the present, the situation of displacement constraints, a comparison of equation (2.16) with the specified limits on the displacements comprises the displacement constraints. Thus, one deals with the *solution* to analysis, and the functional relationship to design variables is studied in terms of the inverse of **K**. The process of matrix inversion produces a matrix which is a non-linear function of the design variables.

Consideration of stress constraints follows directly from this. A familiar approach to the calculation of stresses in finite-element analysis is through the establishment of element or system stress matrices, which relate the stresses to the displacements. Adopting, for convenience, the latter form

$$\boldsymbol{\sigma} = \mathbf{S}\boldsymbol{\Delta} \tag{2.17}$$

where $\boldsymbol{\sigma}$ lists all stress values of interest. By substitution of equation (2.16),

$$\boldsymbol{\sigma} = \mathbf{S}\mathbf{K}^{-1}\mathbf{P}. \tag{2.18}$$

Hence, because of the dependence on the inverse of **K**, the stresses are also, in general, a non-linear function of the design variables. The scale of analysis by use of the finite-element method requires numerical solution of the above relationships, yielding constraints that are implicit functions of the design variables.

### 2.5  Mathematical statement of the optimum-structural-design problem

In algebraic terms, the minimum-weight design problem represented in the design space of Figure 2.1(d) is stated as 'minimize $W(\mathbf{x})$, which is an objective function of $n$ design variables $\mathbf{x} = [x_1, \ldots, x_n]^\mathrm{T}$, subject to constraints

$$g_i(\mathbf{x}) = 0, \qquad i \in E$$

$$g_i(\mathbf{x}) \geqslant 0, \qquad i \in I$$

representing equality constraints ($E$) and inequality constraints ($I$) on the variables'. If the objective function is not weight, but, rather, a function to be maximized, the problem is identical if one seeks to minimize $-W(\mathbf{x})$.

For simplicity, the geometric representation of the design problem is shown in Figure 2.4 in terms of two design variables. Upper and lower limits may be specified for each designated stress and displacement amplitude for each alternative load condition; so that a large number of constraint surfaces are present. These surfaces intermesh in a complex fashion. The uppermost (*dominant*) constraints form a *composite constraint surface* in which each 'patch' represents a segment of an individual constraint surface. The previously noted side constraints appear as planes parallel to the coordinate directions when they refer to minimum or maximum values of the design variables.

If a design point is located in the space above the composite constraint surface, it is in *free space* and is known as a *feasible-design* or *exterior* point.

Specified lower limit on $x_2$ (side constraint)

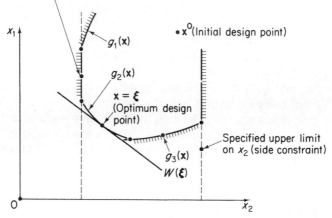

**Figure 2.4**   Two-variable design space; $g_1(\mathbf{x})$, $g_2(\mathbf{x})$ and $g_3(\mathbf{x})$ are behaviour constraints

Conversely, a design point that represents the violation of constraints is *infeasible* or *interior*. The ability to choose a feasible point as an initial design estimate is a precondition on the use of a number of mathematical-programming algorithms.

In geometric terms, Figure 2.4 discloses that the optimum-design problem consists in finding the osculatory point of the weight and constraint surfaces. The designer must first effect an estimate of the design variables $\mathbf{x}^0$. The goal of the chosen mathematical-programming algorithm is then to travel, with least computational effort, from this initial guess to the optimum design point.

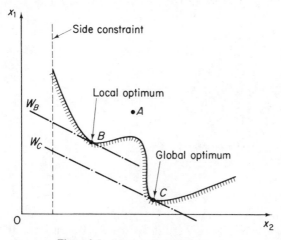

**Figure 2.5**   Local and global optima

Illustration of another representative two-dimensional design space (Figure 2.5) permits description of one of the major pitfalls in the application of mathematical-programming methods—the failure to distinguish between *local* minima and the *global* minimum. If, for example, the procedure leads from the point A to the point B, analytical tests applied at point B will indicate that no further moves are possible without a violation of the constraints. Thus a *local* minimum will have been reached. The absolute, or *global*, minimum is seen to exist a point C, however.

A local minimum coincides with the global minimum when the constraint surface is convex, but, unfortunately, constraint surfaces in structural design are not always convex. Kavlie and Moe[7] and Moses and Onada[8] have given examples of structural-design problems with concave constraint surfaces, and Moe discusses this point further in Chapter 9.

The analytical test for a local minimum, alluded to above, is the Kuhn–Tucker optimality condition. This test forms a useful procedure for deciding when to terminate the search for an optimum. A Kuhn–Tucker point is a point at which there is no feasible descent direction. The mathematical statement of this condition is as follows.

Consider the case where $q$ constraints are active at a design point. At this point $\bar{x}$, the objective function is designated as $W(\bar{x})$. Denote the normal (gradient) vector to the $i$th active constraint as $g_i(\bar{x})$. The Kuhn–Tucker optimality condition states that the necessary condition for a local optimum-design point is that the gradient to the objective function be expressible as the negative of a linear combination of the constraint gradients, i.e.

$$\nabla W(\bar{x}) = - \sum_{i=1}^{q} \lambda_i \nabla g_i(\bar{x}) = - \nabla g(\bar{x})\lambda \qquad (2.19)$$

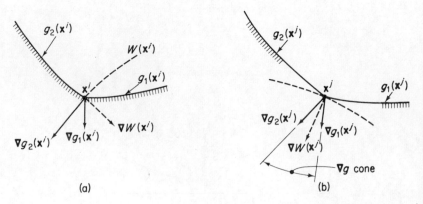

(a)                                   (b)

**Figure 2.6** Kuhn–Tucker condition : (a) design point $x_j^i$ is not a minimum and (b) design point $x^j$ is a minimum

where the $\lambda_i$ ($i = 1, \ldots, q$) are non-negative multipliers. The $\lambda_i$ are also termed Lagrange multipliers, since the development of equation (2.19) may be accomplished by the application of the Lagrange-multiplier approach to constrained minimization.

A graphical representation of the Kuhn–Tucker condition can also be constructed (Figure 2.6). The negatives of the constraint-surface normals can be seen to comprise a cone in design space; equation (2.19) requires that the normal to the objective function be within this cone in order for the optimality condition to be met.

## References

1. R. L. Fox, *Optimization Methods for Engineering Design*, Addison–Wesley, Reading, Mass., 1971.
2. M. Aoki, *Introduction to Optimization Techniques*, Macmillan, New York, 1971.
3. T. Au and T. Stelson, *Introduction to Systems Engineering: Deterministic Models*, Addison–Wesley, Reading, Mass., 1969.
4. O. C. Zienkiewicz, *The Finite Element Method in Engineering Science*, McGraw-Hill, London, 1971.
5. J. L. Meek, *Matrix Structural Analysis*, McGraw-Hill, New York, 1971.
6. F. Beaufait, W. Rowan, P. Hoadley and R. Hackett, *Computer Methods of Structural Analysis*, Prentice–Hall, Englewood Cliffs, N.J., 1970.
7. D. Kavlie and J. Moe, 'Automated design of frame structures', *J. Struct. Div.*, *ASCE*, **97**, No. ST1, 33–62 (1971).
8. F. Moses and S. Onada, 'Minimum weight design of structures with application to elastic grillages', *Int. J. Num. Methods Engrg.*, **1**, No. 4, 311–331 (1969).

# Chapter 3

# Fully Stressed Design

*Richard H. Gallagher*

## 3.1 Introduction

A fully stressed design, abbreviated here as f.s.d., is a design in which each structural member sustains a limiting allowable stress under at least one of the specified loading conditions. Analysis is restricted to the selection of member sizes for a fixed structural geometry and specified materials, and no consideration is given to displacement limitations (this restriction is removed in the work described in Chapter 4). If maximum- or minimum-member-size limitations are taken into account, the affected members may not be fully stressed.

The fully stressed design is one of the traditional concepts of optimal structures, and retains its significance owing to its familiarity and intuitive appeal. Other reasons for the significance of the concept include:

(a) The structural proportions calculated in an f.s.d. analysis procedure can be proved to be optimal under certain circumstances.
(b) The f.s.d. analysis procedure is inexpensive to apply in comparison with methods based on mathematical-programming concepts.
(c) An f.s.d. is a useful starting point for optimum-design procedures based on non-linear-programming concepts[1,2].

This chapter first examines the general character of f.s.d. analysis to define its relationship to the more general type of optimum-design analysis (e.g. one based on mathematical-programming concepts) and to identify certain of its limitations. The alternative forms of f.s.d. analysis procedures are studied. Ways of determining if an f.s.d. is an optimum design are then discussed, followed by a description of conditions which must be met to achieve convergence when applying an algorithm for f.s.d. Finally, a review of published experience in f.s.d. is given.

## 3.2 Characteristics of fully stressed design

It is useful to begin the treatment of f.s.d. with a discussion of its distinguishing features. These features identify various limitations of the concept without resorting to analytical or numerical evidence, and indicate the relationship to optimum design as it is seen in the context of mathematical programming.

A foremost consideration in f.s.d. is the absence of an objective function whose extreme value is sought. Hence one cannot expect that a figure of merit, such as weight or volume, will be minimized, or that stiffness will be maximized. It was proved nearly seventy years ago[3,4] that, for a truss under a single load condition, the minimum weight/stiffness ratio is represented by a determinate, and therefore fully stressed, design. This consideration does not extend, however, to indeterminate structures with alternate load conditions.

This is because a fully stressed design is not unique. Since every statically determinate structure can be proportioned directly to yield an f.s.d., if one allows the members of an initially described statically indeterminate structure to assume zero size, *each subsidiary statically determinate form of the structure is an alternative f.s.d.* A numerical demonstration of this situation is given by Barta[5], who studies a twice-redundant eight-bar truss with *twelve* subsidiary statically determinate forms.

There is no assurance that an algorithm for the calculation of an f.s.d. will converge to the *minimum-weight* f.s.d. This is because, as has been noted and will be demonstrated in the next section, the analytical statement of the goal of minimum weight (the merit function) is absent from the algorithm for the calculation of the f.s.d. Numerical evidence of this behaviour can be found in Reference 6.

These aspects of fully stressed design can be shown graphically in design space (Figure 3.1.), using a three-bar truss as an example. Symmetry is imposed on the sloping members to produce a two-dimensional situation in

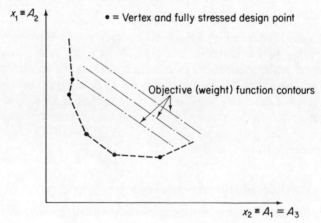

**Figure 3.1** Fully stressed designs in design space; the axes refer to the design variables of the three-bar truss of Figure 2.2 with $\phi = \beta$ and $A_1 = A_3$

which the design variables are $x_1$ and $x_2$. In general, the problem may involve alternative load conditions, and each condition produces a constraint surface relative to both the upper and lower stress limits of each member. As shown schematically in Figure 3.1, this produces many vertices, and each vertex resulting from constraints on $x_1$ as well as $x_2$ is a fully stressed design.

Thus there are many vertices in design space. Those vertices that join constraints from at least one of *each* of the members represent fully stressed designs. Other vertices do not include a stress-limiting constraint from each member, and, although such a point cannot be a fully stressed design, it might well represent the optimum design. Finally, the optimum design might exist on a constraint surface between vertices.

## 3.3 Analytical basis

There is neither a unique nor universally accepted method for the calculation of an f.s.d. The simplest approach consists of a cyclic analysis procedure in which the results from a given cycle are used to scale the members to the fully stressed state. The scaled sizes are then employed in the next analysis cycle. To express this procedure in algebraic form, we denote the design variable and stress state of the $i$th member in the $k$th cycle as $x_i^k$ and $\sigma_i^k$, respectively. The allowable stress is designated by $\bar{\sigma}_i$ (a further distinction might be made between allowed states of tension and compression, but it is disregarded here for simplicity). Hence the design variable to be employed in the $(k + 1)$th cycle is calculated from

$$x_i^{k+1} = x_i^k \frac{\sigma_i^k}{\bar{\sigma}_i}. \tag{3.1a}$$

This approach is termed the *stress-ratio* method. The process is continued to convergence unless, as will be described in Section 3.4, convergence cannot be achieved because of relationships between the form of the structure and the number of load conditions.

Recognizing that an increase in the size of a member draws more load to that member and that a decrease in the size relieves the load, convergence may be hastened by a process of 'overrelaxation'. To accomplish this, equation (3.1a) can be revised to the form

$$x_i^{k+1} = x_i^k \left(\frac{\sigma^k}{\bar{\sigma}}\right)_i^{\beta} \tag{3.1b}$$

where $\beta$ is the overrelaxation factor greater than unity.

Clearly, a more realistic approach than the stress-ratio method would account for the influence of *all* the member sizes on the stress in a given member. We can develop such relationships by considering the common (in truss design) situation in which the stress in a member is defined by the simple product of two terms; the member force $F_i$ and $b_i$ the member 'compliance'.

$$\sigma_i = b_i F_i \tag{3.2}$$

Taking the derivative of $\sigma_i$ with respect to $x_j$, we have

$$\frac{\partial \sigma_i}{\partial x_j} = F_i \frac{\partial b_i}{\partial x_j} + b_i \frac{\partial F_i}{\partial x_j}. \tag{3.3}$$

This relationship shows that the change in stress of a member arises from two sources: (a) the change in the member compliance at constant force and (b) a change in force at constant compliance. Using equation (3.3), Reference 6 demonstrates that the stress-ratio method [equation (3.1)] for trusses may be derived from equation (3.3) by treating the reciprocal of the member area as the design variable (i.e. $x_i = 1/A_i$) by disregarding the second term in equation (3.3), and by observing that, for trusses, $b_i = 1/A_i$.

We return now to equation (3.3) to establish more sophisticated f.s.d. algorithms than the stress-ratio method. From physical laws, the compliance for an axial member $b_i = 1/A_i$, and for flexural members $b_i = 1/Z_i$. (Note that this simple approach excludes the consideration of more complex cases, such as combined bending and axial behaviour, where no single term characterizes the compliance.) Also, the member size, rather than the reciprocal of the member size, is chosen as the design variable ($x_i = A_i$ or $Z_i$).

$$\frac{\partial \sigma_i}{\partial x_i} = -\frac{F_i}{x_i^2} + \frac{1}{x_i} \frac{\partial F_i}{\partial x_i} \tag{3.4}$$

and, for a finite change $\delta x_j$, with $\sigma_i = F_i/x_i$, in an $n$-member system,

$$\delta \sigma_i = -\frac{\sigma_i}{x_i} \delta x_i + \frac{1}{x_i} \sum_{j=1}^{n} \frac{\partial F_i}{\partial x_j} \delta x_j. \tag{3.5}$$

Now, if the change takes place from the $k$th design point to the $(k + 1)$th design point, $\delta x_j = x_j^{k+1} - x_j^k$ and $\delta \sigma_i = \sigma_i^{k+1} - \sigma_i^k$, so that equation (3.5) becomes

$$\sigma_i^{k+1} - \sigma_i^k = -\left(\frac{\sigma}{x}\right)_i^k (x_i^{k+1} - x_i^k) + \frac{1}{x_i^k} \sum_{j=1}^{n} \left(\frac{\partial F_i}{\partial x_j}\right)^k (x_j^{k+1} - x_j^k) \tag{3.6}$$

and, for all members, with $\boldsymbol{\sigma}^{k+1}$ set equal to the allowable stress $\bar{\boldsymbol{\sigma}}$:

$$\bar{\boldsymbol{\sigma}} - \boldsymbol{\sigma}^k = [\mathbf{G}(\boldsymbol{\sigma}) + \mathbf{H}(\mathbf{F})](\mathbf{x}^{k+1} - \mathbf{x}^k) \tag{3.7}$$

where

$$\mathbf{G}(\boldsymbol{\sigma}) = \left[ \begin{matrix} \diagdown \\ & \dfrac{\sigma_i}{x_i} \\ & & \diagdown \end{matrix} \right]^k, \qquad i = 1, \ldots, n, \text{ (a diagonal matrix)} \qquad (3.7a)$$

and

$$\mathbf{H}(\mathbf{F}) = \left[ \dfrac{\partial F_i}{\partial x_j} \right]^k, \qquad i, j = 1, \ldots, n \qquad (3.7b)$$

is a square matrix, evaluated at design point $\mathbf{x}^k$. Equation (3.7) can be solved for the design variables $\mathbf{x}^{k+1}$ to give a new estimate of the design variables for a fully stressed design.

It is difficult to evaluate the matrix $\mathbf{H}(\mathbf{F})$, and for this reason it is preferable to supplant equation (3.4) with the direct differentiation of stress with respect to the design variables

$$\delta\sigma_i = \sum_{j=1}^{n} \dfrac{\partial \sigma_i}{\partial x_j} \delta x_j = \dfrac{\partial \sigma_i}{\partial x_1} \delta x_1 + \cdots + \dfrac{\partial \sigma_i}{\partial x_j} \delta x_j + \cdots + \dfrac{\partial \sigma_i}{\partial x_n} \delta x_n. \qquad (3.8)$$

The terms $\partial\sigma_i/\partial x_j$ are no simpler to evaluate than $\partial F_i/\partial x_j$, but they do represent, in aggregate, the complete set of coefficients of the matrix to be inverted. Using the prior definitions of $\delta\sigma_i$ and $\delta x_j$, we have:

$$\bar{\boldsymbol{\sigma}} - \boldsymbol{\sigma}^k = \mathbf{H}(\boldsymbol{\sigma})(\mathbf{x}^{k+1} - \mathbf{x}^k) \qquad (3.9)$$

where

$$\mathbf{H}(\boldsymbol{\sigma}) = \left[ \dfrac{\partial \sigma_i}{\partial x_j} \right]^k, \qquad i, j = 1, \ldots, n \qquad (3.9a)$$

is a square matrix, evaluated at the design point $\mathbf{x}^k$, each row of which is the gradient to the surface in design space, corresponding to the stress $\sigma_i$. Solving for $\mathbf{x}^{k+1}$, we have

$$\mathbf{x}^{k+1} = \mathbf{x}^k + [\mathbf{H}(\boldsymbol{\sigma})]^{-1}(\bar{\boldsymbol{\sigma}} - \boldsymbol{\sigma}^k). \qquad (3.10)$$

Equations (3.8)–(3.10) represent application of Newton's method for the solution of non-linear algebraic equations. Certain implications of this point will be taken up in Section 3.5.

The relationship between this approach and mathematical-programming concepts can be established as follows. The constraints $g_i(\mathbf{x})$ in fully stressed design consist entirely of equality constraints on stress, i.e.

$$g_i(\mathbf{x}) = \bar{\sigma}_i - \sigma_i = 0 \qquad (3.11)$$

where $\sigma_i$ here represents the vector of member stresses as calculated in analysis. We now expand $g_i(\mathbf{x})$ in a Taylor series at the $k$th design point:

$$g_i(\mathbf{x} + \delta\mathbf{x}) = g_i(\mathbf{x}^k) + \sum_{j=1}^{n} \frac{\partial g_i(\mathbf{x}^k)}{\partial x_j} \delta x_j + \cdots \qquad (3.12)$$

and, retaining only the terms indicated on the right-hand side of equation (3.12), and with reference to equation (3.11),

$$g_i(\mathbf{x}^k) = -\sum_{j=1}^{n} \frac{\partial g_i(\mathbf{x}^k)}{\partial x_j} \delta x_j \qquad (3.13)$$

Also, by virtue of equation (3.11), since $\bar{\boldsymbol{\sigma}}$ is constant,

$$\frac{\partial g_i(\mathbf{x}^k)}{\partial x_j} = -\frac{\partial \sigma_i}{\partial x_j} = -H_{ij}(\boldsymbol{\sigma}) \qquad (3.14)$$

so that, for a finite change in design variable, and with substitution of equation (3.11), equation (3.13) becomes

$$\bar{\sigma}_i - \sigma_i = \sum_{j=1}^{n} H_{ij}(\boldsymbol{\sigma}) \, \delta x_j. \qquad (3.15)$$

Noting, finally, that, if equation (3.15) is used iteratively as the algorithm for fully stressed design, then $\sigma_i = \sigma_i^k$ and $\delta x_j = x_j^{k+1} - x_j^k$, and, writing this expression for the full system,

$$\bar{\boldsymbol{\sigma}} - \boldsymbol{\sigma}^k = \mathbf{H}(\boldsymbol{\sigma})(\mathbf{x}^{k+1} - \mathbf{x}^k) \qquad (3.16)$$

which is identical to equation (3.9) and to the format of a portion of the linear-programming problem, which is discussed in detail in Chapter 6. The portion here consists of the constraint equations for stress. Displacement constraints are absent, as are side-constraint conditions and the merit function. Since the constraint equations are non-linear in the design variables, linear programming, *per se*, is not applicable. The *iterative* application of linear programming is a feasible scheme, however. This idea, introduced into structural design by Moses[7], stems from the 'cutting-plane' method[8] of mathematical programming and is developed in some detail in Chapter 7.

There are a number of practical considerations which must be accounted for when developing programs which employ these algorithms. These include: (a) the choice of the form of design variables for flexural members, (b) the manner of treatment of statically determinate portions of the total structure, (c) the identification of 'subsidiary' structural forms as defined by zero-size members, (d) devices for hastening the convergence of the iterative process and (e) criteria for discontinuing the iterative design process.

For some optimization processes the cross-sectional area may be used as the design parameter for a flexural member when a relationship between the area and the section modulus can be established. This substitution would be made in the interest of having consistent formulations of the constraint and merit-function equations. In f.s.d., however, there is no merit function, and it is convenient to retain the section modulus (or its reciprocal) as the design variable in flexural situations. Indeed, Moses and Onada[9] have reported that attempts to use the cross-sectional area as the design variable have caused difficulties in achieving convergence.

Cornell[10] has observed that the use of the reciprocal of the apparent design variable (i.e. the use of $1/A$ or $1/Z$) is advantageous in the more sophisticated f.s.d. analysis procedures. The rationale for this form of design variable derives from the mathematical-programming format of the design problem, where the inverse form linearizes the constraint equations at the expense of non-linearization of the merit function, enabling the use of more advantageous mathematical-programming algorithms.

Certain structures are composed of determinate and indeterminate segments. Members in the determinate segment are sized once and for all in the first cycle and will, of course, remain unchanged during successive iterations. These members can be identified by monitoring the iterative sequence, and they should be removed from the sequence, once they have been identified, for the sake of computational efficiency.

It has been shown[11,12] that, for a single-load condition, the determinate form of a truss with specified joints possesses a weight that is equal to or less than the lightest indeterminate truss with the same joints. This suggests that, for the more general (multiple-load condition) design problem, the same circumstance may exist. Indeed, Sved and Ginos[13] demonstrated that a three-bar truss that had been studied intensively in earlier design investigations was, in fact, lighter in the determinate, two-bar form. Thus it appears that there is a need for algorithms that incorporate the automatic removal of members as they approach zero size so as to include a consideration of 'subsidiary' structural forms.

One method of dealing with this situation is to set a member size to zero automatically when it reaches an arbitrarily defined small size. Alternatively, an organized procedure for the examination of all possible subsidiary forms, including the candidate determinate forms, might be devised as suggested by Sved and Ginos[13]. A procedure of this type has been defined in the context of linear-programming concepts by Dorn, Gomory and Greenberg[14].

More recently, Sheu and Schmit[15] have presented a method for the automated selection of a minimum-weight truss from a subset of configurations obtained by omitting various member configurations from a primary configuration. A feasible-direction algorithm, which is in the class of algorithms described in Chapter 8, was used to find an 'upper-bound' solution, and 'lower-bound' solutions were obtained by an application of the dual simplex algorithm (see Chapter 6).

A device for hastening the convergence of the iterative process has already been cited for the stress-ratio method in the form of the overrelaxation scheme of equation (3.1b). Melosh[16] describes more generally applicable extrapolation schemes. For example, when the iterative process has stabilized to the extent that the active constraint for a member does not change in three successive cycles with design parameter values of $x_i^k$, $x_i^{k+1}$ and $x_i^{k+2}$, a quadratic expression is fitted through these three values. The extrapolated design variable $x_i^{k+3}$ is then given by

$$x_i^{k+3} = x_i^{k+1} + 2\frac{(x_i^k - x_i^{k+1})(x_i^{k+2} - x_i^{k+1})}{(x_i^k - 2x_i^{k+1} + x_i^{k+2})}. \tag{3.17}$$

It is not essential that extrapolation be based on a quadratic fit, and perhaps a preferable scheme is to employ a representation in the form of a decaying exponential.

Finally, Melosh[16] has also considered criteria for discontinuing the iterative design process. Two schemes are described. In one, after the completion of an iterative cycle, the two members with the highest and lowest ratios of 'actual' to allowable stress are identified, and these ratios are used to construct a parameter that is indicative of convergence. It is also recommended that iteration should be discontinued after a specified number of cycles to forestall cases in which convergence does not occur within an economically acceptable number of iterations.

### 3.4   Illustrative example

The significance of some of the alternatives and considerations discussed in Section 3.3 is illustrated by numerical results for the f.s.d. of the three-bar truss shown in Figure 3.2. This problem, solved previously in analytical form by Schmidt[17] in one of the earliest careful studies of the implications of f.s.d., is subjected to two alternative load conditions, $P_1$ acting alone and $P_2$ acting alone; $P_1 = P_2 = P$.

Figure 3.2 shows the variation of calculated cross-sectional area $A_2$ against the number of iterations for various methods. The simple stress-ratio method [Equation (3.1a)] was first employed to calculate the f.s.d. with starting values of $A_1 = A_2 = A_3 = 1.0 \text{ in}^2$; the results are shown by the solid line. It is apparent that the process leads to the 'exact' solution[11], but at a relatively slow rate.

Next, with the same starting point, the more sophisticated algorithm represented by equation (3.16) was applied, leading immediately to a divergent calculation. The difficulty here can be explained by the previously noted fact that equation (3.16) corresponds to Newton's method, which is prone to divergence when the starting point represents a poor estimate of the exact solution.

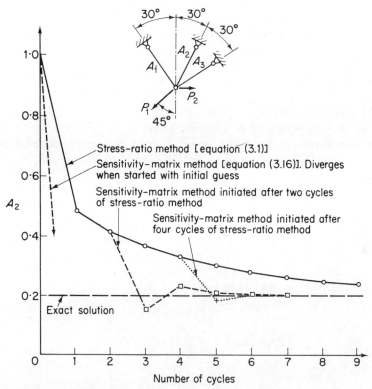

**Figure 3.2**   Numerical experience for all fully stressed design algorithms

With the above consideration in mind, a 'mixed-mode' approach was attempted, i.e. the stress-ratio scheme was applied for a few cycles, after which the algorithm represented by equation (3.16) was invoked. This scheme proved to be quite effective. As Figure 3.2 shows, with the use of just two cycles of the stress-ratio method, the estimate is brought sufficiently close to the exact solution to permit a convergent solution with only three cycles of the equation (3.16) algorithm.

These results simply indicate the potential value of a 'mixed-mode' approach. It must be kept in mind that the software costs and complexities and the specific circumstances of a given problem (e.g. the availability of a good initial estimate) may result in a preference for a single-mode scheme.

### 3.5   Conditions for a non-convergent solution

A simple example[18] demonstrates that a certain relationship between the number of members and redundants in a problem and the number of load conditions must be satisfied for the f.s.d. analysis procedure to converge. Consider the three-bar truss (Figure 2.2), acted upon by a vertical load $P_y$ with $P_x = 0$. ($\phi = \beta$ in this case.)

If member $L_2$ is subjected to a stress $\sigma$, the elongation $v_2 = (\sigma/E)L_2$. If $L_1$ and $L_3$ (of the same material as $L_2$) were also stressed to $\sigma$, the elongations would be $v_1 = v_3 = (\sigma/E)(L_2/\cos\phi)$. Compatibility of displacement of the junction demands, however, that $v_1 = v_3 = v_2 \cos\phi$, from which it follows that $\sigma_1 = \sigma_2 \cos^2\phi$. Thus the stress in the inclined bars *must* be less than the stress in the vertical member, and a 'fully stressed' design is not possible for a single choice of material. On the other hand, an f.s.d. is possible if unequal allowable stresses can be specified for different elements.

This condition on the relationship between the number of redundant forces and the number of load conditions for convergence of the iterative fully stressed design approach is readily generalized[19,20]. In this general case, the structure has $n$ members and $r$ redundants. Thus there are $n - r$ independent element equilibrium equations for each load condition. With $p$ load conditions, the requirement that the number of independent equations exceeds the $n$ unknowns can be written as

$$p(n - r) \geqslant n \qquad (3.18)$$

or, to define the required number of load conditions

$$p \geqslant \frac{n}{(n - r)}. \qquad (3.19)$$

In the previous simple example, $p = 1, n = 3$ and $r = 1$; so that

$$p = 1 < \frac{3}{3 - 1} = \frac{3}{2} = \frac{n}{n - r}$$

and the requirement is violated. In design space this means that the constraints are so contrived that a vertex does not exist.

Equation (3.19) indicates that more than one load condition is necessary for any redundant structure if the determination of member sizes is to be accomplished on a truly independent basis. On the other hand, the geometric arrangement of the truss may be such that a uniform strain is produced throughout the system. The 'radial' or 'fan' trusses of Figure 3.3 are illustrative of this consideration, and each would be fully stressed under the indicated single load, provided that the limiting stress is the same for each member. Also, by introducing prestressing or initial-fit parameters in a design problem, in the manner discussed by Hofmeister and Felton[21], deficiencies indicated by a comparison of $p$ and $n/(n - r)$ can be eliminated and an f.s.d. achieved.

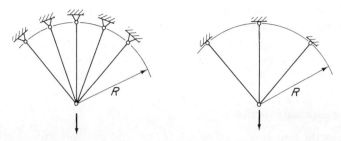

**Figure 3.3** Truss arrangements that permit fully stressed design for a single load

## 3.6 Optimum versus fully stressed design

An analytical procedure that determines whether or not a f.s.d. is the optimum design is desirable, provided that it is relatively inexpensive to apply. The basic test for optimality in any design problem is the classic Kuhn–Tucker optimality condition[22], discussed previously in Section 2.5. Razani[23] adapted this condition to the problem of truss design as follows:

Once an f.s.d. has been computed, the active constraints—one for each of the $n$ members—are identified. Each is of the form of equation (3.11). Restating equation (2.19), the Kuhn–Tucker optimality condition stipulates that, at the optimal design, the gradient to the objective function is equal to a linear combination of all the active constraint gradients. Using $\mathbf{A}$, the vector of truss-member cross-sectional areas to designate the design variables, rather than the symbol $\bar{\mathbf{x}}$, and with $n$ supplanting $q$ in equation (2.19), we have

$$\nabla W(\mathbf{A}) = - \sum_{i=1}^{n} \lambda_i \nabla g_i(\mathbf{A}) = -\lambda^{\mathrm{T}} \nabla g(\mathbf{A}). \tag{3.20}$$

If the objective function is weight, then, as shown in equations (2.4)–(2.7) for a truss, $\nabla W(\mathbf{A}) = (\rho \mathbf{L})$. Also, from equation (3.14), $\nabla g(\mathbf{A}) = -\partial \sigma / \partial \mathbf{A}$. Hence equation (3.20) becomes

$$(\rho \mathbf{L}) = \lambda \frac{\partial \sigma}{\partial \mathbf{A}} \tag{3.21a}$$

and, solving for $\lambda$,

$$\lambda = \left( \frac{\partial \sigma}{\partial \mathbf{A}} \right)^{-1} (\rho \mathbf{L}). \tag{3.21b}$$

By virtue of the requirements for a linear combination, in order for the f.s.d. to be optimal, the result of the product on the right-hand side of equation (3.21b) must be positive for all $\lambda_i$. This product is easily computed, since the inverse of $\partial\boldsymbol{\sigma}/\partial\mathbf{x}$ will have been computed in the final step of the f.s.d. process represented by equation (3.16).

Although the Kuhn–Tucker approach is a perfectly valid criterion for optimality, it does not help with the selection of improved designs for non-optimal cases. The methods discussed in Chapter 8 have this as their goal.

## 3.7 Application experiences

Despite the time-honoured nature of f.s.d. and an unquestioned history of utilization, few organized studies of f.s.d. analysis have been reported in the literature.

Razani[23], Melosh[16] and Pope[24] have studied the percentage difference in weight between fully stressed and minimum-weight designs, and their results indicate that this difference is relatively small. The comparisons are for the lowest weight of f.s.d. for a given configuration, but, as noted earlier and demonstrated also by Barta[5], wide differences may exist within a given class of structure for alternative f.s.d.s. One of the most interesting aspects of the numerical results obtained by Melosh, who dealt with graduated structures with successively greater numbers of analysis degrees of freedom and design parameters, is that the number of cycles to convergence was not significantly affected by the size of the problem.

Lansing and co-workers[25] and Dwyer and co-workers[26] have performed f.s.d. analyses for wing structures, using the stress-ratio method. These studies, however, are mainly directed towards an identification of the influence on the f.s.d. of alternatives in the interpretation of analysis results.

Still other studies, by Wright and Feng[2] and by Tocher and Karnes[27], are principally directed towards the application of mathematical-programming methods to the structural design of practical structures, but also include comments about f.s.d. or comparisons with f.s.d. results.

## 3.8 Closure

The fully stressed design approach is often the most efficient and economical method of calculating low-weight design proportions for complex structures. If the relationship between the number of members, redundants and load conditions permits, and if the mode of structural action is simple, solutions can be obtained in relatively few cycles. F.S.D. is also valuable as a starting point for applications of mathematical-programming methods to optimum structural design.

The fully stressed design is not unique, and will not always correspond to the optimum design. Large differences in the objective function may exist between the best and worst f.s.d.s. One may test for optimality, but a failure of the test will not disclose the gap between the f.s.d. and optimality, nor will it disclose the relative proportions of members for an optimum design. Available numerical evidence, however, indicates that the gap between the best f.s.d. and the optimum design is often quite small for many classes of real problems not selected for their pathological behaviour.

Significant questions remain unanswered in the practical calculation of fully stressed designs. There are a number of possible algorithms for computations to produce an f.s.d., and it is possible that the most effective approach is to employ a combination of these. The method adopted should strive to account for such factors as determinate segments of the whole structure and the automatic removal of individual members (zero-size members) for the examination of subsidiary structural forms.

## References

1. R. A. Gellatly and R. H. Gallagher, 'A procedure for automated minimum weight structural design. Part I, theoretical basis', *Aeronautical Quart.*, **XVII**, August, 216–230 (1966).
2. P. M. Wright and C. C. Feng, 'Optimum design of plane frames using a multi-mode scheme', *Trans. Engrg. Inst. of Canada* (1971).
3. A. G. M. Michell, 'The limits of economy of material in framed structures', *Phil. Mag. (Series 6)*, **8**, 589–597 (1904).
4. F. H. Cilley, 'The exact design of statically indeterminate frameworks, an exposition of its possibility, but futility', *Trans. ASCE*, **42**, June, 353–407 (1900).
5. J. Barta, 'On the minimum weight of certain redundant structures', *Acta Tech. Acad. Sci. Hungar*, **18**, 67–76 (1957).
6. K. Reinschmidt, C. A. Cornell and J. F. Brotchie, 'Iterative design and structural optimization', *J. Struct. Div.*, *ASCE*, **92**, No. ST6, December (1966).
7. F. Moses, 'Optimum structural design using linear programming', *J. Struct. Div.*, *ASCE*, **90**, No. ST6, 89–104 (1964).
8. H. J. Kelley, 'The cutting plane method for solving convex programs', *SIAM J.*, **8**, No. 4, 703–712 (1960).
9. F. Moses and S. Onada, 'Minimum weight design of structures with application to elastic grillages', *Int. J. Num. Methods Engrg.*, **1**, No. 4, 311–331 (1969).
10. C. A. Cornell, 'Examples of optimization in structural design', in *An Introduction to Structural Optimization*, S.M. Study No. 1 (Ed. M. Z. Cohen), University of Waterloo Press, 1969.
11. R. L. Barnett and P. C. Herrmann, 'High performance structures', NASA CR-1038, May 1968.
12. J. B. B. Owen, *The Analysis and Design of Light Structures*, Arnold, London, 1965.
13. G. Sved and Z. Ginos, 'Structural optimization under multiple loading', *Int. J. Mech. Sci.*, **10**, 803–805 (1968).
14. W. S. Dorn, R. E. Gomory and H. J. Greenberg, 'Automatic design of optimal structures', *J. Mecanique*, **3** (1964).
15. C. Y. Sheu and L. A. Schmit, 'Minimum weight design of elastic redundant trusses under multiple static load conditions', *AIAA J.*, **10**, February, 155–162 (1972).
16. R. J. Melosh, 'Structural analysis, frailty evaluation and design. Vol. 1. Safer theoretical basis', AFFDL TR 70-15, Vol. 1, July 1970.
17. L. C. Schmidt, 'Fully stressed design of elastic redundant trusses under alternative loading systems', *Australian Journal of Applied Science*, **9**, 337–348 (1958).

18. E. Traum and W. Zalewski, 'Conceptual rather than "exact" structural design', *Civil Engineering Magazine*, July (1968).
19. P. Dayaratnam and S. Patnaik, 'Feasibility of fully stressed design', *AIAA J.*, 7, April, 773–774 (1969).
20. S. Patniak and P. Dayaratnam, 'Behavior and design of pin connected structures', *Int. J. Num. Methods Engrg.*, 2, 577–595 (1970).
21. L. D. Hofmeister and L. D. Felton, 'Prestressing in structural synthesis', *AIAA J.*, 8, February, 363–364 (1970).
22. H. W. Kuhn and A. W. Tucker, 'Nonlinear programming', in *Proceedings of 2nd Berkeley Symposium on Mathematical Statistics and Probability*, University of California Press, 1951, pp. 481–490.
23. R. Razani, 'The behavior of the fully-stressed design of structures and its relationship to minimum weight design', *AIAA J.*, 3, December, 2262–2268 (1965).
24. G. Pope, 'The design of optimum structures of specified basis configuration', *Int. J. Mech. Sci.*, 10, 251–265 (1968).
25. W. Lansing, W. Dwyer, R. Emerton and E. Ranalli, 'Application of fully stressed design procedures to wing and empennage structures', *J. Aircraft*, 8, September, 683–688 (1971).
26. W. Dwyer, J. Rosenbaum, M. Shulman and H. Pardo, 'Fully stressed design of airframe redundant structures', in *Proceedings of the 2nd Conference on Matrix Methods in Structural Mechanics, Dayton, Ohio, October 15–17, 1968*.
27. J. L. Tocher and R. N. Karnes, 'The impact of automated structural optimization on actual design', AIAA Paper 71-361, in *Proceedings of AIAA/ASME 12th Structures Conference, Anaheim, Calif., April 1971*.

*Chapter 4*

# Optimality-criterion-based Algorithms

*R. A. Gellatly and L. Berke*

## 4.1 Introduction

As the preceding chapter has shown, fully stressed designs are not necessarily minimum-weight designs for indeterminate structures under multiple-load conditions. There is no explicit reference to weight as a merit function in the usual methods for the determination of a fully stressed design. In certain cases, using the procedure of Reference 1, a fully stressed design can be found for which optimality with respect to weight can be verified. In other cases, a fully stressed design may be heavier than the least-weight design, and, in still other cases, no fully stressed design can be found.

In general, operational experience with fully stressed design methods does indicate, in the vast majority of problems not selected for their pathological behaviour, that the resultant design—for stress limits only—is indeed either the optimum or close to it. It is also relatively easy to select problems for which the fully stressed design is remote from the optimum, but this is usually achieved by selecting unrealistic and disparate allowable stresses for members which provide parallel load paths in a redundant structure.

Using a simple stress-ratio redesign method, convergence on the fully stressed design will occur in one step for a statically determinate structure and will require a finite number of iterations for an indeterminate structure. The number of iterations required for convergence is another paradoxical problem that is apparently controlled by characteristics that are poorly understood at present. One factor which is relatively clear is that the number of iterations is not, in general, linked closely to the size and number of variables in the problem. It is this last fact that makes the use of fully stressed design so attractive. Each redesign step is simple, and, in many cases, the convergence is extremely rapid. The fact that a true least-weight structure is not generated becomes of less importance when compared with the ease with which the improved design is reached.

Clearly, the basis for this type of optimization, which is only applicable to strength considerations, is the *a priori* specification of a set of conditions to be satisfied by the optimum design. That these criteria may not be rigorously applicable at the optimum is of little consequence. An *a priori* approach

33

is in direct contrast to a search algorithm in mathematical programming, where the determination of an optimum is predicated on a strictly *post hoc* basis; for example, the search is terminated only when further improvement in merit is impossible.

It might be concluded, from a consideration of the known characteristics of the fully stressed design procedure, that the rapid convergence and the relative independence of the number of iterations from problem size was largely attributable to the *a priori* specification of optimality criteria. The stress-ratio method attempts to satisfy these arbitrary criteria and hence converges directly on the prescribed point in the design space without the need for exploration of convexity or any other use of the constraints.

With this recognition, a new avenue of approach to structural optimization, through the use of 'optimality criteria', becomes possible. This requires the definition of additional criteria for each response phenomena, coupled with some procedure which permits the simultaneous consideration of multiple criteria.

The approach presented here stems from work by Barnett and Hermann[2] on the optimum design of determinate structures for a single displacement constraint. Berke[3] reevaluated this work and proposed its application to an indeterminate structure with a generalized constraint. From this starting point, the optimality criteria have been further generalized for indeterminate structures under a multiplicity of displacement constraints. A full exposition of these ideas is contained in Reference 4. Illustrative examples, additional to those described here, are presented in Reference 5.

## 4.2   Theoretical basis

A structure is represented by an assembly of discrete elements for which, initially, one stress is sufficient to describe the response behaviour of each element. For more complex elements, involving multiple stress components, the same overall derivation will apply, but it will need to be modified slightly to reflect the additional stress components. This will be demonstrated later. Single stress elements may be axial-force members or shear panels. For an axial member, the appropriate design variable is selected to be its cross-sectional area, whereas, for a plate or panel, the thickness is the chosen variable.

The starting point for the development of this approach is the consideration of a structure under the action of a single loading system $\mathbf{P}$, which may consist of many load components. Application of external loading results in internal forces $\mathbf{F}$ defined as the product of stress and cross-sectional area for axial members and stress and thickness for plates. The internal forces in each member $\mathbf{F}^P$ arise as the result of the application of the loading system $\mathbf{P}$. Associated with the loading system will be displacements $\boldsymbol{\delta}$ at all nodes of the structure. The value of a generalized displacement, which, as a special case, is a single nodal displacement, can be determined through use of a generalized-virtual-load method.

A virtual loading system $\mathbf{Q}$, corresponding to the generalized displacement system in node point and direction, is applied to the structure consisting of $n$ elements. The virtual work $\Pi$ arising between the virtual system $\mathbf{Q}$ and the real strains and displacements is given by

$$\Pi = \sum_{i=1}^{n} \frac{F_i^P F_i^Q L_i}{A_i E_i} = \mathbf{Q}^T \boldsymbol{\delta} \qquad (4.1)$$

where $F_i^Q$ are the forces due to the virtual system and $A_i$, $E_i$ and $L_i$ are as defined in prior chapters, except that the geometric dimension $L_i$ is given by

$$L_i = \frac{V_i}{A_i} \qquad (4.2)$$

where $V_i$ is the volume of the element. For plate elements, $L_i$ is the surface area, and, for shear panels, the appropriate value for $E_i$ is the shear modulus. For the special case of a single displacement, the corresponding virtual loading system $\mathbf{Q}$ consists of a single unit load, and hence $\Pi$, as calculated by equation (4.1), is the value of displacement.

It is assumed now that the variables in the structure may be divided into two groups—active and passive members. Active members are those whose cross-sectional dimension $A_i$ may be varied to achieve an optimized displacement-limited design, whereas passive members will remain unchanged. The basis for the allocation of the individual member into the two groupings will be discussed later, and has no influence on the derivation.

With this division, equation (4.1) can be rewritten as

$$\Pi = \sum_{i=1}^{m} \frac{F_i^P F_i^Q L_i}{A_i E_i} + \sum_{i=m+1}^{n} \frac{F_i^P F_i^Q L_i}{A_i E_i} \qquad (4.3)$$

where the first group contains $m$ passive members.

Since, in subsequent manipulation, the passive members are unaltered, the first term of equation (4.3) is replaced by $\Pi_0$, i.e.

$$\Pi = \Pi_0 + \sum_{i=m+1}^{n} \frac{F_i^P F_i^Q L_i}{A_i E_i}. \qquad (4.4)$$

The weight of the structure $W$ is given by

$$W = \sum_{i=1}^{n} V_i \rho_i = \sum_{i=1}^{n} A_i L_i \rho_i \qquad (4.5)$$

where $\rho_i$ is material density. Introducing the active- and passive-member grouping, this becomes

$$W = W_0 + \sum_{i=m+1}^{n} A_i L_i \rho_i \tag{4.6}$$

where $W_0$ is the weight of the passive members.

A constraint on the allowable values of the virtual work of the load system **Q** can be prescribed as $\Pi^*$, given by

$$\Pi^* = \sum Q_j \delta_j = \mathbf{Q}^T \boldsymbol{\delta} \tag{4.7}$$

where the summation is taken over all terms of the generalized system.

The minimum-weight structure for which the generalized displacement will have the specified value $\Pi^*$ can now be determined by finding the stationary value of $W$ subject to the equality constraint

$$\Pi = \Pi^*. \tag{4.8}$$

A Lagrange-multiplier approach is ideally suited to optimization problems involving equality constraints. Using a Lagrange multiplier $\lambda$, the problem can be expressed as the minimization of the function

$$\overline{W} = \left( W_0 + \sum_{i=m+1}^{n} A_i L_i \rho_i \right) + \lambda \left( \sum_{i=m+1}^{n} \frac{F_i^P F_i^Q L_i}{A_i E_i} + \Pi_0 - \Pi^* \right). \tag{4.9}$$

For a minimum

$$\frac{\partial \overline{W}}{\partial A_j} = 0 = L_j \rho_j - \lambda \frac{F_j^P F_j^Q L_j}{A_j^2 E_j} + \lambda \sum_{k=m+1}^{n} \left( \frac{\partial F_k^P}{\partial A_j} F_k^Q + \frac{\partial F_k^Q}{\partial A_j} F_k^P \right) \frac{L_k}{A_k E_k}. \tag{4.10}$$

It has been shown (Reference 6) that the terms in the summation are actually identically zero. For determinate structures, the $F_k$ are independent of the $A_j$, and the derivatives are therefore zero. For indeterminate structures, the derivative terms form a self-equilibrating internal load system. The virtual work of this system, represented by the summation, is zero, by the principle of virtual displacements. Equation (4.10) therefore becomes

$$0 = L_j \rho_j - \lambda \frac{F_j^P F_j^Q L_j}{A_j^2 E_j}. \tag{4.11}$$

From equation (4.11):

$$A_j = \sqrt{\lambda} \sqrt{\left( \frac{F_j^P F_j^Q}{E_j \rho_j} \right)}. \tag{4.12}$$

Substituting equation (4.12) into equation (4.4), and noting equation (4.8):

$$\Pi^* = \Pi_0 + \frac{1}{\sqrt{\lambda}} \sum_{i=m+1}^{n} L_i \sqrt{\left(\frac{F_i^P F_i^Q \rho_i}{E_i}\right)}. \tag{4.13}$$

Rewriting equation (4.13):

$$\sqrt{\lambda} = \frac{1}{(\Pi^* - \Pi_0)} \sum_{i=m+1}^{n} L_i \sqrt{\left(\frac{F_i^P F_i^Q \rho_i}{E_i}\right)}. \tag{4.14}$$

Finally, substituting equation (4.14) into equation (4.12):

$$A_j = \frac{1}{(\Pi^* - \Pi_0)} \sum_{i=m+1}^{n} L_i \sqrt{\left(\frac{F_i^P F_i^Q \rho_i}{E_i}\right)} \sqrt{\left(\frac{F_j^P F_j^Q}{E_j \rho_j}\right)}. \tag{4.15}$$

Equation (4.15) now represents the criterion which the variables $A_j$ must satisfy at the optimum for the desired displacement-constrained minimum-weight design.

Equation (4.15) can be rearranged as:

$$A_j = \sqrt{\left(\frac{F_j^P F_j^Q}{E_j \rho_j}\right)} \times \frac{1}{(\Pi^* - \Pi_0)} \sum_{i=m+1}^{n} L_i \sqrt{\left(\frac{F_i^P F_i^Q \rho_i}{E_i}\right)}. \tag{4.16}$$

In finite-element analyses, the more usual form of output from a computer is stress rather than element force. Introducing elemental stresses, equation (4.16) becomes:

$$A_j = A_j \sqrt{\left(\frac{\sigma_j^P \sigma_j^Q}{E_j \rho_j}\right)} \times \frac{1}{(\Pi^* - \Pi_0)} \sum_{i=m+1}^{n} (A_i L_i \rho_i) \sqrt{\left(\frac{\sigma_i^P \sigma_i^Q}{E_i \rho_i}\right)}. \tag{4.17}$$

For a statically determinate structure, there is no redistribution of internal forces with variation in member size. Hence equation (4.16) will generate the active member sizes directly for a minimum-weight structure with its critical displacement equal to the specified value $\Pi^*$. If any members have been selected as passive, their fixed values will be reflected in the computation of the term $\Pi_0$. This procedure is then analogous to a constrained minimization with one main constraint (the displacement limit) and a number of side constraints.

It has long been known that, for a statically determinate structure subject to strength (stress) limitations only, the minimum-weight design has all members either fully stressed or at their minimum allowable size. If such a member satisfies the inequality displacement limitation, no redesign is possible or required. If the displacement constraint is violated, certain (active) members must be increased in size. With reference to equation (4.3), if the product $F_i^P F_i^Q$ is negative, an increase in the member size (in the denominator) will decrease the negative component of $\Pi$ arising from this member and hence further increase the total value of $\Pi$. On the other hand, reduction of the area is not possible, since it will already have its minimum size to satisfy the stress or fabrication requirements. Thus, in effect, no variation in such a member size can be permitted, and this provides one criterion for the selection of a passive member in a statically determinate structure, i.e. $F_i^P F_i^Q$ is negative. As a logical corollary to this definition, it also follows that any member for which an external condition would be governing (i.e. requiring a larger size than that demanded by the displacement constraint condition) must also be treated as a passive member.

As indicated previously, equation (4.17) is also valid for statically indeterminate structures. It can be used as a recursion expression for the iterative determination of $A_j$, in a manner exactly analogous to the use of the simple stress-ratio expression for determining a fully stressed design. Equation (4.17) can then be written

$$A_j^{k+1} = A_j^k \sqrt{\left(\frac{\sigma_j^P \sigma_j^Q}{E_j \rho_j}\right)^k} \times \frac{1}{(\Pi^* - \Pi_0)} \sum_{i=m+1}^{n} (A_i L_i \rho_i)^k \sqrt{\left(\frac{\sigma_i^P \sigma_i^Q}{E_i \rho_i}\right)^k}. \qquad (4.18)$$

where the superscripts $k$ and $k+1$ refer to iterations. From some arbitrary starting point, iterative application of equation (4.18) will then lead to convergence, in a finite number of steps, to the statically indeterminate structure of minimum weight with its critical displacement equal to $\Pi^*$. As in the case of the statically determinate structure, certain members may be treated as being passive and non-participatory in this process based upon external criteria.

### 4.3  Discussion of procedure

Before a practical use can be made of the recursion expression [equation (4.18)] for the design of minimum-weight structures with displacement constraints, a number of points must be clarified.

First, in using a virtual unit-load method, the internal force or stress system corresponding to the virtual external load $\mathbf{Q}$ need only be statically equivalent to $\mathbf{Q}$. For a determinate structure, only one such system exists, but, in an indeterminate structure, $\mathbf{F}^Q$ is not unique, and a multiple choice exists for the statically equivalent system. Since the selection of any one particular statically equivalent system means that $\mathbf{F}^Q$ will be arbitrarily set to zero in some (redundant) members, this is tantamount to an arbitrary selection of active and passive members for redesign. This is also equivalent to introducing side constraints on the design variables, which, in turn, can result only in an increase in the attainable minimum weight. Therefore, when some members are to be excluded from design changes, the correct approach is to use the actual distribution of $\mathbf{F}^Q$, which arises from the application of the virtual load $\mathbf{Q}$ to the real structure. This permits all members, potentially, to participate in redesign. It also has the virtue of not requiring a separate analysis of a degenerate structure under the virtual load, but simply of including $\mathbf{Q}$ as an additional loading case in the finite-element analysis. The division between active and passive members is of greater complexity, as are the associated problems of introducing multiple loads along with displacement and other types of constraints.

The recursion expression [equation (4.18)] will cause convergence at a vertex in the design space formed by the intersection of the main constraint (displacement condition) and the side constraints (passive-member sizes). It is clear that the source and nature of these side constraints is irrelevant to the iteration procedure. This, then, provides the mechanism for the inclusion of additional criteria, whereby other forms of constraint conditions (e.g. multiple displacement constraints, stress limits, fabrication considerations, etc.) can be permitted to specify certain member sizes, while other members are sized by the primary constraint. The effect of this type of combination of criteria will be to force the design simultaneously towards all potential individual minima, resulting in convergence in a vertex formed by many active constraints. It is, therefore, necessary to establish rules or criteria for the selection of active and passive members consistent with this philosophy.

One criterion for the selection of passive members, similar to the condition used for statically determinate structures, can be observed immediately from an examination of equation (4.18). If $F_i^P F_i^Q$ is negative for a member, its inclusion as an active member would require its square root to be found, introducing imaginary numbers which are computationally and physically unacceptable. Thus one criterion for a passive member is:

$$\sigma_i^P \sigma_i^Q < 0. \tag{4.19}$$

Similarly to the corollary for the statically determinate case, if the size of a member is defined dominantly by other than a displacement constraint, that member must be treated as passive in the redesign to satisfy displacement constraints. That is, if the size required for a member to satisfy a stress or fabricational constraint is greater than the size required to satisfy a displacement constraint, that member should be considered passive when using equation (4.18).

The extension of these principles to multiple loads and displacement constraints can be accomplished in a manner similar to that used in determining the appropriate stress ratio for a fully stressed design. In the stress-ratio method, the ratio of the actual to the allowable stress is computed for an element for each loading case, and the largest ratio selected for the redesign. With displacement constraints, the largest area generated using equation (4.18) is taken from the combined set resulting from all loads and displacement constraints. Thus, if four loading cases are specified and three displacements are constrained, a total of twelve possible areas for one member are generated by equation (4.18) and the largest is selected as the dominant value. A second iteration of equation (4.18) is then performed, in which members designed in the first iteration by a specific load/displacement case are treated as passive, except when considering that particular case.

The procedure for redesign with displacement constraints, which has evolved from an intensive exploration of the numerical aspects of the problem, automatically partitions members into active and passive categories for a given redesign phase. Tests on the sign of $\sigma_i^P \sigma_i^Q$ are incorporated, as are checks to determine whether stress and fabrication limits dominate for any member and to ensure that that member is treated as being passive.

An additional feature of note is the method of treating negative displacement limits. Frequently displacement limits are expressed in the form $-\Pi^* \leqslant \Pi \leqslant +\Pi^*$. A negative virtual load system $-Q$ is required to satisfy the negative limit. This has the effect of reversing all the signs on the virtual stresses and thereby interchanging the active and passive members selected by the sign of $\sigma_i^P \sigma_i^Q$. The effect of this is to cause some elements to participate in the satisfaction of the positive displacement, while the others (passive in the first case) then participate in the redesign for the satisfaction of the negative displacement. The fact that these two constraint conditions are apparently in direct conflict does not introduce, in practice, any complications. Other conditions of stress limits, minimum sizes and multiple loads and displacements intrude and eliminate one of the contradictory constraints.

In summary, the redesign process is as follows. As a first stage, a stress-ratio redesign is effected for each member with due regard to multiple loads and minimum member sizes. For every limited displacement specified, a unit load vector is generated and the corresponding stresses $\sigma_i^Q$ are computed for all members, together with the actual stresses $\sigma_i^P$. A matrix of the products $\sigma_i^P \sigma_i^Q$ is computed for all combinations of loads and limited displacements (all **P** and **Q**) for each member.

Using equation (4.18), new values of $A_j$ are computed for every member. The partitioning into active and passive members at this state is controlled wholly by the test on the sign of $\sigma_i^P \sigma_i^Q$. In the presence of both negative and positive limits on the restrained displacements, new areas will be found for each member.

The largest area is selected for each member from all the loading/displacement combinations, and a record is made of the particular combination which controlled the design of each area. The areas so generated are compared with those based on stresses or minimum sizes, and the larger values selected for each member.

Again, using equation (4.18), a new evaluation of areas is made for each load/displacement case, but now the passive members for each case consist not only of those elements for which $\sigma_i^P \sigma_i^Q$ is negative, but also of those which were, at the end of the first iteration, critically designed by stress limits, minimum size or by a load/displacement combination other than that being currently considered. At the end of this redesign, a comparison is again made between the stress-designed and the displacement-designed sizes. This cycle is repeated, up to a maximum of three times, until no transfer occurs between the members of the group designed by stresses and those designed by displacements. The resultant design is then reanalysed and scaled until it becomes critical. This process can be repeated as many times as is necessary until a minimum-weight design is achieved.

This derivation has been presented for a structure composed of elements for which only one stress was defined. In the case of membrane plates, three components of stress $\sigma_x$, $\sigma_y$ and $\tau_{xy}$ are generated for each element. For the calculation of stress limits, a single reference stress is calculated based on the von Mises criterion.

$$\sigma_{\text{ref}} = (\sigma_x^2 + \sigma_y^2 - \sigma_x\sigma_y + 3\tau_{xy}^2)^{\frac{1}{2}}. \tag{4.20}$$

A limiting value is then specified for $\sigma_{\text{ref}}$. For displacement constraints, equations (4.17) and (4.18) are modified slightly to reflect the additional terms. Wherever the term $\sigma_i^P \sigma_i^Q / E_i \rho_i$ appears, it is replaced by the term

$$\frac{1}{\rho_i}\left[\frac{(\sigma_x^P - v\sigma_y^P)\sigma_x^Q + (\sigma_y^P - v\sigma_x^P)\sigma_y^Q}{E} + \frac{\tau_{xy}^P\tau_{xy}^Q}{G}\right]_i. \tag{4.21}$$

A similar modification is introduced in the calculation of $\Pi_0$. The test on the sign of $\sigma_i^P \sigma_i^Q$ is replaced by a test on expression (4.21).

## 4.4 Computer program

A large-scale computer program, based on a finite-element analysis procedure, has been developed to demonstrate the practicality of the optimality-criteria approach to the design of least-weight structures.

The element library is the basis of a finite-element program. At present, four simple elements are provided in the optimization program. These are the axial-force member, the triangular-membrane plate, the quadrilateral shear panel and the special half-web shear panel used for symmetric-wing analysis (Figure 4.1). The triangle is a constant-stress element and generates the three components of stress $\sigma_x$, $\sigma_y$ and $\tau_{xy}$ at its centroid. The use of the half-web element is necessary when performing the analysis and design of thin-symmetric-wing-type structures. In these structures the inclusion of all degrees of freedom in both upper and lower halves usually leads to severe conditioning problems associated with the numerical disparities of out-of-plane and in-plane stiffness characteristics. The solution to this difficulty lies in considering symmetric (in-plane) and antisymmetric (out-of-plane) behaviour separately and only treating half the structure. This is no problem for elements lying completely in either half, but a special half-web element is

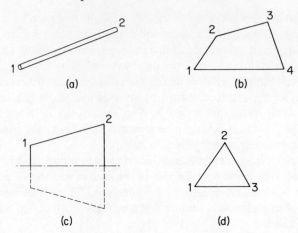

**Figure 4.1**   Finite elements: (a) axial-force member, (b) quadrilateral shear panel, (c) half-web element and (d) triangular-membrane plate

required to cross the centreline. This trapezoidal shear web is defined by only two node points.

The program uses a simple stress-ratio method to account for strength constraints and the recursion relationship [equation (4.18)] to account for displacement constraints. Fabricational constraints are included, and the program performs the partitioning between active and passive members automatically, according to the criteria defined previously. The results presented here illustrate typical results obtained using the program operationally and indicate some of the convergence characteristics of the iterative process.

## 4.5   Examples of application

Three example problems are presented as illustrations of the efficiency of this approach to structural optimization. The first example is a standard structure frequently used for checking out optimization programs, whereas the other examples are based on the use of the optimization program as a practical tool in the early developmental phases of two project designs.

### 4.5.1   Twenty-five-bar transmission tower

This structure, shown in Figure 4.2, has been used as a demonstration problem on a number of previous occasions[7,8]. The tower, representative of a structure carrying transmission lines, has twenty-five axial-force members. There are two loading cases [Table 4.1(a)]. In spite of the directional nature of these loads, the structure is required to be double symmetric about the

Table 4.1   Twenty-five-bar transmission tower

(a) Applied loading systems

| Load condition | Node | Direction | | |
|---|---|---|---|---|
| | | $x$ | $y$ | $z$ |
| 1 | 1 | 1000 | 10,000 | −5000 |
| | 2 | 0 | 10,000 | −5000 |
| | 3 | 500 | 0 | 0 |
| | 6 | 500 | 0 | 0 |
| 2 | 5 | 0 | 20,000 | −5000 |
| | 6 | 0 | −20,000 | −5000 |

(b) Allowable compressive stresses

| Member | Allowable stress (lbf/in$^2$) | Member | Allowable stress (lbf/in$^2$) |
|---|---|---|---|
| 1 | −35,092 | 12, 13 | −35,092 |
| 2, 3, 4, 5 | −11,590 | 14, 15, 16, 17 | −6759 |
| 6, 7, 8, 9 | −17,305 | 18, 19, 20, 21 | −6969 |
| 10, 11 | −35,092 | 22, 23, 24, 25 | −11,082 |

(c) Iteration history

| Iteration | 1 | 2 | 3 | 4 | 5 | 6 | 7 |
|---|---|---|---|---|---|---|---|
| Weight (lb) | 734·38 | 555·72 | 549·08 | 546·54 | 545·92 | 545·45 | 545·36 |

(d) Final design

| Member | Area (in$^2$) | Member | Area (in$^2$) |
|---|---|---|---|
| 1 | 0·01 | 12, 13 | 0·01 |
| 2, 3, 4, 5 | 2·0069 | 14, 15, 16, 17 | 0·6876 |
| 6, 7, 8, 9 | 2·9631 | 18, 19, 20, 21 | 1·6784 |
| 10, 11 | 0·01 | 22, 23, 24, 25 | 2·6638 |

$x$ and $y$ axes. The maximum displacements of the upper nodes 1 and 2 are $\pm 0.35$ in in the $x$ and $y$ directions. The bar members are all designed for a 35,000 lbf/in$^2$ tensile stress and a compressive stress based on buckling allowables for thin-wall circular tubes [Table 4.1(b)]. The minimum member size is 0·01in$^2$.

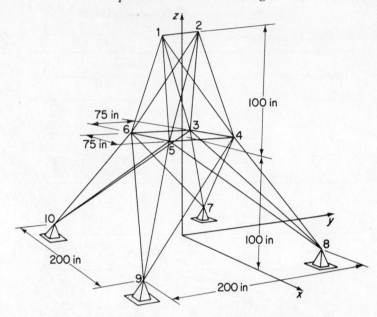

**Figure 4.2**    Twenty-five-bar transmission tower

In some optimizations, symmetry is achieved through the use of additional loading cases. This is not necessary with the present program, which has a built-in capability for linking member sizes to preserve symmetry.

Slight differences have been found between the final designs generated in previous studies. These differences are principally associated with variations in tolerances and specified allowable stresses. An average weight is 550 lb, and, using a numerical search method, over one hundred analyses may be required.

Using the current program, convergence occurred in seven iterations to a weight of 545·4 lb. Details of the history are given in Figure 4.3.

### 4.5.2    Wing centre section

Several fixed obstructions at the wing/fuselage junction of the aircraft prevented the use of main wing spars that were continuous through the fuselage.

The restrictions imposed by the overall aerodynamic envelope of the aircraft resulted in the design indicated in Figure 4.4. There are three main spars in the outer wing that are attached to a longitudinal torsion box that transfers wing loads to the three transverse beams crossing the fuselage.

This structure was first optimized purely on a strength basis, using a simple axial-force-member/shear-panel idealization. The model of the

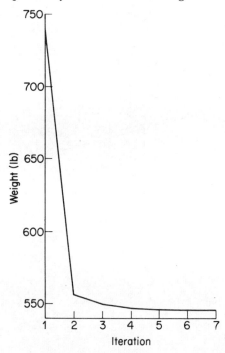

**Figure 4.3**  Iteration history for twenty-five-bar transmission tower

symmetric half structure has seventy degrees of freedom and eighty-nine elements with the boundary conditions shown in Figure 4.4.

On a strength basis, the structural weight was 56·6 lb, but the vertical deflections at the tip of the stub wing (nodes 27–32) lay between 1·1 and 1·45 in. Since these displacements were considered to be excessive from an aeroelastic standpoint, the structure was reoptimized with the vertical displacements at nodes 28, 29 and 32 limited to 0·2 in. This type of restraint may be unrealistic with regard to torsional response, but this situation is discussed in Section 4.5.3. With the imposed displacement limitation, the weight of the structure increased to 216·0 lb, with significant redistribution of the principal load-carrying material compared with the stress-limited design.

Thus, for an increase in stiffness of approximately 7·0, the weight has increased only by a factor of 3·83, indicating the value of the optimization procedure in allocating structural material in an efficient manner.

From the iteration history (Figure 4.5), it can be seen that the stress-limited design required twenty iterations for convergence, although it was within 1 per cent of the final weight by iteration 12, whereas the introduction of displacement limits yielded the least-weight design at iteration 7.

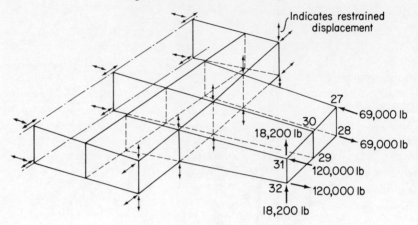

**Figure 4.4**   Wing centre section

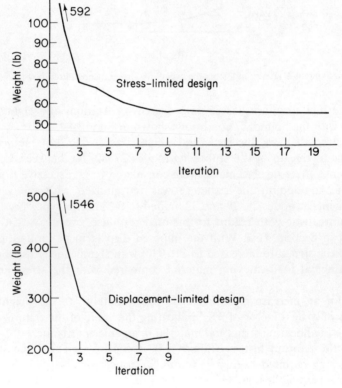

**Figure 4.5**   Iteration histories for wing centre section

### 4.5.3  *Wing carry-through box*

In a further illustration of the capabilities of this type of optimization, the centre section of a wing structure is again used as the subject model. The wing shown in Figure 4.6 is a relatively simple unswept four-cell box structure. The structure has 210 degrees of freedom and over two hundred elements are used. The rectangular cover panels are each made up from two triangular-membrane elements as indicated, and a program-linking option is used to ensure that each pair of constituent triangles has the same thickness. Two loading cases are imposed at the tip section, one consisting principally of bending with limited torsion, and the other consisting of torsion with a small bending component.

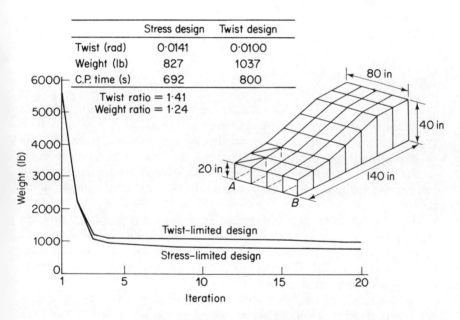

|  | Stress design | Twist design |
|---|---|---|
| Twist (rad) | 0·0141 | 0·0100 |
| Weight (lb) | 827 | 1037 |
| C.P. time (s) | 692 | 800 |

Twist ratio = 1·41
Weight ratio = 1·24

**Figure 4.6**  Wing carry-through box

The box was optimized with stress and fabricational constraints only, yielding a weight of 827 lb.

In this structure the amount of twist was of some importance to the design efficiency. For the stress-limited design, the effective twist of the tip cross-section, as measured by the difference in vertical displacement of nodes at A and B in Figure 4.6, was 0·0141 rad. Instead of applying a single unit-load system to restrain a single displacement, a generalized loading system consisting of a vertically upward unit load at A and a downward unit load at B were imposed using a special program option. The value of this generalized displacement was then restrained to have a value corresponding to a twist of 0·0100 rad, and the structure was reoptimized. The new weight was 1037 lb, and the results are shown in Figure 4.6.

The ratio of the torsional stiffness of the two designs is 1·41:1, compared with a weight ratio of 1·24:1, indicating that the optimized design is a considerable improvement over that achievable by simple scaling of the stress-limited design.

The computational costs for the two structures are of interest in this case. In both approaches the iteration process was allowed to run for twenty cycles. For the simplest stress-ratio redesign method, a total central-processing-unit (c.p.u.) time of 692 s was required. The introduction of the traditionally more complex displacement limits only raised this time to 800 s, a modest increase of only 14 per cent for a large gain in capability.

### 4.6 Closure

The approach to the weight minimization of fixed-geometry structures with displacement and stress constraints based on the use of optimality criteria appears to offer considerable advantages over mathematical-programming-based methods. For comparable problems, the present method reaches a similar or better design in considerably fewer iterations than most numerical search methods and involves considerably reduced computational costs.

The results presented for the specific examples are very encouraging and indicate that some, if not all, of the difficulties encountered in large-scale optimization problems can be eliminated through this type of approach. Certain problems may still remain to be resolved, particularly with regard to convergence characteristics.

The extension of these procedures beyond stress- and displacement-constraint conditions can be accomplished within the general framework discussed in this chapter. The criteria for the automatic partitioning between active and passive members is generally valid for other types of response phenomena. It is then only necessary to derive the criteria that must be satisfied (exactly or approximately) at the optimum design in a form analogous to equation (4.15), containing provision for both variant and invariant member sizes. In such an extension, it must be recognized that, while the convergence can be expected to be more rapid than with a non-linear-programming approach, there will still be a considerable level of computational complexity owing to the fundamental nature of the analysis processes associated with the relevant response phenomena.

**References**

1. R. Razani, 'The behavior of the fully-stressed design of structures and its relationship to minimum weight design', *AIAA J.*, **3**, No. 12, December, 2262–2268 (1965).
2. R. L. Barnett and P. C. Herrmann, 'High performance structures', NASA CR-1038, May 1968.
3. L. Berke, 'An efficient approach to the minimum weight design of deflection limited structures', USAF AFFDL-TM-70-4-FDTR, May 1970.
4. R. A. Gellatly and L. Berke, 'Optimal structural design', USAF AFFDL-TR-70-165, February 1971.
5. R. A. Gellatly, L. Berke and W. Gibson, 'The use of optimality criteria in automated structural design', in *Proceedings of the 3rd Air Force Conference on Matrix Methods in Structural Mechanics, October 1971.*
6. L. Berke, 'Convergence behavior of optimality criteria based iterative procedures', USAF AFFDL-TM-72-1-FBR, January 1972.
7. P. V. Marcal and R. A. Gellatly, 'Application of the created response surface technique to structural optimization', in *Proceedings of the 2nd Air Force Conference on Matrix Methods in Structural Mechanics, October 1968*, AFFDL-TR-68-150.
8. V. B. Venkayya, 'Design of optimum structures', *Computers and Structures*, **1**, No. 1/2, 265–309 (1971).

*Chapter 5*

# Mathematical-programming Methods— A Critical Review

*R. Fletcher*

## 5.1 Introduction

The aim of this chapter is to explain and critically review what are, at present, considered to be the important concepts in the design of algorithms for optimization. Its scope is the minimization of differentiable merit functions, subject to constraints on the design variables in which the constraint functions are differentiable. The amount of detail which can be given here is limited, but many of the individual topics are amplified considerably in later chapters, and for those topics which are not the reader is referred to mathematical-programming texts such as References 1–4 and to the review by Powell[5]. It should also be noted that *dynamic programming, integer programming* and *stochastic programming* are excluded from the present review, although aspects of each of these are discussed in the context of optimum design in Chapters 10–13.

This chapter is a development of an earlier review[6]. Interesting research and new algorithms have since been reported, however, and have been incorporated in this chapter. The coverage of concepts in linearly and non-linearly constrained optimization has also been extended. In explaining and reviewing concepts in the design of optimization algorithms, it has been necessary to be somewhat selective to avoid obscuring the fundamental ideas. However, there are a number of well tried methods which, together with some promising new ideas, represent key optimization concepts. Section 5.2 discusses unconstrained optimization in this way, with similar coverage being given to constrained optimization problems in Sections 5.3–5.6. I have been particularly concerned to present implemented algorithms for which a reasonable amount of experimental evidence is available. I am equally concerned about the reliability of algorithms, and whether there is proof or good reason to think that convergence of the algorithm will occur, and whether it will be at an ultimately rapid rate. Also presented in this chapter, in Section 5.7, is a decision tree which is intended to enable selection of a suitable algorithm for the solution of any one problem.

Before proceeding into the above subject matter, certain basic aspects of the optimization problems in question will be defined. Thus the purpose is to

$$\text{minimize } W(\mathbf{x}), \qquad \mathbf{x} = \begin{pmatrix} x_1 \\ x_2 \\ \vdots \\ x_n \end{pmatrix} \tag{5.1a}$$

which is an *objective function* of $n$ variables,

$$\text{subject to} \qquad \begin{aligned} g_1(\mathbf{x}) &= 0 \quad i \in E \\ g_i(\mathbf{x}) &\geqslant 0 \quad i \in I \end{aligned} \tag{5.1b}$$
$$\tag{5.1c}$$

representing *equality constraints* $(E)$ and *inequality constraints* $(I)$ on the variables. $\nabla$ will refer to the gradient operator and $\mathbf{G}$ to the matrix $[\partial^2 W/(\partial x_i \partial x_j)]$.

A special case of equation (5.1), and, in particular, a subproblem which is solved in many methods, is the problem of finding the minimum of a function of one variable. As a subproblem this usually involves finding the minimum point along a line in the space of the variables and is called the *linear-search* subproblem. In particular, if $\mathbf{x}^k$ is a point and $\mathbf{d}^k$ is a direction vector, then

$$\mathbf{x}^{k+1} = \mathbf{x}^k + \alpha^k \mathbf{d}^k \tag{5.2}$$

where the superscript $k$ refers to the $k$th iterate and $\alpha^k$ is chosen so as to minimize $W(\mathbf{x}^k + \alpha \mathbf{d}^k)$ with respect to $\alpha$. It will be assumed that algorithms for this step are available, a good survey being given in Reference 1.

In general, the problem posed in equation (5.2) cannot be solved in a finite number of operations, and a linear-search subproblem often attempts to find only a crude estimate of the minimum, such as that obtained by interpolating a quadratic or cubic function to known information. In fact, attempting to locate the exact minimum is not desirable, because it may not exist at all, even in a well behaved problem. More recently (see Reference 5), methods have been developed in which a linear search is carried out with the lesser aim of finding a point at which a sufficiently large decrease in the objective function is obtained. Methods using partial linear searches of this type are promising, although further experience is desirable.

Although the problem described by expressions (5.1) is posed in terms of the *global minimum* of the function, the methods given here can only be relied on to find a point $\xi$ which is a local minimum, i.e. around $\xi$ there is a region in which $W(\xi)$ is the minimum feasible point. In fact, the methods could terminate at a *stationary point* (for unconstrained problems), which is a point at which $\nabla W = 0$. However, the fact that $\xi$ is found as the limit of a sequence of points $\mathbf{x}^k$ for which $W^k$ [i.e. $W(\mathbf{x}^k)$] decreases monotonically is usually sufficient to ensure that $\xi$ is at least a local minimum. The equivalent concept for constrained problems is the *Kuhn–Tucker point,* which is a point at which there is no feasible descent direction. If all the constraints are equalities, a necessary condition for $\xi$ to be a Kuhn–Tucker point is that Lagrange multipliers $\lambda$ exist such that

$$\nabla W = N\lambda \qquad (5.3)$$

where the columns of $N$ are the vectors $\nabla g_i$, $i \in E$, and $\nabla W$ and $N$ are evaluated at $\xi$. The generalization of this to inequality problems will be considered further in Section 5.3 [see also Chapter 2, equation (2.19)].

## 5.2 Unconstrained optimization

### 5.2.1 Ad-hoc methods

Much of the early research into unconstrained optimization was concerned with the development of *ad-hoc* methods based on the study of small problems. Although in general these methods have been superseded, some of these methods are still useful, particularly when the function values may be subject to large errors, in which case more sophisticated methods are unreliable. A typical example is the *simplex method* (see Nelder and Mead[7]), which has also been extended to constrained problems by Paviani and Himmelblau[8]. The *DSC method* (see Reference 1) is probably the most efficient of the *ad-hoc* methods (Fletcher[9]). It is one of the few methods that deal satisfactorily with problems of more than three or four variables.

When first derivatives are available, the choice of $-\nabla W$ as the search direction gives the well known *method of steepest descent.* Although the method can be proved to be convergent, the rate of convergence has been shown to be unacceptable both theoretically and in practice. However, the ideas used in proving the convergence of this method do have implications for the more sophisticated methods to be described.

### 5.2.2 Conjugate-direction methods

It has become clear that methods are only generally successful if they attempt in some way to retain information about the second derivatives of the objective function. When the matrix of second derivatives cannot be evaluated, an indirect way of accumulating information about second derivatives is to set out a finite sequence of operations that finds the minimum of a quadratic function, yet which can be applied iteratively to general problems.

An efficient way of doing this is to use the concept of *conjugate directions,* a set of vectors $d^i$ being defined as mutually conjugate with regard to a positive definite symmetric matrix $G$ if

$$(d^i)^T G d^j = 0 \qquad (5.4)$$

for all $i, j$, $i \neq j$. In applications to quadratic functions **G** is the Hessian of the quadratic function, in which case two important theorems can be stated.

*Theorem 1* If a linear search is carried out successively along a set of mutually conjugate directions, the function is minimized in the space spanned by those directions (see Reference 10).

*Theorem 2* If **y** and **z** are the minimum points in two parallel subspaces, the direction **z** − **y** is conjugate to any vector which lies in either subspace. Both theorems are illustrated by Figure 5.1 for a two-dimensional quadratic function, the directions **z** − **y** and **y** − **x** being conjugate.

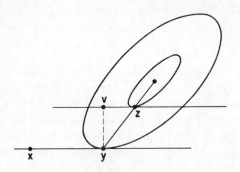

Reproduced by permission from R. Fletcher, 'Methods for the solution of optimization problems', *Study No. 5*, *Computer-aided Engineering*, Solid Mechanics Division, University of Waterloo, Ontario

**Figure 5.1**   Conjugacy properties for a quadratic equation

Smith[11] proposed a method for minimization without using derivatives based on these theorems, in which a set of conjugate directions is built up. It is assumed that a set of independent directions $\mathbf{p}^1, \mathbf{p}^2, \ldots, \mathbf{p}^n$ and an initial approximation $\mathbf{x}^1$ are given. The first conjugate direction $\mathbf{d}^1 = \mathbf{p}^1$, and $\mathbf{x}^2$ is obtained by a linear search [equation (5.2)] along $\mathbf{d}^1$. For $i = 2, 3, \ldots, n$, each point $\mathbf{x}^i$ is the minimum point in the subspace spanned by $\mathbf{p}^1, \ldots, \mathbf{p}^{i-1}$ and $\mathbf{x}^{i+1}$ is defined as follows.

(i) Displace $\mathbf{x}^i$ to the point $\mathbf{v} = \mathbf{x}^i + h^i \mathbf{p}^i$ where $h^i \neq 0$ is a constant supplied by the user. $\mathbf{v}$ can be considered as an arbitrary point in a parallel subspace (Figure 5.1).

(ii) Minimize from $\mathbf{v}$ successively along $\mathbf{s}^1, \mathbf{s}^2, \ldots, \mathbf{s}^{i-1}$ giving the minimum point $\mathbf{z}^i$ in the parallel subspace, by virtue of Theorem 1.

(iii) Define $\mathbf{s}^i = \mathbf{z}^i - \mathbf{x}^i$ which is conjugate to $\mathbf{s}^1, \mathbf{s}^2, \ldots, \mathbf{s}^{i-1}$ by Theorem 2, and choose $\mathbf{x}^{i+1}$ as the minimum point along $\mathbf{s}^i$ through $\mathbf{x}^i$.

The point $\mathbf{x}^{n+1}$ determined by this process is the vector that minimizes the quadratic function. The conjugacy properties are obtained entirely by the use of linear searches. Smith suggested that, for general functions, $x^{n+1}$ should become $x^1$ and that the complete cycle is repeated iteratively. However, except for problems of about four or fewer variables, the method has not worked well in practice. It seems that this is because the directions with high suffices ($p^n$ say) are used very little in comparison to directions such as $p^1$, so that many linear searches along $p^1$ are, essentially, wasted because they are at a considerable distance from the solution in the direction $p^n$.

Powell[12] has suggested a modification whereby all directions are used equally, replacing the displacement of step (i) by a linear search along all the directions $p^i, p^{i+1}, \ldots, p^n$, which would not otherwise be introduced at the $i$th stage (except for $p^i$). A particularly nice way of arranging the storage enables the approximate conjugate directions to be carried forward as independent directions for the next cycle. In practice it was found that, when the complete cycle of Powell's method was repeated iteratively, the would-be conjugate directions tended to be linearly dependent. A more sophisticated iteration was therefore suggested, in which the introduction of a new conjugate direction in step (iii) might be bypassed in order to preserve linear independence. Powell noted when using this program that as $n$ was increased, step (iii) was bypassed to an increasing degree, with an adverse effect on rates of convergence (see also Reference 9). More recently, Zangwill[13] has proposed a modification, but a paper by Rhead[14] suggests the modification to be inferior.

When methods that use derivatives are considered, ideas that relate conjugacy properties to the direction of steepest descent are attractive. If $\delta^k = x^{k+1} - x^k = \alpha^k d^k$ is a typical correction, and $\gamma^k = \nabla W^{k+1} - \nabla W^k$ is the corresponding change in gradients,

$$\gamma^k = G\delta^k \qquad (5.5)$$

holds for a quadratic function. Thus, for any vector $d^j$ to be conjugate to $d^k$ ($k \neq j$), the condition represented by equation (5.4) is equivalent to the orthogonality condition

$$(\gamma^k)^T d^j = 0. \qquad (5.6)$$

In the *method of conjugate gradients* the first direction of search is the steepest-descent vector; so $d^1 = -\nabla W^1$ and, for $k = 2, 3, \ldots, n$, $d^k$ is chosen as the component of $-\nabla W^k$ that is orthogonal to $\gamma^1, \gamma^2, \ldots, \gamma^{k-1}$. Fletcher and Reeves[10] show that the theory of the method of conjugate gradients for linear equations implies that this is achieved when $d^k$ is determined by the simple relationship

$$d^k = -\nabla W^k + d^{k-1}[(\nabla W^k)^T \nabla W^k]/[(\nabla W^{k-1})^T \nabla W^{k-1}]. \qquad (5.7)$$

Although for general functions it would be possible to apply equation (5.7) for all $k \geqslant 1$, it has been found better to restart the cycle every $n$ iterations by setting $\mathbf{d}^{pn+1} = -\nabla W^{pn+1}$ for all $p$. Experience with the method suggests that, while it is not greatly inferior to the class of quasi-Newton methods described in Section 5.2.3 in the number of function evaluations required to solve realistic problems, it is an order of magnitude better as regards the number of housekeeping operations or the amount of computer storage required. If either of these considerations is significant when using a quasi-Newton method, the method of conjugate gradients is to be preferred.

Another conjugate-direction method using linear searches is Zoutendijk's *method of feasible directions*[15] in which the conditions (5.6) are applied explicitly at each iteration to obtain a search direction that is the solution of a linear program. The performance of the method on general functions is thought to be similar to that of the method of conjugate gradients, except that the housekeeping and storage requirements are only comparable with those for quasi-Newton methods.

### 5.2.3   Newton-like methods

When second derivatives can be computed, the correction $\boldsymbol{\delta}$ for which $\mathbf{x} + \boldsymbol{\delta}$ minimizes a quadratic function can be written directly as

$$\boldsymbol{\delta} = -\mathbf{G}^{-1}\mathbf{g} \tag{5.8}$$

and the use of this equation iteratively, as

$$\mathbf{x}^{k+1} = \mathbf{x}^k - (\mathbf{G}^k)^{-1}\nabla W^k \tag{5.9}$$

is known as *Newton's method*. In these circumstances, equation (5.9) can also be derived by truncating the Taylor series for $\nabla W(\mathbf{x})$ expanded about $\mathbf{x}^k$. When the iteration converges, it does so at a rate which is second order, a very desirable feature. At each iteration $\boldsymbol{\delta}$ is determined as the solution of a set of linear equations, which, although time consuming when compared with the housekeeping requirements of the quasi-Newton methods to be described, is as efficient as can be expected when the information in the second-derivative matrix is used to its full extent.

Some modifications have to be made to Newton's method if it is not to be divergent remote from the minimum, and the usual philosophy is to attempt to ensure that $W(\mathbf{x})$ is reduced on each iteration. This can be done by modifying $\boldsymbol{\delta}$ so that the condition $\boldsymbol{\delta}^T\nabla W < 0$ is achieved, so that the existence of a correction $\alpha\boldsymbol{\delta}$ for which $W(\mathbf{x}^k + \alpha\boldsymbol{\delta}) < W(\mathbf{x}^k)$ is guaranteed. Note that, although these modifications prevent divergence, they do not ensure convergence to a stationary point. Some implementations of Newton's method have introduced a linear search to give the optimum reduction in the direction $-(\mathbf{G}^k)^{-1}\nabla W^k$. However this is unnecessarily sophisticated, because it involves evaluating the function a number of times on each iteration, whereas the ultimate second-order rate of convergence requires only one evaluation per iteration. A partial linear search in which equation (5.9) is used if it causes a sufficient decrease in $W$ is preferable. Powell[5] reviews ways of carrying out both these modifications.

When only first derivatives are available, Newton's method cannot be implemented directly, but the spirit of this type of method is contained in the class of *quasi-Newton methods*. In these methods a symmetric matrix $\mathbf{H}^k$ is used as an approximation to $\mathbf{G}^{-1}$. After each iteration $\mathbf{H}^{k+1}$ is calculated by correcting $\mathbf{H}^k$ so as to take into account the information about second derivatives that has become available on that iteration. In particular, if $W$ were quadratic, equation (5.5) would be exact, and, by analogy with equation (5.5), the matrix $\mathbf{H}^{k+1}$ is constructed so as to satisfy

$$\mathbf{H}^{k+1}\gamma^k = \delta^k. \tag{5.10}$$

It has been usual to use $\mathbf{H}^k$ to determine a direction of search

$$\mathbf{d}^k = -\mathbf{H}^k\mathbf{g}^k \tag{5.11}$$

by analogy with equation (5.8), and each iteration uses the linear search [equation (5.2)] to determine $\mathbf{x}^{k+1}$.

The best known implementation of a quasi-Newton method has been the *DFP. method*[16,17], in which the correction formula

$$\mathbf{H}^{k+1} = \mathbf{H}^k + \frac{\delta^k(\delta^k)^{\mathrm{T}}}{(\delta^k)^{\mathrm{T}}\gamma^k} - \frac{\mathbf{H}^k\gamma^k(\mathbf{H}^k\gamma^k)^{\mathrm{T}}}{(\gamma^k)^{\mathrm{T}}\mathbf{H}^k\gamma^k} \tag{5.12}$$

is used. $\mathbf{H}^1$ is chosen to be an arbitrary positive definite matrix ($\mathbf{H}^1 = \mathbf{I}$ is usually used), in which case all subsequent $\mathbf{H}^k$ can be shown to be positive definite. The DFP method terminates when the function is quadratic. In practice, the quasi-Newton methods seem to be more efficient than conjugate-gradient methods, in terms of the number of evaluations of the objective function and its derivatives that are required. The housekeeping requirements are about $3n^2$ multiplications per iteration, which is usually tolerable. An Algol procedure is given in References 18 and 19.

More recently, a whole family of such formulae has been discovered (Broyden[20]), and Powell[5] reviews the situation. However, a very recent and startling discovery due to Dixon[21] is that, when they are used in a quasi-Newton method with exact linear searches, all formulae in the family give exactly the same points $\{\mathbf{x}^k\}$ when applied to *any* function. Although the line search must, in practice, be approximate, this result suggests that there will be little to choose between the methods using different formulae from the family if reasonably accurate line searches are carried out. Some results of Fletcher[22], however, indicate that quasi-Newton methods are most efficient if the line search is carried out as crudely as possible, although there have been indications that such an approach may not be as robust on ill conditioned problems such as those that arise when using penalty functions. A generalization of the DFP method to solve unconstrained optimization problems without derivatives has been made by Stewart[23], and this method is probably better than Powell's[12] conjugate-direction method. An Algol program due to Lill[24] is available.

### 5.2.4   Restricted-step methods

The occasional difficulties that arise with Newton or quasi-Newton methods seem to be problems of reliability when the approximation is remote from the minimum. In these circumstances, the prediction given by the Newton-like method is irrelevant, because the Taylor series from which it is derived is dominated by the higher terms; so that the truncation is not realistic. A much more reasonable way of using information about second derivatives of the objective function (assuming that the first derivatives are given) is to assume that the quadratic function $q^k(\mathbf{x})$ which is determined by this information is only valid approximation to $W(\mathbf{x})$ in a restricted region $\Omega$ about $\mathbf{x}^k$. [$q^k(\mathbf{x})$ is formally defined by

$$q^k(\mathbf{x}^k) = W^k, \qquad \nabla q^k(\mathbf{x}^k) = \nabla W^k, \qquad \nabla^2 q^k(\mathbf{x}) = \mathbf{\Gamma}^k \qquad (5.13\text{a,b,c})$$

where $\mathbf{\Gamma}^k$ is either the Hessian matrix $\mathbf{G}^k$ or an approximation to it.] Thus the new point $\mathbf{x}^{k+1}$ is taken as the vector that solves the problem

$$\min_{\mathbf{x}} q^k(\mathbf{x}), \qquad \mathbf{x} \in \Omega. \qquad (5.14\text{a,b})$$

The first application of, essentially, these ideas was to sums-of-squares problems as described in Section 5.2.5, but a similar application to problems involving general functions was made by Goldfeld, Quandt and Trotter[25]. However, the more recent *method of hypercubes* due to Fletcher[26] will be described, as it follows more closely the ideas that are illustrated in this section. An iteration of the method of hypercubes consists of the following steps:

(a) Given $\mathbf{x}^k$, $W^k$, $\nabla W^k$, $\mathbf{\Gamma}^k$ and a parameter $h$.
(b) Define the quadratic function $q^k(\mathbf{x})$ by equations (5.13).
(c) Let $\mathbf{y}$ solve expression (5.14) where $\Omega = \{\mathbf{x}, \|\mathbf{x} - \mathbf{x}^k\|_\infty \leqslant h\}$ is a hypercube with centre $\mathbf{x}^k$ and side $2h$. (Note that $\|\mathbf{a}\|_\infty = \max |a_j|$.)
(d) If $W(\mathbf{y})$ and $q^k(\mathbf{y})$ do not agree satisfactorily, halve $h$ and go to step (c).                                                                              (5.15)
(e) Let $\mathbf{x}^{k+1} = \mathbf{y}$.
(f) Consider whether to double $h$ for the next iteration if $\Omega$ is unduly restrictive, or whether to leave $h$ unchanged.

A typical iteration of the method is illustrated in Figure 5.2. $\mathbf{y}$ is the point predicted by equation (5.15c) on the square $\Omega$. However $W(\mathbf{y})$ and $q^k(\mathbf{y})$ do not agree satisfactorily; so $h$ is halved. The predicted minimum on the smaller square $\Omega'$ is then examined, and is found to be satisfactory, and

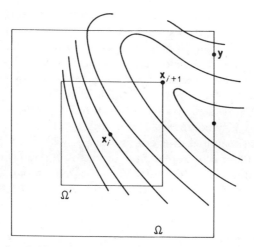

**Figure 5.2**   An iteration of the method of hypercubes

hence becomes $x^{k+1}$. The precise implementation of the method of hypercubes depends on the particular way of choosing $\Gamma^k$ and on the tests that
are implied by steps (5.15d) and (5.15f), and the reader is referred to Reference
26. The method generalizes readily to solve linearly constrained problems,
because these additional constraints are just added to the quadratic-
programming problem in step (5.15c). The method of Reference 25 differs
mainly in that $\Omega$ is defined by $\{x, \|x - x^k\|_2 \leqslant h\}$ $[\|a\|_2 = (a^Ta)^{\frac{1}{2}}]$, with the
result that the eigensolution of $G$ is required to solve the equivalent of
equation (5.15c). The details of the way in which $h$ is changed are also different, and the method does not generalize conveniently to solve linearly
constrained problems.

Restricted-step methods have proved very satisfactory from a practical
viewpoint, and perform as well as, if not better than, the equivalent quasi-
Newton method, even when the latter work well. The main advantage of
the restricted-step methods is that, if the size of $\Omega$ is varied in a sensible
way, convergence of the methods is guaranteed, so that there are no doubts
about reliability. The disadvantage of the methods is that the housekeeping
operations are more expensive than with the Newton-like methods. This
difference is not very great when second derivatives are used, so that restricted-step methods are to be preferred because of their more satisfactory
convergence. However, when only first derivatives can be calculated, the
discrepancy is more marked, and, until this feature is rectified, restricted-step
methods will have only a limited use in these circumstances.

There is another class of methods which will be referred to as *trajectory methods*, and which have a similar philosophy to the restricted-step methods. Consider the parameter $h$ in step (5.15c) on any one iteration, and let $h$ be varied from zero upwards, in which case the solution $\mathbf{y}$ describes a trajectory starting at $\mathbf{x}^k$, with an initial direction determined by the steepest-descent vector $\nabla W^k$. The trajectory terminates at the point $\mathbf{y} = \mathbf{x}^k - (\mathbf{\Gamma}^k)^{-1}\nabla W^k$, which would be predicted by a Newton-like method. In the trajectory methods, an easily computable trajectory is set up between the same two points so that its initial direction is along $-\nabla W^k$. Step (5.15c) can now be replaced by the much more simple problem of minimizing $q^k(\mathbf{x})$ along the trajectory, subject to $\|\mathbf{y} - \mathbf{x}^k\|_2 \leqslant h$. The conditions on the trajectory enable convergence to be proved, and they do not affect the possibility of being able to achieve ultimate superlinear convergence. Such methods seem to be a satisfactory way of implementing the spirit of a restricted-step method without the penalties as regards housekeeping. A method of this type, which only requires $O(n^2)$ housekeeping operations per iteration, is suggested by Powell[27]. Unfortunately, trajectory methods are not conveniently generalized to solve linearly constrained problems.

### 5.2.5 Sums-of-squares problems

Problems will now be considered in which the objective function $W(\mathbf{x})$ is a sum of squares of $p$ other functions $f_j(\mathbf{x})$, $j = 1, 2, \ldots, p$, where $p \geqslant n$, i.e.

$$W(\mathbf{x}) = \sum_{j=1}^{p} f_j(\mathbf{x})^2 = \mathbf{f}^\mathsf{T}\mathbf{f}. \qquad (5.16)$$

Derivatives of $f_j$ are expressed by the Jacobian matrix $\mathbf{J}$ ($J_{jk} = \partial f_j/\partial x_k$), from which the gradient of $W$ is given by

$$\nabla W = 2\mathbf{J}^\mathsf{T}\mathbf{f}. \qquad (5.17)$$

The Hessian of $W$ can be obtained by a further differentiation as

$$\mathbf{G} = 2\mathbf{J}^\mathsf{T}\mathbf{J} + 2\sum_{j} f_j\mathbf{K}^j \qquad (5.18)$$

where $\mathbf{K}^j$ is the matrix of second derivatives of $f_j$. Sums-of-squares problems can be treated without using the fact that the function has this special form. However, it is usually possible to make use of the fact that $W$ is a sum of squares to reduce considerably the time required to solve the problem. The usual way of doing this is to make the approximation

$$\mathbf{G} \approx 2\mathbf{J}^\mathsf{T}\mathbf{J} \qquad (5.19)$$

in equation (5.18), assuming either that the $f_j$ are all small at the solution, or that they are nearly linear so that the $\mathbf{K}^j$ are small. Thus many of the advantages of second derivatives can be obtained while only evaluating first derivatives. There may, of course, be cases in which approximation (5.19) is not adequate, and in which the ultimate rate of convergence obtained by using approximation (5.19) is poor. In these cases a general minimization method should be used.

When approximation (5.19) is used in Newton's method [equation (5.9)], the result is the *generalized-least-squares* or *Gauss–Newton Method*. Because $2(J^k)^T J^k$ does not tend to $G(\xi)$ as $x^k \to \xi$, a second-order rate of convergence is not obtained; so the case for carrying out a linear search along the direction $-[2(J^k)^T J^k]^{-1} \nabla W^k$ is stronger, and this feature is usually found in computer programs. Powell[28] has described a version of this method that does not require derivatives. Although these methods usually work, they can fail to converge, and this is most likely to happen when $p$ is equal to, or not much greater than, $n$ (see below). Approximation (5.19) can be used very satisfactorily in restricted-step methods, in which case convergence can be guaranteed. Marquardt[29] shows that, if the subproblem

$$2[(J^k)^T J^k + \lambda I]\delta = -\nabla W^k \qquad (5.20)$$

is solved, $x^k + \delta$ solves the subproblem posed by expression (5.14), with $\Gamma^k = 2(J^k)^T J^k$ in the definition of $q^k(x)$, and where $\Omega$ is defined by the hypersphere $\{x\}$, $\|x - x^k\|_2 \leqslant h$, where $h = \|\delta\|_2$. Thus a restricted-step method can be based on equation (5.20), differing from method (5.15) in that $\lambda$ and not $h$ is used to control $\Omega$. A suitable algorithm is given in Reference 30, and, although the use of $\lambda$ rather than $h$ is less convenient in some ways, it does permit formulation in terms of the readily solvable subproblem of equation (5.20). The method only generalizes to linearly constrained problems at the expense of solving a quadratic-programming subproblem, and a least-squares version of the method of hypercubes would probably be preferable in these circumstances.

If $\lambda$ were allowed to vary between $+\infty$ and zero on any iteration, the point $y = x^k + \delta$, where $\delta$ is obtained from the solution of equation (5.20), traces out a trajectory which lies between $x^k$ and the Gauss–Newton prediction $x^k - [2(J^k)^T J^k]^{-1} \nabla W^k$, and whose initial direction is along the direction $-\nabla W^k$. Thus trajectory methods spring to mind, and the possibility of replacing the Marquardt trajectory by a more readily computable one has been considered in the *spiral method* of Jones[31] and in a method by Powell[32] which can be used when $J$ is not directly available.

Finally, the important special case of equation (5.16) in which $p = n$ will be considered, where the sum-of-squares function has been constructed so as to solve exactly the set of non-linear equations

$$f(x) = 0. \qquad (15.21)$$

In this case the Gauss–Newton method reduces to calculating the correction

$\delta$ from

$$\delta = -\mathbf{J}^{-1}\mathbf{f} \tag{5.22}$$

and is identical to the *Newton–Raphson method* for solving non-linear equations.

Powell[33] gives an example that shows that the above method can fail to converge for a well behaved problem, even when linear searches are used, and this demonstrates that the Newton–Raphson method [equation (5.22)] is not adequate as it stands, and that even the Gauss–Newton method should be used with care. I have known a number of systems of non-linear equations for which the Newton–Raphson method failed to converge, so the phenomenon would appear not to be just a theoretical quirk. The problems are best treated, therefore, by restricted-step or trajectory methods. Powell[33] gives a trajectory method which does not require derivatives. A version of Powell's method for the solution of non-linear equations in which $\mathbf{J}$ is sparse is given by Reid[34].

There are also a number of Newton–Raphson-type methods for solving non-linear constrained problems by treating the Lagrangian equation in the form of equation (5.20). I regard these methods with suspicion for the same reasons.

### 5.3  Linear and quadratic programming

A fundamental optimization problem is *linear programming* (l.p.), in which the objective and constraint functions are all linear. In this case the solution must lie at a point at which $n$ independent constraints intersect, and this is usually the *vertex* of a polyhedral feasible region formed by the inequality constraints. As there are only a finite number of such vertices, the solution technique can be finite.

A well known technique for l.p. is the *revised simplex method*, and this will be explained briefly. It is usual to transform the problem represented by equation (5.1) so that it fits into a traditional formulation for linear-programming problems, but, for insight, this explanation will be in terms of the untransformed problem. Figure 5.3 shows a three-variable problem with four inequality constraints (1–4) that are represented by the planes that form the boundary of the feasible region. The method works systematically from one vertex to another, so it will be assumed that a feasible vertex such as $\mathbf{x}$ in Figure 5.3 has been found. If $\mathbf{N}$ is the square matrix whose columns are the normal vectors of the constraints which intersect at $\mathbf{x}$, the rows of $\mathbf{N}^{-1}$ give the directions of the *edges* which leave $\mathbf{x}$, and which will be denoted by $\mathbf{d}^i$, $i = 1, 2, \ldots, n$. If scalars $\lambda_i$ given by the components of the gradient along the $\mathbf{d}^i$ ($\lambda_i = \nabla W^T \mathbf{d}^i$) are calculated, $\mathbf{d}^i$ will be 'downhill' (i.e. in a direction along which $F$ decreases) if $\lambda_i$ is negative.

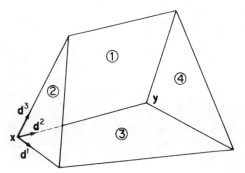

**Figure 5.3**   An iteration of the revised simplex method

Consequently, if all the $\lambda_i$ are non-negative, $\mathbf{x}$ is a solution, because there is no feasible downhill direction at $\mathbf{x}$. Otherwise the revised simplex method chooses the direction $\mathbf{d}^i$ corresponding to the most negative $\lambda_i$ and looks along $\mathbf{d}^i$ for the first new vertex. In Figure 5.3, if $\mathbf{d}^2$ is chosen as the edge along which to move, $\mathbf{y}$ is the new vertex. The procedure can now be repeated anew. The $\lambda_i$ are known variously as Kuhn–Tucker multipliers, generalized Lagrange multipliers, reduced costs, and so on.

An important feature of the revised simplex method is that matrices $\mathbf{N}$ corresponding to neighbouring vertices such as $\mathbf{x}$ and $\mathbf{y}$ differ in only one column, with the result that $\mathbf{N}^{-1}$ corresponding to $\mathbf{y}$ can be calculated from $\mathbf{N}^{-1}$ corresponding to $\mathbf{x}$ in only about $n^2$ multiplications. This calculation is called the *pivot* step. Although the number of possible vertices increases rapidly with $n$, the way in which the vertices are chosen means that, in practice, only a small proportion of all possible vertices are visited. In fact, $2n$ is a typical number of iterations to solve a problem, showing that a small multiple of $n^3$ computer operations are required by this method. Computer subroutines for the revised simplex method are given in Reference 35. The application of linear programming to structural analysis and the design problems to which it applies directly is taken up in the next chapter.

Ideas similar to those used in linear programming are at the root of most generalizations of unconstrained minimization methods to treat constrained problems. These generalizations also borrow from the theory of Lagrange multipliers. It has already been noted that equation (5.3) must hold ($\mathbf{N}$ being relevant only to constraints that are satisfied by $\mathbf{x}$ as equations) and that the parameters $\lambda_i$ must be non-negative if $i \in I$. These are usually known as the *Kuhn–Tucker conditions*.

The Kuhn–Tucker conditions are closely involved with the problem in which the objective function is quadratic and the constraints linear; this is known as *quadratic programming*. Although quadratic programming is more complicated than linear programming, the problem can still be solved in a finite number of operations. There are a number of algorithms for this problem, but two efficient general algorithms for which subroutines are available are given by Beale[36] and Fletcher[37]. A typical figure for the number of operations in solving quadratic-programming problems is $\sim 4n^3$ multiplications (see Reference 26).

A special case of this problem that often arises in data fitting is *least-squares quadratic programming* in which the objective function is a sum of squares of linear functions. It is particularly important in such problems that attention be paid by the method to ill conditioning, and a suitable algorithm has been suggested by Bartels, Golub and Saunders[38].

### 5.4 General minimization subject to linear constraints

Methods for solving linearly constrained problems are extensions of those for unconstrained optimization, and rely heavily on concepts that have been described in earlier sections, for instance linear searches, conjugacy, approximations to inverse Hessians and so on. However, there are a number of additional ideas that are required to handle the linear constraints, and these are, essentially, extensions of the ideas for solving linear- or quadratic-programming problems.

First of all, it is important to know how to deal with linear *equality* constraints before considering the inequality problem; so it will be assumed that the function $W$ is being minimized subject merely to the $k \leqslant n$ constraints

$$N^T x = d \qquad (5.23)$$

where $N$ is an $n \times k$ matrix such that its columns $\nabla g_1, \nabla g_2, \ldots, \nabla g_k$ are the normal vectors of the equality constraints. Note that these are constant vectors for linear constraints.

One way of handling the problem is to make an affine transformation of variables from $x$ to a new set $y \in E^n$, so that the constraints represented by equation (5.23) can be satisfied by setting $y_1 = y_2 = \cdots = y_k = 0$. The function will then be minimized with respect to the variables $y_{k+1}, \ldots, y_n$ so that the problem has been transformed into the unconstrained minimization of a function of $n - k$ variables. To do this $y$ is divided into two partitions, $y_1 \in E^k$ and $y_2 \in E^{n-k}$, where

$$y_1 = N^T x - d \qquad (5.24)$$

and

$$y_2 = M^T x - e. \qquad (5.25)$$

$M$ is an $n \times (n - k)$ matrix and is chosen in different ways by different methods. Equations (5.24) and (5.25) can be written collectively as

$$y = [N:M]^T x - \binom{d}{e} \qquad (5.26)$$

where $[N:M]$ is a non-singular $n \times n$ matrix. Clearly, given any value of the variables $y_2$, and because $y_1 = 0$, and given the matrix $[N:M]^{-1}$, the variables $x$, and hence $W(x)$, can be calculated. Thus the problem has been reduced to the unconstrained minimization of a function $W(y_2)$ of $n - k$ variables.

The methods which use this analysis are the Wolfe *reduced-gradient method*[39], methods which eliminate variables and, indirectly, Rosen's *gradient-projection method*[40]. The relation between these methods is described by Fletcher[41], who also shows how, using the matrix $[N:M]^{-1}$, the gradient vector $\nabla_{y_2} W$ can be calculated from $\nabla_x W$, enabling gradient methods to be used to minimize $W(y_2)$. It is also shown how estimates of the Lagrange multipliers $\lambda \in E^k$ can be obtained using the same information. Of course, methods such as those described in References 39 and 40 originally used gradient information in steepest-descent-like methods, whereas the importance of incorporating second-order information is now appreciated. However, there is no difficulty in devising conjugate-gradient versions of the above techniques, or alternatively estimates of inverse Hessian matrices in the $y_2$ variables can be built up, giving quasi-Newton-like methods. An example of this is the *variable-reduction method* of McCormick[42].

An alternative approach is to tackle the constraints represented by equation (5.23) by using a Lagrange-multiplier analysis, and this is outlined in Reference 41. It leads to Newton-like methods in which the direction for the linear search is given by equation (5.11), where $H$ is now a positive semidefinite matrix of rank $n - k$ defined by

$$H = G^{-1} - G^{-1} N (N^T G^{-1} N)^{-1} N^T G^{-1}. \qquad (5.27)$$

If the Hessian $G$ is explicitly available, the algorithm to use is fairly straightforward (see Fletcher[43]). If only first derivatives are available, quasi-Newton methods have to be considered. One method is given by Goldfarb and Lapidus[44], who show that an estimate of $H$ as defined by equation (5.27) can be updated by formula (5.12) at each iteration. An alternative approach is that of Murtagh and Sargent[45,46], who update approximations to construct $H$ as indicated by equation (5.27).

So far, only equality-constraint problems have been considered. The generalization to inequalities is most conveniently made through the *active-set strategy*. It will now be assumed that all the constraints are inequalities, although there are no difficulties in generalizing to mixed problems. In this approach, a feasible point $\mathbf{x}^k$ is obtained, and the indices of the inequality constraints which are satisfied exactly as equalities are collected and become the *active set*. The function is then minimized subject to these equality constraints as above, neglecting the other constraints. If the active set corresponds to the constraints which are active at the solution, this iteration will, hopefully, converge to the solution of the inequality problem. However, in general this will not happen; so there have to be rules for adding or deleting indices to the active set. The rule for adding constraints is obvious, in that, if the point $\mathbf{x}^{k+1}$, which would be obtained by a linear search [equation (5.2)], is no longer feasible, a smaller value of $\alpha^k$ is chosen in equation (5.2), so that $\mathbf{x}^{k+1}$ is the minimum point subject to feasibility being maintained. The index added to the set is then that of the constraint which limited the size of step. The choice of rule for removing a constraint is not so obvious, and many different ones have been tried. They usually involve examining estimates of the Lagrange multipliers $\lambda$, deleting indices of constraints for which $\lambda_i < 0$ under certain circumstances, in accordance with the Kuhn–Tucker conditions of Section 5.3. The difficulty is to avoid the phenomenon of *zigzagging*, in which a constraint repeatedly enters and leaves the active set, and which is a cause of slow convergence. A review of the different rules is given in Reference 41.

An important feature of methods using the active-set strategy is that the various matrices, $[\mathbf{N}:\mathbf{M}]^{-1}$, $\mathbf{H}$, etc., can be modified easily when changes in the active set are made. In fact it is always possible to recalculate the new matrices by making a change of low rank, so that the housekeeping is only of $O(n^2)$ operations per iteration.

Finally, there are some methods involving the solution of linear or quadratic programs at each iteration. These methods are not, in general, as satisfactory as those based on the active-set strategy, either for reasons of housekeeping or poor rates of convergence. A review is given in Reference 41. Of the methods based on solving a linear program, Zoutendijk's *method of feasible directions*[15] is probably the most satisfactory. However, the method of hypercubes described in Section 5.2.4 can be implemented at no extra cost when linear constraints are present. Although the housekeeping requirements are $O(n^3)$ operations per iteration, the reliability of the method may make it preferable in some applications, especially when the function is expensive to calculate.

### 5.5 Direct methods for non-linear constraints

As with methods for linearly constrained problems, one may consider algorithms based on the active-set strategy for solving non-linearly constrained problems. Again, two subproblems can be distinguished, the one of minimizing the function subject to non-linear *equality* constraints, and the other of adding and deleting constraints in an active set. The major difficulty is in the first of the two subproblems, because it is no longer generally possible to find a transformation of variables from $\mathbf{x}$ to $\mathbf{y} \in E^n$ which achieves the desired effect. If a linear transformation is made which is valid locally, the linear search direction will be tangential to the manifold of points which satisfy the equality constraints. Thus, given a feasible point $\mathbf{x}^k$, any $\mathbf{x}^{k+1}$ generated by equation (5.2) will, in general, no longer be feasible. Algorithms circumvent this by projecting $\mathbf{x}^{k+1}$ onto the constraint manifold after each linear search, but this is a process fraught with difficulty. It involves the solution of a set of non-linear equations, a difficult problem in itself, and there are many difficulties that can arise. The situation is illustrated in Figure 5.4.

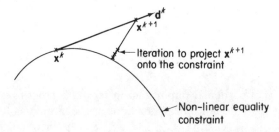

**Figure 5.4**   Direct method for non-linear equality constraints

The problem of when to add or delete constraints in the active set is tackled in a similar way to the approach of Section 5.4, although further difficulties arise because of the non-linearities. Nevertheless, this sort of approach has been tried variously with some success. The *generalized-reduced-gradient method* of Abadie and Carpentier[47], a method due to Murtagh and Sargent[48] and some work generalizing the Goldfarb–Lapidus method by Davies[49] and more recently by Myers[50] all present encouraging results on small test problems. However, I still feel that the methods are not sufficiently reliable to be recommended without reservation.

Other approaches are, again, those in which a linear or quadratic sub-problem is solved at each iteration. This can be done in various ways. One is to linearize the objective and constraint functions by

$$W(\mathbf{x}) \simeq W(\mathbf{x}^k) + (\nabla W^k)^T(\mathbf{x} - \mathbf{x}^k) \qquad (5.28)$$

and solve the linear program obtained. The new point is used to construct further linearizations, which are added to the linear program which is again solved, and so on. This is the *cutting-plane method* (see Wolfe[39], for example); each linearization of a constraint cuts off a part of the feasible region in which the solution cannot lie. The computation is carried out using the dual formation of the linear program, which readily permits the addition of extra equations. However, the method only converges at first order and is fraught with numerical difficulties.

Another method of a similar nature is the *method of approximation programming* of Griffith and Stewart[51], in which a new linearization is made at each point, but a restriction on the size of the change in x is made. The rate of convergence is again linear. These methods are best used only when the solution is known to lie at a vertex. Chapter 7 gives the details of one such method in the context of structural design.

This is equally true in the alternative *decomposition method* (see Wolfe[39], for example), in which the variables are linearized by using

$$\mathbf{x} \simeq \sum_k \alpha^k \mathbf{x}^k, \qquad \alpha^k \geqslant 0, \qquad \sum_k \alpha^k = 1 \qquad (5.29)$$

so that functions can be replaced by the linearizations

$$W(\mathbf{x}) = \sum_k \alpha^k f(\mathbf{x}^k). \qquad (5.30)$$

This method is of particular importance in *separable programming*, in which it is assumed that the non-linear functions are separable as functions of a single variable, i.e. $W(\mathbf{x}) \equiv \sum_j f_j(x_j)$. The functions are then replaced by piecewise linear approximations. However, the conditions under which it will work are somewhat restrictive, and I have had mixed reports of its success.

Finally, it is possible to make a quadratic approximation to the function, linear approximations to the constraints and to solve a quadratic program at each iteration. This is the basis of Wilson's Solver method (see Beale[52]), and is an approach with some promise. However, I have seen no satisfactory numerical evidence of its application to non-linear problems.

### 5.6 Penalty functions

A more reliable, although possible slower, technique for solving a non-linearly constrained problem is to convert it into a sequence of unconstrained minimization problems using a *penalty function*. The most obvious penalty function for solving an equality-constrained problem is

$$\phi(\mathbf{x}, r) = W(\mathbf{x}) + \frac{1}{r} \sum_{i \in E} g_i^2(\mathbf{x}). \qquad (5.31)$$

Typically, values of $r = 1, 0\cdot1, 0\cdot01, 0\cdot001, \ldots$ are chosen, and the vector $\boldsymbol{\xi}_r$ which minimizes $\phi(\mathbf{x}, r)$ is found for each value of $r$ in turn. As $r \to 0$, the sequence $\{\boldsymbol{\xi}_r\}$ converges to the vector that minimizes $W(\mathbf{x})$ subject to $g_i(\mathbf{x}) = 0$ $(i \in E)$, under fairly mild conditions on the functions $W$ and $g$. The effect of solving a simple one-dimensional problem is illustrated in

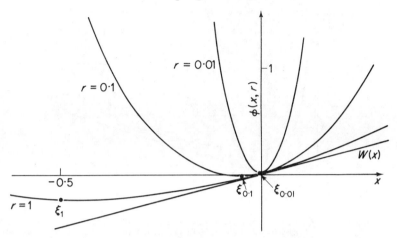

Reproduced by permission from R. Fletcher, 'Methods for the solution of optimization problems', *Study No. 5, Computer-aided Engineering*, Solid Mechanics Division, University of Waterloo, Ontario

**Figure 5.5**   Interior penalty function (equality constraint)

Figure 5.5. A similar penalty function for inequality constraints is

$$\phi(\mathbf{x}, r) = W(\mathbf{x}) + \frac{1}{r} \sum_{i \in I} \{\min[0, g_i(\mathbf{x})]\}^2. \tag{5.32}$$

However this penalty function has the property that every intermediate design $\xi_r$ is infeasible, which is disadvantageous. This type of penalty function is termed an *exterior penalty function*. Another penalty function, which does not suffer from this difficulty, is

$$\phi(\mathbf{x}, r) = W(\mathbf{x}) + r \sum_{i \in I} 1/g_i(\mathbf{x}). \tag{5.33}$$

This function is an example of an *interior penalty function*. Use of this penalty function is illustrated in Figure 5.6. Note particularly that $\phi$ is only defined on the feasible region; so that it is not possible to use an unconstrained minimization routine without modifying it in some way. Other choices of penalty functions are discussed in this book in Chapter 8, which explores in some detail the application of the penalty-function concept.

The chief advantage of using penalty functions lies in their reliability and the ease with which an unconstrained minimization routine can be converted to solve constrained problems. The chief disadvantage is that a considerable amount of time is required to solve the unconstrained minimization subproblem many times. To alleviate this difficulty, various devices for acceleration have been suggested, notably by Fiacco and McCormick[53] in Sumt, which uses a penalty function combining equations (5.31) and (5.33).

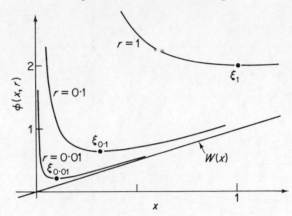

Reproduced by permission from R. Fletcher, 'Methods for the solution of optimization problems', *Study No. 5, Computer-aided Engineering*, Solid Mechanics Division, University of Waterloo, Ontario

**Figure 5.6**   Exterior penalty function

However, my experience is that a saving of, at most, about 50 per cent can be gained by the use of acceleration techniques.

Apart from the obvious disadvantage of having to solve an unconstrained optimization problem many times, another difficulty with penalty functions concerns the severe ill conditioning of the minimization problems in which arbitrarily large curvature can occur and which makes the problem of minimizing the functions more acute (see Figures 5.4 and 5.5). For equality-constrained problems, Powell[54] has suggested a penalty function that avoids the ill conditioning aspect; Fletcher[55] gives a well conditioned penalty function whose minimum is at the solution of the equality-constraint problem and which therefore avoids the sequence of unconstrained optimization problems. Current research shows that Powell's method is probably more reliable if first derivatives or no derivatives are available, whereas Fletcher's method is efficient if second derivatives can be calculated (see Fletcher[56]). For inequality problems, Osborne and Ryan[57] give a generalization of Powell's method, and there is reason to hope for a suitable generalization of Fletcher's method. Murray[58] also gives a modified penalty-function technique, in which variation of the parameter $r$ is carried out in parallel with the optimization, so that, again, the sequence of unconstrained problems is avoided.

### 5.7   Choosing an algorithm

Figure 5.7 presents a decision tree that enables an algorithm to be chosen which is suitable for solving any particular optimization problem. The choice of methods represented is somewhat subjective, and, when viewed from the standpoint of an existing subroutine library, might not be completely appropriate. When a number of satisfactory methods exist, however, the matter has been resolved by the choice of the most readily available method.

The decision tree is necessarily crude, and does not account for any special knowledge that the user might have. This special knowledge should be used to as large an extent as possible in formulating the problem and in choosing an algorithm. For example, linear equality constraints in the formulation can sometimes be used to eliminate variables, and so reduce the dimension of the problem; although this is not always preferable, or even possible, linear programming being an example. On the other hand, I have known cases (see Reference 59) in which sparse non-linear equality constraints have been used satisfactorily to eliminate variables. Another device is that constraints can sometimes be eliminated by making a non-linear transformation of variables. Typically, a variable $x$ subject to a constraint $x > 0$ can be replaced by the variable $y = \log x$. When minimizing with respect to $y$, the constraint $x > 0$ will automatically be imposed. Further possibilities are outlined in Reference 1. Yet again, in certain sums-of-squares problems, it often happens that $f$ is linear in some of the variables, and this can be taken into account when posing the problem. The variables supplied to the optimization routine should be the *non-linear* variables only, because, every time $W$ is calculated, the linear variables can be found by a linear least-squares calculation. This device reduces the number of variables and hence the complexity of the optimization problem. Care is required when evaluating derivatives in the modified formulation. Also, when solving linear-programming problems, it is often possible to solve the *dual* problem much more efficiently.

Users are also expected to pose their problems with care, giving attention to such problems as scaling, and avoiding cancellation, which can be of crucial importance. It is also recommended that any formula for evaluating derivatives should first be checked by differences in the function values. The choice of method is also subject to the local availability of computer subroutines, and to this end all methods for which Algol or Fortran programs are freely available are given in Figure 5.7 with the author's name in *capital letters*.

In closing, some general remarks on the various classes of method will be made. Methods for general minimization without derivatives are a class of methods ripe for improvement. Although Stewart's method[23] seems to be the best, evaluating the gradient vector by differences is not very attractive, and I think it likely that still better methods can be devised. People researching into this problem should bear in mind the desirability of being able to extend a method to deal with linear constraints. The chief disadvantage of

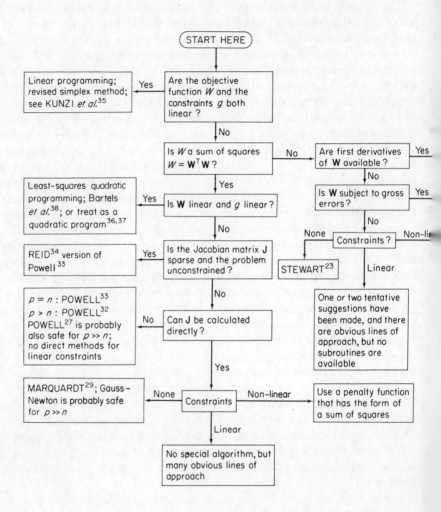

START HERE

Linear programming; revised simplex method; see KUNZI *et al.*[35] — Yes — Are the objective function $W$ and the constraints $g$ both linear?

No

Is $W$ a sum of squares $W = \mathbf{W}^\mathsf{T}\mathbf{W}$? — No — Are first derivatives of $W$ available? — Yes

Yes

No

Least-squares quadratic programming; Bartels *et al.*[38]; or treat as a quadratic program[36,37] — Yes — Is $\mathbf{W}$ linear and $g$ linear? 

Is $\mathbf{W}$ subject to gross errors? — Yes

No

None

REID[34] version of Powell[33] — Yes — Is the Jacobian matrix $\mathbf{J}$ sparse and the problem unconstrained? 

Constraints? — Non-li

STEWART[23]     Linear

No

$p = n$ : POWELL[33]
$p > n$ : POWELL[32]
POWELL[27] is probably also safe for $p \gg n$; no direct methods for linear constraints — No — Can $\mathbf{J}$ be calculated directly?

One or two tentative suggestions have been made, and there are obvious lines of approach, but no subroutines are available

Yes

MARQUARDT[29]; Gauss–Newton is probably safe for $p \gg n$ — None — Constraints — Non-linear — Use a penalty function that has the form of a sum of squares

Linear

No special algorithm, but many obvious lines of approach

**Figure 5.7**  Decision tree; all methods for which Algol or Fortran programs are freely available are given with the author's name in capital letters

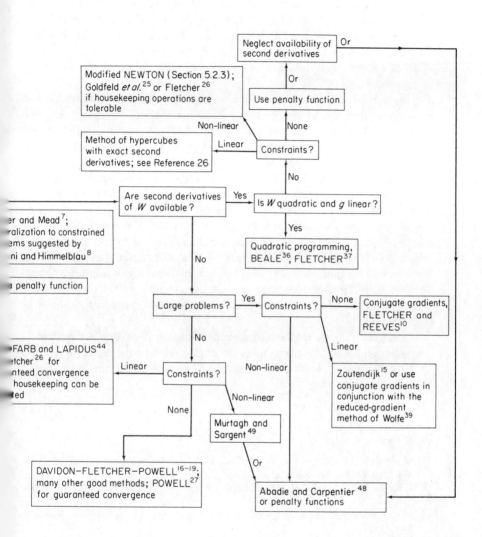

The flow-chart boxes read as follows:

Neglect availability of second derivatives — Or

Modified NEWTON (Section 5.2.3); Goldfeld *et al.*[25] or Fletcher[26] if housekeeping operations are tolerable

Or — Use penalty function

Method of hypercubes with exact second derivatives; see Reference 26

Non-linear / Linear / None

Constraints?

No

Are second derivatives of W available? — Yes — Is W quadratic and g linear?

...r and Mead[7]; ...ralization to constrained ...ems suggested by ...ni and Himmelblau[8]

Yes

Quadratic programming, BEALE[36], FLETCHER[37]

No

...a penalty function

Large problems? — Yes — Constraints? — None — Conjugate gradients, FLETCHER and REEVES[10]

No

Linear

...FARB and LAPIDUS[44] ...tcher[26] for ...nteed convergence ...housekeeping can be ...ed

Linear — Constraints? — Non-linear

Non-linear

Zoutendijk[15] or use conjugate gradients in conjunction with the reduced-gradient method of Wolfe[39]

None

Murtagh and Sargent[49]

DAVIDON–FLETCHER–POWELL[16-19]; many other good methods; POWELL[27] for guaranteed convergence

Or

Abadie and Carpentier[48] or penalty functions

Reproduced by permission from R. Fletcher, 'Methods for the solution of optimization problems', *Study No. 5, Computer-aided Engineering*, Solid Mechanics Division, University of Waterloo, Ontario

these methods is that, as a class, they are the least reliable. There is a lot to be gained by evaluating first derivatives if it is at all possible. First-derivative methods have been much studied, and there are many good methods. Here the choice has been what to leave out, and not what to put in. If $W$ and $\nabla W$ are very readily calculated, it may not be desirable to evaluate second derivatives, and a first-derivative method may be most suitable. Equally, sums-of-squares methods are generally satisfactory, given the suitability of approximation (5.19).

Figure 5.7 discloses that there are very few algorithms for treating linear constraints directly. There are many good ideas about for doing this, and a number of gaps in the tree could be filled in very readily. In particular, provision for simple upper and lower bounds is often very easy to build into an algorithm for unconstrained optimization. Of course, linearly constrained problems can be solved using penalty functions, but, in my experience, this is very inefficient (see Reference 26) and a direct method should be used where possible. There is also a lack of direct methods for tackling non-linearly constrained problems, but this is because of the real difficulty in solving such problems. I have mentioned one or two direct methods for which good results have been quoted, but I do not regard these methods as the ultimate weapon, because they involve solving a set of non-linear equations at each iteration to ensure that the constraints are satisfied. This is a process which can fail to converge, and the reliability of the methods may be questionable.

Finally, of course, mathematical programming is a very active field of research, and new developments are continually being made. It is very likely that Figure 5.7 will have to be revised as new algorithms prove themselves to be preferable.

### References

1. M. J. Box, D. Davies and W. H. Swann, *Non-Linear Optimization Techniques*, ICI Monograph No. 5, Oliver and Boyd, London, 1969.
2. J. Kowalik and M. R. Osborne, *Methods for Unconstrained Optimization Problems*, Elsevier, New York, 1968.
3. R. L. Fox, *Optimization Methods for Engineering Design*, Addison-Wesley, Reading, Mass., 1971.
4. M. Aoki, *Introduction to Optimization Techniques*, Macmillan, New York, 1971.
5. M. J. D. Powell, 'Recent advances in unconstrained optimization', AERE Report TP.430, 1970, presented at the 7th International Mathematical Programming Symposium, The Hague, proceedings to be partly published in *Mathematical Programming*.
6. R. Fletcher, 'Methods for the solution of optimization problems', AERE Report TP.432, 1970, and in *Proceedings of the Symposium on Computer Aided Engineering, University of Waterloo* (Ed. G. L. M. Gladwell), 1971.

7. J. A. Nelder and R. Mead, 'A simplex method for function minimization', *Computer Jour.*, **7**, 308 (1965).

8. D. A. Paviani and D. M. Himmelblau, 'Constrained non-linear optimization by heuristic programming', *Operations Res.*, **17**, 872 (1969).

9. R. Fletcher, 'Function minimization without evaluating derivatives—a review', *Computer Jour.*, **8**, 33 (1965).

10. R. Fletcher and C. M. Reeves, 'Function minimization by conjugate gradients', *Computer Jour.*, **7**, 149 (1964). (Includes an Algol procedure.)

11. C. S. Smith, 'The automatic computation of maximum likelihood estimates', NCB Sc. Dept., Report SC 846/MR/40, 1962.

12. M. J. D. Powell, 'An efficient method of finding the minimum of a function of several variables without calculating derivatives', *Computer Jour.*, **7**, 155 (1964). (AERE Library subroutine VAO4A.)

13. W. I. Zangwill, 'Minimizing a function without calculating derivatives', *Computer Jour.*, **10**, 293 (1967).

14. D. G. Rhead, 'Some experiments on Zangwill's method for unconstrained minimization', University of London Institute of Science working paper ICSI 319, 1971.

15. G. Zoutendijk, *Methods of Feasible Directions*, Elsevier, Amsterdam, 1960.

16. W. C. Davidon, 'Variable metric method for minimization', Argonne Nat. Lab., ANL-5990 Rev., 1959.

17. R. Fletcher and M. J. D. Powell, 'A rapidly convergent descent method for minimization', *Computer Jour.*, **6**, 163 (1963). (An Algol procedure given in References 18 and 19.)

18. M. Wells, 'Algorithm 251, function minimization', *Comm. ACM*, **8**, 169 (1965).

19. R. Fletcher, 'Certification of Algorithm 251', *Comm. ACM*, **9**, 686 (1966).

20. C. G. Broyden, 'Quasi-Newton methods and their application to function minimization', *Maths. Comp.*, **21**, 368 (1967).

21. L. C. W. Dixon, 'Variable metric algorithms: necessary and sufficient conditions for identical behavior on non-quadratic functions', Hatfield Polytechnic Numerical Optimization Centre, Report 26, 1971.

22. R. Fletcher, 'A new approach to variable metric algorithms', *Computer Jour.*, **13**, 317 (1970).

23. G. W. Stewart III, 'A modification of Davidon's minimization method to accept difference approximations to derivatives', *Jour. ACM*, **14**, 72 (1967). (Algol procedure given in Reference 24.)

24. S. A. Lill, 'Algorithm 46: a modified Davidon method for finding the minimum of a function using difference approximations for derivatives', *Computer Jour.*, **13**, 111 (1970).

25. S. M. Goldfeld, R. E. Quandt and H. F. Trotter, 'Maximization by quadratic hill climbing', *Econometrica*, **34**, 541 (1966).

26. R. Fletcher, 'An efficient, globally convergent, algorithm for unconstrained and linearly constrained optimization problems', AERE Report TP.431, presented at the 7th International Mathematical Programming Symposium, The Hague, 1970, proceedings to be partly published in *Mathematical Programming*.

27. M. J. D. Powell, 'A Fortran subroutine for unconstrained minimization, requiring first derivatives of the objective function', UKAEA Research Group Report AERE R.6469, 1970, available from HMSO.

28. M. J. D. Powell, 'A method of minimizing a sum of squares of non-linear functions without calculating derivatives', *Computer Jour.*, **7**, 303 (1965). (AERE Library subroutine VAO2A.)

29. D. W. Marquardt, 'An algorithm for least squares estimation of non-linear parameters', *Jour. SIAM*, **11**, 431 (1963). (Fortran subroutine available, see Reference 30.)

30. R. Fletcher, 'A modified Marquardt subroutine for non-linear least squares', AERE Report R. 6799, 1971.

31. A. Jones, 'Spiral—a new algorithm for non-linear parameter estimation using least squares', *Computer Jour.*, **13**, 301 (1970).

32. M. J. D. Powell, AERE library subroutine VA05A, report in preparation, 1970.

33. M. J. D. Powell, 'A hybrid method for non-linear equations', and 'A Fortran subroutine for solving systems of non-linear algebraic equations', in *Numerical Methods for Non-linear Algebraic Equations* (Ed. P. Rabinowitz), Gordon and Breach, London, 1970.

34. J. K. Reid, 'A Fortran subroutine for the solution of a sparse system of non-linear equations', UKAEA Research Group Report (in preparation), AERE Library subroutine NSO2A, 1970.

35. H. P. Kunzi, H. G. Tzschach and C. A. Zehnder, *Numerical Methods of Mathematical Optimization*, Academic Press, New York, 1968.

36. E. M. L. Beale, 'On quadratic programming', *Naval Res. Logistics Quarterly*, **6**, 227 (1959). (See the subroutine in Reference 35.)

37. R. Fletcher, 'A Fortran subroutine for general quadratic programming', UKAEA Research Group Report, AERE Report R. 6370, 1970 (available from HMSO).

38. R. H. Bartels, G. H. Golub and M. A. Saunders, 'Numerical techniques in mathematical programming', presented at the 7th International Mathematical Programming Symposium, The Hague, 1970, proceedings to be partly published in *Mathematical Programming*.

39. P. Wolfe, 'Methods of non-linear programming', in *Non-linear Programming* (Ed. J. Abadie), North-Holland, Amsterdam, 1967.

40. J. B. Rosen, 'The gradient projection method for non-linear programming: Part I, linear constraints', *SIAM J.*, **8**, 181 (1960).

41. R. Fletcher, 'Minimizing general functions subject to linear constraints', AERE Report T.P. 453, presented at the Dundee Optimization Conference, June 1971, proceedings to be published (Ed. F. A. Lootsma).

42. G. P. McCormick, 'The variable reduction method for non-linear programming', *Management Sci.*, **17**, 146 (1970).

43. R. Fletcher, 'Generalized inverses for non-linear equations and optimization', in *Numerical Methods for Non-Linear Algebraic Equations* (Ed. P. Rabinowitz), Gordon and Breach, London, 1970.

44. D. Goldfarb and L. Lapidus, 'Conjugate gradient method for non-linear programming problems with linear constraints', *I & EC Fundamentals*, **1**, 142 (1966). (AERE Library subroutine VEO1A.)

45. B. A. Murtagh and R. W. H. Sargent, 'A constrained minimization method with quadratic convergence', in *Optimization* (Ed. R. Fletcher), Academic Press, London, 1969.

46. B. A. Murtagh and R. W. H. Sargent, 'Computational experience with quadratically convergent minimization methods', *Computer Jour.*, **13**, 185 (1970).

47. J. Abadie and J. Carpentier, 'Generalization of the Wolfe reduced gradient method to the case of non-linear constraints', in *Optimization* (Ed. R. Fletcher), Academic Press, London, 1969.

48. B. A. Murtagh and R. W. H. Sargent, 'Projection methods for non-linear programming', presented at the 7th International Mathematical Programming Symposium, The Hague, 1970, proceedings to be partly published in *Mathematical Programming*.

49. D. Davies, 'Some practical methods of optimization', in *Integer and Non-linear Programming* (Ed. J. Abadie), North-Holland, Amsterdam, 1970.

50. G. E. Myers, 'Numerical experience with accelerated gradient projection', presented at the Dundee Optimization Conference, June 1971, proceedings to be published (Ed. F. A. Lootsma).

51. R. E. Griffith and R. A. Stewart, 'A nonlinear programming technique for the optimization of continuous processing systems', *Management Science*, **7**, 379–392 (1961).

52. E. M. L. Beale, 'Numerical methods', in *Nonlinear Programming* (Ed. J. Abadie), North-Holland, Amsterdam, 1967.

53. A. V. Fiacco and G. P. McCormick, *Non-linear Programming*, Wiley, New York, 1968.

54. M. J. D. Powell, 'A method for non-linear constraints in minimization problems', in *Optimization* (Ed. R. Fletcher), Academic Press, London, 1969.

55. R. Fletcher, 'A class of methods for non-linear programming with termination and convergence properties', in *Integer and Nonlinear Programming* (Ed. J. Abadie), North-Holland, Amsterdam, 1970.

56. R. Fletcher, 'A class of methods for nonlinear programming—III—rates of convergence', AERE Report T.P. 449, presented at the Dundee Optimization Conference, June 1971, proceedings to be published (Ed. F. A. Lootsma).

57. M. R. Osborne and D. M. Ryan, 'A hybrid method for non-linear programming', presented at the Dundee Optimization Conference, June 1971, proceedings to be published (Ed. F. A. Lootsma).

58. W. Murray, 'An algorithm for constrained minimization', in *Optimization* (Ed. R. Fletcher), Academic Press, London, 1969.

59. H. W. Dommel and W. F. Tinney, 'Optimal power flow solutions', *I.E.E.E. Trans. P.A.S.*, **87**, 1866 (1968).

*Chapter 6*

# Linear Programming in Structural Analysis and Design*

*R. K. Livesley*

## 6.1 Introduction

Although many optimization problems in engineering design are not strictly linear, in practice the designer often has the choice of an approximate optimization based on a simple linear model (i.e. a linear-programming solution) or a more sophisticated approach giving rise to a non-linear problem. In this choice the time and money consumed by the optimization procedure are relevant factors. Since a linear-programming problem is far easier to solve than a non-linear one, an approximate linearized optimum may be 'better', in an overall-cost sense, than the 'exact' solution. Even when a non-linear problem has to be solved, it is often possible to use an iterative procedure in which linearization in the neighbourhood of a succession of trial solutions effectively converts the problem into a series of linear-programming problems whose solutions converge to the exact answer; this point is taken up in the next chapter.

The first systematic method of solving the general linear-programming problem was developed by Dantzig and his associates in the late 1940s[1]. The first successful solution of a linear-programming problem on a digital computer was carried out in 1952[2]. Since that time linear-programming models have been used in many fields, including economics, management, transportation and production planning. Some of these applications have given rise to systems of inequalities with many thousands of variables, and considerable effort has gone into the construction of efficient techniques for their solution. This has produced both an extensive literature and a large number of commercially available computer-program packages. It is largely the existence of these programs that makes linear programming an important practical technique.

The most prevalent applications of linear programming in structural engineering have occurred in limit analysis and design, where the designer looks for upper- and lower-bound solutions. Like the first papers on linear

* Throughout this chapter *italic* bold-faced symbols refer to elements and *roman* bold-faced symbols refer to complete structures.

79

programming, a formal statement of the basic upper- and lower-bound theorems of limit analysis dates from the late 1940s[3]. At about the same time Charnes and Greenberg[4] showed that the 'equilibrium' (lower-bound) and 'mechanism' (upper-bound) methods of limit analysis were, in the case of trusses only, dual linear-programming problems. This was proved more generally by Charnes, Lemke and Zienkiewicz[5]. Foulkes[6] demonstrated that the problem of plastic minimum-weight design for a single-load system could also be reduced to a linear-programming problem and showed later[7] that the existence of alternative systems of loading did not significantly alter the design problem.

At the time when these papers were written few computers were available to engineers, and fewer still had reliable linear-programming packages available for general use. Accordingly limit analysis and design developed an 'engineer's theory' in which linear programming was hardly mentioned. The first workers in the field were naturally concerned with persuading practising engineers to use plastic design methods, and rightly felt that papers written in abstract mathematical terms would not help in this aim. Noteworthy among early published work directed at an engineering audience were papers by Heyman[8,9] on minimum-weight design, the method of 'plastic moment distribution' due to Horne[10] and a number of expository papers by Prager[11,12]. These papers tried to communicate basic physical ideas by means of simple examples, with the implication that the reader would go on to do similar calculations by hand, using his engineering intuition to direct the course of the analysis. Most of these early papers used a 'mechanism' approach, since the idea of a collapse mechanism has immediate physical appeal.

Anyone who has tried to do a linear-programming calculation by hand will agree that, for manual limit analysis, the physical, intuitive approach of the engineer is the right one. However, the growth of digital computers during the 1950s stimulated work on computer-oriented methods for structural analysis and design, and it was quickly found that, in these circumstances, a more formal mathematical approach was desirable. Early programs for the minimum-weight design of frames were developed by Heyman and Prager[13] and myself[14]. Although neither of these programs used the standard linear-programming formulation or the standard linear-programming algorithms, they were at least in the spirit of linear programming— systematic general procedures, applicable in principle to any limit-analysis problem.

From a structural point of view, both these early programs were only partial solutions to the overall design problem, since they required the equilibrium equations of a framework to be set up by hand—no trivial task in the case of a complicated structure. However, the development of a general matrix theory of structures (largely for elastic calculations) produced techniques of automatic redundant selection[15,16] and provided a theoretical basis for fully automatic programs which would take in the engineering details of a frame, set up the equilibrium equations automatically and then produce either a load-factor analysis or a minimum-weight design. Such programs have been described by myself[16] and Baty and Wright[17].

Many of the references cited above state that a translation of a problem in limit analysis or design into standard linear-programming form leads to a considerable increase in the number of variables and constraints. This is true, as we shall see in Section 6.4, and was a valid reason for the development of special-purpose algorithms in the early days of computers. Present-day machines, however, have sufficient storage to handle very large linear-programming problems, while commercial stimulus from fields other than engineering has resulted in the development of very efficient and sophisticated programs for solving problems set up in standard form. The choice of a 'standard package' or a 'special-purpose program' is now one that can be made entirely on the grounds of convenience and computing costs.

The existence of efficient computer programs means that a problem stated in standard linear-programming form can be regarded as 'solved', irrespective of whether the originator of the problem knows anything of the solution process. In view of this, is there any virtue in a structural engineer trying to 'understand' the simplex method, the dual simplex method, or any of the other algorithms produced by the analysts?

The answer must surely be that an engineer who sets up a system of equations to be processed by an algorithm of which he knows nothing will get correct answers, but that he is in a vulnerable position. In the first place, he must cast his equations in the form required by the computer program, rather than in the form most natural to his problem. He may find, for example, that the program requires all variables to be positive, so that he is forced to treat one whose sign is not restricted as the difference of two positive variables. In the second place, he may have constraints of a particular form that would, in a specially written program, allow for a considerable saving in computer time or storage. These facts may be minor irritations in a situation where a single large problem has to be solved quickly by whatever means are to hand, but they become important in the development of a design procedure which is to be used over and over again. It is then that the engineer considers writing special-purpose programs of his own.

The structural engineer who reaches this point and turns to the standard texts of linear programming (e.g. References 18, 19 and 20) encounters two difficulties. The first is simply that he has to achieve 'understanding' either in purely mathematical terms, or through accounts presented in the language of economics. The second is that solution algorithms tend to be described as integrated procedures to be applied to problems stated 'in the standard form'. Thus the engineer may feel that, in approaching a linear optimization problem, he has to choose between a standard procedure of formal analysis, which makes no concessions to the special characteristics of his problem, and an *ad hoc* procedure which, although stated in the familiar language of bending-moment diagrams and collapse mechanisms, may not be particularly suitable as the basis of an efficient computer program.

One purpose of this chapter is to show that there can be a middle way. The standard procedures of linear programming are all based on a few quite simple ideas, and if these ideas are really understood they can be used out of their normal context in special-purpose algorithms that provide a more natural (and therefore more efficient) solution of certain problems. We illustrate this later in the chapter by describing an algorithm for rigid–plastic collapse analysis. No account is given of the simplex method or the other standard procedures of linear programming, since these are described well in the texts referred to above.

If a general optimization procedure for structural design is to be of practical value, it must include a systematic method for setting up the governing equations and constraints, as well as means of solving them. This is true whether the method is elastic or plastic, linear or non-linear. A large part of this chapter is, accordingly, concerned with automatic methods for constructing the equations and inequalities associated with a number of general problems in the analysis and design of skeletal structures. We consider:

(1) The rigid–plastic collapse analysis of a frame or truss under a given loading system. This problem is linear, provided that the usual assumptions of simple rigid–plastic limit analysis are made.
(2) The shakedown analysis of a frame or truss. This also requires elastic analyses of the structure.
(3) Minimum-weight design under a single fixed loading system. In the case of a frame, the weight function must be linearized by assuming that the weight of a structural member is proportional to its fully plastic moment.
(4) Minimum-weight design for several different loading systems, but without including the possibility of incremental collapse.

**6.2 The general nature of the solution**

The linear-programming problem has been stated in Section 5.3. The nature of the problem and its solution may be seen from the following example:

Maximize

$$W = x_1 + x_2$$

subject to

$$3x_1 + 5x_2 \leqslant 15$$

$$x_1 + x_2 \geqslant -3$$

$$x \leqslant 4$$

$$-3 \leqslant x_2 \leqslant 2.$$

If we treat the variables $x_1$ and $x_2$ as coordinate axes, as shown in Figure 6.1, we see that the inequalities define a permissible region, shown shaded, within which any 'feasible solution' must lie. Since $W$ is a linear function

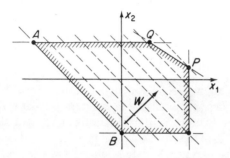

**Figure 6.1**

of the variables, the contours of constant $W$ are straight lines and the maximum value of $W$ is obviously attained at the point P. Thus only two of the inequalities are necessary to define the solution, and these two are satisfied as equalities.

The form of the solution is not altered greatly if some of the constraints are equalities. An equality constraint in this example reduces the permissible region to a line segment. For example, if the first inequality were replaced by the equality $3x_1 + 5x_2 = 15$, the permissible region would be simply the segment PQ in Figure 6.1. The actual solution would, in fact, be unchanged.

As was noted in Chapter 2, the characteristics of the general $n$-variable problem are very similar. If we imagine a linear vector space with the $n$ variables as coordinate axes, the permissible region becomes a convex polyhedron, while the surfaces of constant $W$ form a set of parallel hyperplanes. The solution is always a vertex of the permissible region, at which $n$ of the constraints are satisfied as equalities.

Readers already acquainted with linear-programming literature may be surprised by the omission of the condition that the variables be positive.

We shall see later that structural problems often involve variables that may be positive or negative. Techniques for handling variables of this sort are described later in this chapter.

Our simple example allows us to make general statements about the uniqueness of the solution to a linear-programming problem. Apart from the case where the constraints are mutually exclusive and the permissible region does not exist, it is intuitively obvious that:

(a) If the permissible region is bounded, a solution exists. This solution is unique, and is associated with a particular vertex of the region, except when two or more vertices have the same (maximum or minimum) value of $W$. Thus, if our example required the *minimum* of $W = x_1 + x_2$ subject to the conditions stated, any point on the line AB would be a solution.

(b) If the permissible region is unbounded, a finite solution of the type described in (a) *may* exist. On the other hand, $W$ may be able to increase (or decrease) without limit.

The mathematician will rightly demand formal proofs of existence and uniqueness, and these are available in the literature. However, in the present chapter we shall rely on these simple intuitive ideas, coupled with the usual engineering dogma that a mathematical problem, if it is derived in a logical manner from a real physical situation, must have a sensible solution.

How do we extend our simple graphical ideas when there are more than two variables? Clearly, in the general case, we must make $n$ of the constraints into equalities and solve a system of $n$ simultaneous equations, but how do we select the right constraints? The procedure of solving *all* the sets of $n$ equations which can be chosen from $m$ constraints and then picking from the solutions the one which has the greatest value of $W$ is, to say the least of it, a daunting prospect when $n$ and $m$ may be of the order of several hundreds. The literature of linear programming is largely a record of the search for more efficient systematic procedures, in which only a small fraction of the complete set of vertices need to be examined.

### 6.3  Equilibrium, compatibility and yield

In Section 6.4 we formulate a number of problems of limit analysis and design in linear-programming terms. In this section we present a systematic method of setting up the structural equilibrium equations (or, equivalently, the compatibility equations) that form the basis of the later analysis. These equations are the foundation of both elastic and plastic methods of analysis, and are given here in a form similar to that that I have used elsewhere[21]. To fix our ideas, I shall develop the equations in a form appropriate to plane skeletal structures. Finite-element techniques allow the corresponding equations to be set up for continuum problems, although the constraints imposed by material yield are, in general, non-linear.

### 6.3.1 *Element equilibrium equations and yield constraints*

The finite-element approach to structural analysis considers a structure to be made of *elements* joined together at *nodes*. It assumes that, for each element, the internal stresses and the nodal forces can be expressed adequately as functions of the external loads and a finite number of independent parameters. In a plane continuum problem, for example, we may use triangular elements and assume a state of constant stress in each element, choosing the three independent components of the stress tensor as the parameters. In the straight beam shown in Figure 6.2, the end forces and moments and the in-

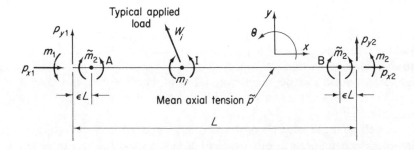

**Figure 6.2**

ternal stresses can be specified uniquely (at least as far as elementary bending theory is concerned) in terms of three linearly independent stress resultants and the applied loads $W_i$. The choice of these basic stress resultants is a matter of convenience—in the analysis given here we choose the mean axial tension $\tilde{p}$ and the moments $\tilde{m}_1$, $\tilde{m}_2$ at two points distant $\varepsilon L$ from the ends, as shown in Figure 6.2. We group the quantities $\tilde{p}$, $\tilde{m}_1$ and $\tilde{m}_2$ into a vector

$$r = \begin{bmatrix} \tilde{p} \\ \tilde{m}_1 \\ \tilde{m}_2 \end{bmatrix}.$$

Simple statics now allows us to express the end-load vectors

$$p_1 = \begin{bmatrix} p_{x1} \\ p_{y1} \\ m_1 \end{bmatrix} \quad \text{and} \quad p_2 = \begin{bmatrix} p_{x2} \\ p_{y2} \\ m_2 \end{bmatrix}$$

in the form

$$p_1 = H_1 r - h_1 \Big\}$$
$$p_2 = H_2 r - h_2 \Big\} \tag{6.1}$$

where $H_1$ and $H_2$ are matrices whose coefficients depend only on the values of $L$ and $\varepsilon$, and $h_1$ and $h_2$ are functions of the applied loads $W_i$. These equations may be transformed into 'system axes' in the usual way:

$$p_1' = H_1' r - h_1' \Big\}$$
$$p_2' = H_2' r - h_2' . \Big\} \tag{6.2}$$

Details of equations (6.1) and (6.2) are given in Appendix 6A.

The assumptions of simple rigid–plastic behaviour imply that the beam shown in Figure 6.2 remains rigid until, at some point, the moment or the axial force reaches its limiting value. This implies that there are 'direct' constraints on the individual variables $p$, $\tilde{m}_1$ and $\tilde{m}_2$, corresponding to axial yield or the appearance of plastic hinges at the points A or B. These constraints may be written $-r^L \leqslant r \leqslant r^U$. (The lower limit on $\tilde{p}$ may actually be set by buckling rather than material yield, but this does not affect our argument.) We shall assume that the elements of $r^L$ and $r^U$ are all non-negative.

As well as these 'direct' constraints, there may be 'indirect' constraints imposed by conditions at other critical points in the element. For example, the bending moment $m_i$ at the point I in Figure 6.2 is limited by the value of the fully plastic moment at point I, and this moment is a linear function of $r$ and the applied loads $W_i$. If there are a number of critical points along the length of the member, we shall have a series of moments ... $m_i$ ..., each one of them being a linear function of $r$ and the applied loads, and each one being limited by a constraint of the form $-m_i^L \leqslant m_i \leqslant m_i^U$. If we combine all these moments into a vector $q$, we may write the statical equations which express $q$ in terms of $r$ in the general form

$$q = q_0 - Gr \tag{6.3}$$

where $q_0$ depends only on the loads ... $W_i$ ... acting on the beam. (It is, in our example, the vector of 'free' bending moments.) Details of the form of equation (6.3) for the case of a straight beam are given in Appendix 6A. The complete set of constraints imposed by yield considerations may now be written

$$-r^L \leqslant r \leqslant r^U$$
$$-q^L \leqslant q \leqslant q^U. \tag{6.4}$$

The same approach may be adopted in the case of a plane finite element with principal stresses $\sigma_1$ and $\sigma_2$ in known constant directions. The Tresca yield criterion gives direct constraints on $\sigma_1$ and $\sigma_2$ and indirect constraints involving $\sigma_1 - \sigma_2$, all constraints being linear. However, if the von Mises condition is applied, or if the directions of the principal stresses are unknown, the constraints become non-linear.

### 6.3.2 The equilibrium equations of a complete structure

The construction of the set of joint equilibrium equations for a structure follows the normal pattern. For example, it is easy to show that the equilibrium equation for joint $j$ in Figure 6.3 is

$$p_j + (h'_2)_a + (h'_1)_b + (h'_1)_c = (H'_2)_a r_a + (H'_1)_b r_b + (H'_1)_c r_c. \qquad (6.5)$$

The set of equations for all joints may be written

$$\mathbf{p_0} = \mathbf{Ar} \qquad (6.6)$$

where $\mathbf{p_0}$ is the vector of all the joint loads $\ldots p_j \ldots$, modified by the various $h'$ vectors as shown in equation (6.5), and $\mathbf{r}$ is the vector of member stress resultants $r_a, r_b, \ldots$ etc. $\mathbf{A}$ is defined as follows. The submatrix of $\mathbf{A}$ associated with joint $j$, element $k$ is $(H'_1)_k$ if element $k$ has $j$ as node 1, $(H'_2)_k$ if element $k$ has $j$ as node 2 and a null matrix otherwise (cf. Reference 21, p. 126).

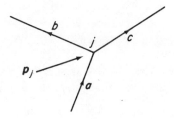

**Figure 6.3**

In the same way all the auxiliary equilibrium equations (6.3) may be combined in the form

$$\mathbf{q_0} = \mathbf{Br} + \mathbf{q} \qquad (6.7)$$

where $\mathbf{q_0}$ and $\mathbf{q}$ are the vectors $(q_0)_a \ldots$ and $(q)_a \ldots$ for all the members. For a rigid-jointed plane frame with $n$ joints (not counting rigidly connected foundations, for which no equations need be set up), $m$ members and $s$ internal loading points, $\mathbf{p_0}$ has $3n$ elements, $\mathbf{r}$ has $3m$ elements and $\mathbf{q_0}$ and $\mathbf{q}$ each have $s$ elements. Equations (6.6) and (6.7) may be written in the form

$$\begin{matrix} 3n \\ s \end{matrix} \begin{bmatrix} \mathbf{p_0} \\ \mathbf{q_0} \end{bmatrix} = \begin{bmatrix} \mathbf{A} & \mathbf{O} \\ \mathbf{B} & \mathbf{I} \end{bmatrix} \begin{bmatrix} \mathbf{r} \\ \mathbf{q} \end{bmatrix}. \qquad (6.8)$$

$$\begin{matrix} 3m & \quad s \end{matrix}$$

These equations hold irrespective of whether deformations are plastic or elastic, provided that the deformations are small enough not to alter the overall geometry of the structure. The matrices **A** and **B** are easily set up automatically.

### 6.3.3 Compatibility equations

The vector of member stress resultants **r** has associated with it a vector of member strains **e**. If we postulate rigid–plastic behaviour we find that corresponding to each moment $\tilde{m}$ there is a possible hinge rotation, the sense of which is shown in Figure 6.4, while corresponding to each $\tilde{p}$ there is an axial extension. In the same way, the vectors **q** and $\mathbf{q}_0$ are linked with a vector $\mathbf{e}_q$ consisting of the hinge rotations at the corresponding intermediate moment points, while to the joint loads $\mathbf{p}_0$ there corresponds a vector of joint displacements and rotations **d**. The assumption of rigid–plastic behaviour implies that each element of **e** (or $\mathbf{e}_q$) is zero unless the corresponding element of **r** (or **q**) has reached one or other of its limiting values. If the deformation is not zero, the sign must be such that positive work is dissipated.

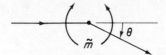

**Figure 6.4**

It is easy to show [by a virtual-work argument applied to equation (6.8)] that the vectors **e**, $\mathbf{e}_q$ and **d** are related by the compatibility equations

$$\begin{array}{c} 3m \\ s \end{array} \begin{bmatrix} \mathbf{e} \\ \mathbf{e}_q \end{bmatrix} = \begin{bmatrix} \mathbf{A}^T & \mathbf{B}^T \\ \mathbf{O} & \mathbf{I} \end{bmatrix} \begin{bmatrix} \mathbf{d} \\ \mathbf{e}_q \end{bmatrix}. \tag{6.9}$$
$$\begin{array}{cc} 3n & s \end{array}$$

### 6.4 Some problems formulated

We now use the equations developed in the previous section to set up a number of different linear-programming problems. We shall take our examples from the field of plane skeletal structures, although the basic ideas have wide application. Since constraints (6.4) include moments and axial forces, we shall not need to distinguish between frames and trusses.

### 6.4.1 Plastic load-factor analysis

If a given loading on a structure corresponds to a particular load vector $\begin{bmatrix} \mathbf{p}_0 \\ \mathbf{q}_0 \end{bmatrix}$, multiplying the loading by a factor $\lambda$ converts the equilibrium equations (6.8) into

$$\lambda \begin{bmatrix} \mathbf{p}_0 \\ \mathbf{q}_0 \end{bmatrix} = \begin{bmatrix} \mathbf{A} & \mathbf{O} \\ \mathbf{B} & \mathbf{I} \end{bmatrix} \begin{bmatrix} \mathbf{r} \\ \mathbf{q} \end{bmatrix}. \tag{6.10}$$

The maximum principle of limit analysis states that the structure will support the factored loading $\lambda \begin{bmatrix} \mathbf{p}_0 \\ \mathbf{q}_0 \end{bmatrix}$ provided that a solution to equations (6.10) exists that does not violate yield constraints (6.4). The collapse-load factor $\lambda_C$ is the greatest value of $\lambda$ for which such a solution can be found. The problem is therefore to maximize $\lambda$ subject to equations (6.10) and constraints (6.4). If we absorb $\mathbf{q}_0$ in $\mathbf{p}_0$ and $\mathbf{q}$ in $\mathbf{r}$, and write the matrix $\begin{bmatrix} \mathbf{A} & \mathbf{O} \\ \mathbf{B} & \mathbf{I} \end{bmatrix}$ as $\mathbf{C}$, our problem is simply to maximize $\lambda$ subject to

$$\lambda \mathbf{p}_0 = \mathbf{C}\mathbf{r} \tag{6.11}$$

$$-\mathbf{r}^L \leqslant \mathbf{r} \leqslant \mathbf{r}^U \tag{6.12}$$

where $\mathbf{r}^L$ and $\mathbf{r}^U$ have components that are all positive. This is the 'bounded-variable problem' first treated by Charnes and Lemke[22]. We describe a direct method of solving this system in Section 6.5.

Equations (6.11) and (6.12) may be put in standard linear-programming form as:
Maximize

$$W = \begin{bmatrix} 1 & \mathbf{O} \end{bmatrix} \begin{bmatrix} \lambda \\ \mathbf{r} \end{bmatrix}$$

subject to

$$\begin{bmatrix} -\mathbf{p}_0 & \mathbf{C} \\ \mathbf{O} & \mathbf{I} \\ \mathbf{O} & -\mathbf{I} \end{bmatrix} \begin{bmatrix} \lambda \\ \mathbf{r} \end{bmatrix} \begin{matrix} = \mathbf{O} \\ \leqslant \mathbf{r}^U. \\ \leqslant \mathbf{r}^L \end{matrix} \tag{6.13}$$

If we impose the condition (common in a number of commercially available linear-programming packages) that all variables be positive or zero, we must write $\mathbf{r} = \mathbf{r}^+ - \mathbf{r}^-$, where the elements of $\mathbf{r}^+$ and $\mathbf{r}^-$ are all positive or zero. The problem then becomes:
Maximize

$$W = \begin{bmatrix} 1 & \mathbf{O} \end{bmatrix} \begin{bmatrix} \lambda \\ \mathbf{r}^+ \\ \mathbf{r}^- \end{bmatrix}$$

subject to

$$\begin{bmatrix} -\mathbf{p}_0 & \mathbf{C} & -\mathbf{C} \\ \mathbf{O} & \mathbf{I} & -\mathbf{I} \\ \mathbf{O} & -\mathbf{I} & \mathbf{I} \end{bmatrix} \begin{bmatrix} \lambda \\ \mathbf{r}^+ \\ \mathbf{r}^+ \end{bmatrix} \begin{aligned} &= \mathbf{O} \\ &\leqslant \mathbf{r}^U \\ &\leqslant \mathbf{r}^L \end{aligned} \qquad (6.14)$$

We now see that a reduction to 'standard form' greatly increases the size of the coefficient matrix that must be handled. For example, if we consider the frame shown in Figure 6.5, the original vectors $\mathbf{p}_0$, $\mathbf{q}_0$, $\mathbf{r}$ and $\mathbf{q}$ have 9, 4,

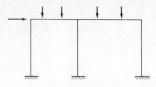

**Figure 6.5**

15 and 4 elements, respectively. Thus the extended vectors $\mathbf{p}_0$ and $\mathbf{r}$ in equation (6.11) have 13 and 19 elements, respectively, and the matrix $\mathbf{C}$ has 13 rows and 19 columns. The matrix in equation (6.14), however, has 51 rows and 39 columns, even without the additional columns for the slack variables of the simplex method. It would be unfair to suggest that this represents a proportionate increase in computer storage and processing time, since many of the coefficients are either 0 or 1, but it does point to the need for efficient handling of zeros in any formal linear-programming system used for structural problems.

### 6.4.2   Duality and the mechanism approach

The well known duality principle of linear programming may be stated in the following form. The two problems:

$$\text{Maximize } W = \mathbf{c}^T\mathbf{u}, \text{ subject to } \mathbf{Au} \leqslant \mathbf{b}, \quad \mathbf{u} \geqslant \mathbf{O}$$

$$\text{Minimize } \overline{W} = \mathbf{b}^T\mathbf{v}, \text{ subject to } \mathbf{A}^T\mathbf{v} \geqslant \mathbf{c}, \quad \mathbf{v} \geqslant \mathbf{O}$$

are said to be dual problems. If one has a finite solution, so has the other, and maximum $W$ = minimum $\overline{W}$.

If we replace the equalities in expression (6.14) by pairs of inequalities, we obtain the problem:

Maximize

$$W = \begin{bmatrix} 1 & \mathbf{O} & \mathbf{O} \end{bmatrix} \begin{bmatrix} \lambda \\ \mathbf{r}^+ \\ \mathbf{r}^- \end{bmatrix}$$

subject to

$$
\begin{bmatrix}
\mathbf{p}_0 & -\mathbf{C} & \mathbf{C} \\
-\mathbf{p}_0 & \mathbf{C} & -\mathbf{C} \\
\mathbf{O} & \mathbf{I} & -\mathbf{I} \\
\mathbf{O} & -\mathbf{I} & \mathbf{I}
\end{bmatrix}
\begin{bmatrix}
\lambda \\
\mathbf{r}^+ \\
\mathbf{r}^-
\end{bmatrix}
\leqslant
\begin{bmatrix}
\mathbf{O} \\
\mathbf{O} \\
\mathbf{r}^U \\
\mathbf{r}^L
\end{bmatrix}
$$

which corresponds to the first problem given above. It is natural to suspect that the dual of this, which attains $\lambda_c$ as a *minimum*, is, in fact, the 'mechanism' formulation of the plastic-collapse problem. It is easy to show that this is indeed the case. We write the dual in the form:
Minimize

$$
\overline{W} = [\mathbf{O} \quad \mathbf{O} \quad (r^U)^T \quad (\mathbf{r}^L)^T]
\begin{bmatrix}
\mathbf{d}^+ \\
\mathbf{d}^- \\
\mathbf{e}^+ \\
\mathbf{e}^-
\end{bmatrix}
$$

subject to

$$
\begin{bmatrix}
\mathbf{p}^T & -\mathbf{p}^T & \mathbf{O} & \mathbf{O} \\
-\mathbf{C}^T & \mathbf{C}^T & \mathbf{I} & -\mathbf{I} \\
\mathbf{C}^T & -\mathbf{C}^T & -\mathbf{I} & \mathbf{I}
\end{bmatrix}
\begin{bmatrix}
\mathbf{d}^+ \\
\mathbf{d}^- \\
\mathbf{e}^+ \\
\mathbf{e}^-
\end{bmatrix}
\geqslant
\begin{bmatrix}
1 \\
\mathbf{O} \\
\mathbf{O}
\end{bmatrix}.
$$

If we define the unrestricted variables $\mathbf{d} = \mathbf{d}^+ - \mathbf{d}^-$ and $\mathbf{e} = \mathbf{e}^+ - \mathbf{e}^-$, the problem becomes:
Minimize

$$
\overline{W} = (\mathbf{r}^U)^T\mathbf{e}^+ + (\mathbf{r}^L)^T\mathbf{e}^- \tag{6.15a}
$$

subject to

$$
\mathbf{p}_0^T\mathbf{d} \geqslant 1 \tag{6.15b}
$$

$$
\mathbf{C}^T\mathbf{d} = \mathbf{e} \tag{6.15c}
$$

We now identify $\mathbf{d}$ and $\mathbf{e}$ with the displacement and deformation vectors introduced in Section 6.3.3, including $\mathbf{e}_q$ as part of both $\mathbf{d}$ and $\mathbf{e}$ in the same way that we made $\mathbf{q}_0$ part of $\mathbf{p}_0$ and $\mathbf{q}$ part of $\mathbf{r}$. If we do this, we see that equation (6.15c) is the condensed form of the compatibility equations (6.9).

Equation (6.15a) is simply an expression for the work done in the plastic hinges or axial plastic strains, while expression (6.15b) is a condition on the work done by the (unfactored) applied loads $\mathbf{p}_0$. Writing $\lambda$ for $\overline{W}$, we may put the problem in the form:

Minimize $\lambda$, subject to

$$\lambda \mathbf{p}_0^T \mathbf{d} \geqslant (\mathbf{r}^U)^T \mathbf{e}^+ + (\mathbf{r}^L)^T \mathbf{e}^- \tag{6.16}$$

where $\mathbf{d}$ and $\mathbf{e}$ are related by the compatibility equations (6.15c). Clearly the final solution will satisfy expression (6.16) as an equality, so that we may state the problem in words as:

Minimize $\lambda$, subject to

work done by applied loads $\lambda \mathbf{p}_0$ = plastic work done in structure
for a compatible system of
deformations.

[Note that, for any particular $e$, $e^+ = e$, $e^- = 0$ if $e > 0$, while $e^- = -e$, $e^+ = 0$ if $e < 0$. Hence we multiply $e$ by the appropriate $r^U$ or $-r^L$ according to whether it is positive or negative.] This statement of the problem is precisely the mechanism approach of elementary plastic theory.

From a formal linear programming point of view, it is irrelevant whether we think in terms of an 'equilibrium' or a 'mechanism' approach, since a simple matrix transposition takes us from one to the other. From a computational point of view, it is held to be more efficient, if the simplex method is used, to apply it to the dual (i.e. mechanism) formulation, since there are usually fewer constraints. For a further discussion of this point the reader is referred to papers by Dorn and Greenberg[23] and Toakley[24].

### 6.4.3  The shakedown problem

Real structures are subjected to varying loads, which may produce either shakedown or incremental collapse (see Reference 25).

These phenomena occur when a number of different loadings applied sequentially each produce a certain amount of yielding, although no one loading is sufficient to generate a full collapse mechanism. If the yielding continues indefinitely with each loading cycle, incremental collapse is said to occur. In effect, a collapse mechanism forms, although not all the plastic flow takes place at the same time. If the yielding eventually ceases so that the behaviour is ultimately entirely elastic, the structure is said to have 'shaken down'.

It is obviously important to be able to find out whether a given design will shake down under variable loads. Perhaps it is more in the spirit of our previous analysis to ask the question 'By what factor do we need to multiply all the loadings in order to just produce incremental collapse?' The answer to this question is provided by the shakedown theorem, which effectively says that, if a state of the structure exists in which it can respond purely elastically to all loadings, it will find that state. More precisely, it states that a structure will shake down under load variations if a set of internal self-equilibrating stresses can be found such that any of the applied loads can be carried by purely elastic behaviour.

From equation (6.11), it is clear that any set of internal stress resultants $\mathbf{r}$ which satisfies

$$\mathbf{Cr} = \mathbf{O} \qquad (6.17)$$

will be self equilibrating (as before, $\mathbf{q}$ is taken to be combined with $\mathbf{r}$). If we carry out elastic analyses of the structure under the various loads, we can compute two vectors, consisting of the greatest and least values of all the internal stress resultants. Let these be $\mathbf{r}^{max}$ and $\mathbf{r}^{min}$. Now, these elastic calculations assume an initial unstressed state; so that, if the initial state is $\mathbf{r}$, the maximum and minimum will be $\mathbf{r}^{max} + \mathbf{r}$ and $\mathbf{r}^{min} + \mathbf{r}$. The structure will shake down if we can find an $\mathbf{r}$ satisfying equation (6.17) such that

$$-\mathbf{r}^{L} \leqslant \mathbf{r}^{min} + \mathbf{r}, \qquad \mathbf{r}^{max} + \mathbf{r} \leqslant \mathbf{r}^{U}.$$

If we now multiply each load by $\lambda$, we multiply $\mathbf{r}^{max}$ and $\mathbf{r}^{min}$ by $\lambda$; so that the calculation of the incremental-collapse load factor leads to the problem:

Maximize $\lambda$, subject to

$$\mathbf{Cr} = \mathbf{O}, \qquad -\mathbf{r}^{L} \leqslant \lambda\mathbf{r}^{min} + \mathbf{r}, \qquad \lambda\mathbf{r}^{max} + \mathbf{r} \leqslant \mathbf{r}^{U}.$$

This may be put in standard linear-programming form as:

Maximize

$$W = \begin{bmatrix} 1 & \mathbf{O} \end{bmatrix} \begin{bmatrix} \lambda \\ \mathbf{r} \end{bmatrix}$$

subject to

$$\begin{bmatrix} \mathbf{O} & \mathbf{C} \\ \mathbf{r}^{max} & \mathbf{I} \\ -\mathbf{r}^{min} & -\mathbf{I} \end{bmatrix} \begin{bmatrix} \lambda \\ \mathbf{r} \end{bmatrix} \begin{matrix} = \mathbf{O} \\ \leqslant \mathbf{r}^{U} \\ \leqslant \mathbf{r}^{L} \end{matrix} \qquad (6.18)$$

which is very similar in form to expression (6.13). As before, $\mathbf{r}$ may be written as $\mathbf{r}^{+} - \mathbf{r}^{-}$ if necessary.

### 6.4.4 Minimum-weight design for a single loading

So far we have been concerned with problems of analysis, in which the limits $\mathbf{r}^{U}$, etc., have been taken as fixed. We now consider the problem of choosing the design of least weight that will just support a given loading.

We imagine that we have certain design parameters $\alpha_1, \ldots, \alpha_k$ at our disposal, which we will write as a vector $\boldsymbol{\alpha}$. In skeletal structures these parameters will normally be cross-sectional areas or fully plastic moments. We assume that the limits $\mathbf{r}^L$ and $\mathbf{r}^U$ are linear functions of the $\alpha$s so that

$$\mathbf{r}^L = \mathbf{S}^L\boldsymbol{\alpha}, \qquad \mathbf{r}^U = \mathbf{S}^U\boldsymbol{\alpha}.$$

The matrices $\mathbf{S}^L$, $\mathbf{S}^U$ will normally have a very simple form. For example, consider the design of the two-bay portal shown in Figure 6.5, assuming that for fabrication reasons both beams are to have fully plastic moments $m_b$ and all the columns are to have fully plastic moments $m_c$. Then $\boldsymbol{\alpha} = \begin{bmatrix} m_b \\ m_c \end{bmatrix}$ and $\mathbf{S}^L$, $\mathbf{S}^U$ are each matrices with 19 rows and 2 columns, each element being 0 or 1.

To keep the problem linear, we assume that the weight of the structure is a linear function of $\boldsymbol{\alpha}$, i.e. $W = \mathbf{l}^T\boldsymbol{\alpha}$ where $\mathbf{l}$ is a given vector. In practice, of course, the $\alpha$s may not be continuously variable (steel sections only come in standard sizes), nor is the weight of a member proportional to its fully plastic moment. These facts must be borne in mind when considering the results of any linear-programming calculation.

Having made these assumptions, our problem is to minimize $W = \mathbf{l}^T\boldsymbol{\alpha}$ subject to the equilibrium equations $\mathbf{p} = \mathbf{C}\mathbf{r}$ and the yield constraints $-\mathbf{S}^L\boldsymbol{\alpha} \leqslant \mathbf{r} \leqslant \mathbf{S}^U\boldsymbol{\alpha}$. This may be written as:

Minimize

$$W = \begin{bmatrix} \mathbf{O} & \mathbf{l}^T \end{bmatrix} \begin{bmatrix} \mathbf{r} \\ \boldsymbol{\alpha} \end{bmatrix}$$

subject to

$$\begin{bmatrix} \mathbf{C} & \mathbf{O} \\ \mathbf{I} & \mathbf{S}^L \\ -\mathbf{I} & \mathbf{S}^U \end{bmatrix} \begin{bmatrix} \mathbf{r} \\ \boldsymbol{\alpha} \end{bmatrix} \begin{matrix} = \mathbf{p}_0 \\ \geqslant \mathbf{O} \\ \geqslant \mathbf{O}. \end{matrix} \qquad (6.19)$$

As before, $\mathbf{r}$ may be replaced by $\mathbf{r}^+ - \mathbf{r}^-$ is required ($\boldsymbol{\alpha}$ is usually essentially positive anyway).

### 6.4.5 *Minimum-weight design for several loading systems*

Finally, consider the case where $\mathbf{p}_0$ is not a fixed loading but may vary between certain limits. Again, keeping within the spirit of linear programming, we assume that $\mathbf{p}_0$ is made up of certain basic loading systems $\mathbf{p}_1, \mathbf{p}_2, \ldots$, so that $\mathbf{p}_0 = \beta_i\mathbf{p}_i = \mathbf{P}\boldsymbol{\beta}$.

If we imagine that the $\beta_i$s are restricted by some system of linear constraints $\mathbf{E\beta} \leqslant \mathbf{f}$, our problem is:
Minimize

$$W = [\mathbf{O} \quad \mathbf{I}^T \quad \mathbf{O}]\begin{bmatrix} \mathbf{r} \\ \boldsymbol{\alpha} \\ \boldsymbol{\beta} \end{bmatrix}$$

subject to

$$\begin{bmatrix} \mathbf{C} & \mathbf{O} & -\mathbf{P} \\ \mathbf{I} & \mathbf{S}^L & \mathbf{O} \\ -\mathbf{I} & \mathbf{S}^U & \mathbf{O} \\ \mathbf{O} & \mathbf{O} & \mathbf{E} \end{bmatrix}\begin{bmatrix} \mathbf{r} \\ \boldsymbol{\alpha} \\ \boldsymbol{\beta} \end{bmatrix} \begin{matrix} = \mathbf{O} \\ \geqslant \mathbf{O} \\ \geqslant \mathbf{O} \\ \leqslant \mathbf{f}. \end{matrix}$$

Once again, we see that a formal linear-programming formulation gives us a large sparse matrix.

This approach produces a design that will not collapse under any single loading within the prescribed limits. It does not, however, ensure that the design will shake down under any sequence of loads within the limits. Design (as opposed to analysis) for shakedown is essentially a non-linear process, and, as such, outside the scope of this chapter. It has been considered by Heyman[26].

### 6.5 A simple procedure for finding the collapse load factor

The standard simplex method of solving the linear-programming problem is described in many texts. In this section we show how the basic ideas used in the simplex procedure may be employed to solve the somewhat simpler problem posed in Section 6.4.1. This problem is:
Maximize $\lambda$ subject to

$$\lambda\mathbf{p} = \mathbf{Cr} \tag{6.20}$$

and the constraints

$$-\mathbf{r}^L \leqslant \mathbf{r} \leqslant \mathbf{r}^U \quad (\mathbf{r}^L, \mathbf{r}^U \geqslant \mathbf{O}) \tag{6.21}$$

where, for simplicity, we drop the suffix zero from $\mathbf{p}$.

Let there be $n$ variables $r_j$ and $m$ loads $p_i$. If $n < m$, the structure will not support an arbitrary set of loads (i.e. it is a mechanism) and the only solution is the trivial one $\lambda = r_j = 0$. If $n = m$ and the equations are linearly independent, the structure is statically determinate and we may solve equation (6.20) for the variables $x_j = r_j/\lambda$, subsequently finding $\lambda$ from the constraints $-r_j^L \leqslant \lambda x_j \leqslant r_j^U$. Our interest is in the case $n > m$.

*Optimum Structural Design*

A basic operation, which we shall call 'procedure A', occurs at a number of places in the analysis. This is the standard Gauss–Jordan operation of transforming a set of equations $\mathbf{b} = \mathbf{Cx}$ into a set $\mathbf{b}^* = \mathbf{C}^*\mathbf{x}$ in such a way that column $j$ of $\mathbf{C}^*$ is a column of zeros, except for a 1 in row $i$. Procedure A consists of two steps:

(1) Equation $i$ is scaled so that $c_{ij} = 1$.
(2) $-c_{kj}$ times this equation is added to the other equations so that $c_{kj}^* = 0$ ($k = 1, \ldots, m, \ k \neq i$).

Equation $i$ is termed the 'pivotal equation', and column $j$ is termed the 'reduced column'.

The first step in the analysis is to transform equation (6.20) into the form $\lambda\mathbf{p}^* = \mathbf{C}^*\mathbf{r}$, where $m$ of the $n$ columns of $\mathbf{C}$ have been replaced by a column of zeros and a single 1, i.e.:

$$\lambda\mathbf{p}^* = \begin{bmatrix} 0 & . & 1 & . & . & 0 & . & & 0 \\ 1 & . & 0 & . & . & 0 & . & & 0 \\ 0 & . & 0 & . & . & 0 & . & & 1 \\ \multicolumn{9}{c}{\dotfill} \\ \multicolumn{9}{c}{\dotfill} \\ 0 & . & 0 & . & . & 1 & . & & 0 \\ 0 & . & 0 & . & . & 0 & . & & 0 \end{bmatrix} \mathbf{r}. \tag{6.22}$$

This is done by the following operations, applied to each scalar equation $i$ of equation (6.20):

(1) The column $j$ containing the coefficient $c_{ij}$ of largest modulus is found.
(2) Procedure A is applied to equation $i$, column $j$.

(The choice of the $c_{ij}$ of largest modulus is not strictly necessary, but helps to keep the equations well conditioned.) Subsequent analysis always leaves the equations in this 'standard form'.

We may note in passing that, if $n = m$, this procedure gives us a solution of the equilibrium equations. This corresponds, as mentioned earlier, to the case of a statically determinate structure.

### 6.5.1 Basic and non-basic variables

We shall find it convenient to divide the elements of $\mathbf{r}$ into two types which (using the normal terminology of linear programming) we shall call 'basic' and 'non-basic' variables.

For our purposes, a *basic* variable is one that satisfies $-r_j^L < r_j < r_j^U$ as a strict inequality, and can therefore be altered by a finite amount without violating its constraints. In the procedure that we shall describe, variables are only altered when they are basic, and correspond to columns of $\mathbf{C}^*$ that have been reduced (i.e. contain only a single 1). A *non-basic* variable is one which is at one or other of its limits, except for an initial stage when its value may be zero. Non-basic variables have constant values and correspond to the columns of $\mathbf{C}^*$ that have no specific arrangement. By reducing the equilibrium equations to the form of equation (6.22), we have, effectively, chosen $m$ of the $r_j$s to be basic and $n - m$ to be non-basic. Note that each basic variable only appears in one equation.

The remainder of the computational process consists of two alternating steps:

(1) Increasing $\lambda$, altering only the basic variables, until one of them reaches one of its limiting values.
(2) Interchanging a basic and a non-basic variable so that a further increase in $\lambda$ becomes possible.

When step (2) cannot be carried out, the maximum value of $\lambda$ has been reached. Initially we put $\lambda = 0$ and all $r_j = 0$.

### 6.5.2 Increasing $\lambda$

Imagine that at a particular stage we have a value of $\lambda$ equal to $\lambda^{(0)}$, and $m$ basic variables $r_{i'}^{(0)}$ satisfying $-r_{i'}^{L} < r_{i'}^{(0)} < r_{i'}^{U}$. (The suffix $i'$ denotes the basic variable associated with equation $i$.)

Consider now the effect of an increase $\delta\lambda$ in the load factor. Since each basic variable is strictly within its limits, equalities (6.22) may be satisfied by keeping all the *non-basic* variables constant and letting the *basic* variables vary. For equation $i$ we have

$$p_i^* \delta\lambda = \delta r_{i'},$$

by virtue of the form of $\mathbf{C}^*$. Thus if $p_i^*$ is positive, $r_{i'}$ increases, while if $p_i^*$ is negative, $r_{i'}$ decreases. The limits on $r_{i'}$ mean that

$$\delta\lambda \leqslant (r_{i'}^{U} - r_{i'}^{(0)})/p_i^* \quad \text{if } p_i^* > 0$$
$$\delta\lambda \leqslant (-r_{i'}^{L} - r_{i'}^{(0)})/p_i^* \quad \text{if } p_i^* < 0.$$

The smallest value of $\delta\lambda$ taken over all the equations determines the permissible increase in $\lambda$. Let the critical equation be $k$ and the permissible increase $\delta\lambda_k$. Then we have

$$\lambda^{(1)} = \lambda^{(0)} + \delta\lambda_k, \quad r_{i'}^{(1)} = r_{i'}^{(0)} + p_i^* \delta\lambda_k \quad \text{(all basic variables)}$$

with $r_{k'}^{(1)} = -r_{k'}^{L}$ or $r_{k'}^{U}$, according to the sign of $p_k^*$, and all other basic variables still within their limits. The non-basic variables remain at their previous values.

98 *Optimum Structural Design*

### 6.5.3 Altering the set of basic variables

It is apparent that to achieve any further increase in $\lambda$ we must find some other variable to take over the role of basic variable for equation $k$. Now equation $k$ has the form

$$\lambda^{(1)} p_k^* = c_{kj}^* r_j^{(1)} + r_k^{(1)}$$

where $j$ is a dummy suffix denoting all the non-basic variables. (All the basic variables, except for $r_{k'}$, have zero coefficients in this equation.) We need an $r_j$ which can be varied in a permissible manner in such a way that $\lambda$ will increase. Each non-basic $r_j$ may be in one of three states:

(1) $r_j^{(1)} = 0.$      This is the initial state. Both positive and negative variations of $r_j$ are (in general) permissible, and one of these will certainly increase $\lambda$, provided that $c_{kj}^* \neq 0$.

(2) $r_j^{(1)} = -r_j^L.$      Only an increase in $r_j$ is permissible. This will increase $\lambda$ provided that $c_{kj}^* \neq 0$ and is of the same sign as $p_k^*$.

(3) $r_j^{(1)} = r_j^U.$      Only a decrease in $r_j$ is permissible. This will increase $\lambda$, provided that $c_{kj}^* \neq 0$ and is of different sign to $p_k^*$.

It seems reasonable, if there is a choice, to choose the $r_j$ which produces the largest rate of increase in $\lambda$. Hence we find the value of the index $j$ which maximizes the quantity $z_j$, where

$$z_j = |c_{kj}^*| \qquad \text{if } r_j^{(1)} = 0,$$

$$z_j = c_{kj}^* \operatorname{sign} p_k^* \qquad \text{if } r_j^{(1)} = -r_j^L,$$

$$z_j = -c_{kj}^* \operatorname{sign} p_k^* \qquad \text{if } r_j^{(1)} = r_j^U.$$

If this value is $l$, and if $z_l > 0$, then $r_l$ is chosen as the new basic variable. We accordingly reduce column $l$ to a column of zeros with a 1 in row $k$ by applying procedure A to equation $k$, column $l$. This eliminates the coefficients of $r_l$ in all the other equations and generates a column of coefficients for $r_{k'}$. The equations are now in 'standard form' again, so that we may return to the procedure described in Section 6.5.2 and consider the next increase in $\lambda$. In this increase, of course, the variable we have called $r_{k'}$, which is now non-basic, remains at the limit which it has just reached. If $z_l \leqslant 0$, no further increase in $\lambda$ is possible and the solution has been reached.

A difficulty arises if $z_l$ is positive but very small (i.e. of round-off order). This implies that $c_{kj}^*$ is very small, so that an application of procedure A multiplies the elements of $\mathbf{C}^*$ by a large number. This ill-conditioning phenomenon occurs when part of a structure is still statically indeterminate at collapse. It may be avoided by terminating the solution process when $z_l$ is of round-off order (say $z_l \leqslant 10^{-6}$ on a computer working to a precision of seven significant figures) rather than when $z_l \leqslant 0$. This feature of the analysis is best understood in terms of the collapse-mode interpretation discussed in the next section. It is shown there that the final pivotal row $k$ of $\mathbf{C}^*$ is equivalent to the deformations associated with the collapse mechanism. A coefficient in this row which is of round-off order corresponds to a point where a moment (or axial force) has reached a limiting value, but for which no significant plastic strain occurs during collapse.

### 6.5.4 *Determination of the collapse mode*

A structural designer may also wish to know the mode of collapse. This involves finding a pair of non-zero vectors $\mathbf{e}$ and $\mathbf{d}$ satisfying the compatibility equations (6.15c) and the yield conditions

$$e_j > 0, \qquad r_j = r_j^U$$

$$e_j < 0, \qquad r_j = -r_j^L$$

$$e_j = 0, \qquad -r_j^L \leqslant r_j \leqslant r_j^U.$$

The magnitude of the vector $\mathbf{e}$ may, of course, be chosen arbitrarily.

These vectors may be determined by a very simple extension of the procedure described in the previous section. We write equation (6.20) as

$$\lambda\mathbf{p} = \lambda\mathbf{Up} = \mathbf{Cr} \qquad (6.23)$$

where $\mathbf{U}$ is a unit matrix. When we apply procedure A to $\mathbf{p}$ and $\mathbf{C}$ we also operate on the rows of $\mathbf{U}$, so that instead of equation (6.22) we obtain

$$\lambda\mathbf{p}^* = \lambda\mathbf{U}^*\mathbf{p} = \mathbf{C}^*\mathbf{r} \qquad (6.24)$$

where the form of $\mathbf{C}^*$ is identical with that given in equation (6.22). This requires no extra computational effort, since each column of $\mathbf{U}^*$ is simply the set of equation multipliers computed in step (2) of procedure A, divided by the appropriate pivotal coefficient.

With this extension of procedure A the computation proceeds exactly as before. When the maximum value of $\lambda$ has been attained, the pivotal equation has the form

$$\lambda\mathbf{p}_k^* = \lambda(\mathbf{u}_k^*)^T\mathbf{p} = (\mathbf{c}_k^*)^T\mathbf{r} \qquad (6.25)$$

where $(\mathbf{u}_k^*)^T$ and $(\mathbf{c}_k^*)^T$ are the final pivotal rows of $\mathbf{U}^*$ and $\mathbf{C}^*$. A comparison between equation (6.25) and the virtual-work equation for the collapse mechanism:

$$\lambda\mathbf{d}^T\mathbf{p} = \mathbf{e}^T\mathbf{r} \qquad (6.26)$$

suggests that $\mathbf{u}_k^*$ and $\mathbf{c}_k^*$ are identical with the vectors $\mathbf{d}$ and $\mathbf{e}$, and this is indeed the case. A proof of this statement is given in Appendix 6B. All that is necessary, therefore, to obtain the mode shape is to multiply the rows $(\mathbf{u}_k^*)^T$ and $(\mathbf{c}_k^*)^T$ by sign $p_k^*$ (to make the work input to the structure positive) and to scale them in a manner suitable for plotting or printing.

*Optimum Structural Design*

The above interpretation of the rows of $\mathbf{U}^*$ and $\mathbf{C}^*$ as a mode shape holds at any stage of the calculation, regardless of whether a further increase of $\lambda$ is possible. At intermediate values of $\lambda$, however, it will be found that the mode has non-zero $e_j$s associated with non-basic variables which still have their initial zero values, which is physically incorrect. In fact the whole procedure may be interpreted in a 'mechanism' rather than an 'equilibrium' sense by imagining fictitious hinges to be present initially at these points. These fictitious hinges are gradually removed during the analysis, though some may still be present (with zero rotation) if part of the structure is statically indeterminate at collapse.

The procedure described above already has considerable advantages over a normal linear-programming formulation as far as storage is concerned. A further economy is achieved by avoiding the storage of the reduced columns of $\mathbf{C}^*$, which each contain only a single 1. If we write the original equation (6.8) in the form

$$
\begin{array}{c}
m\text{ equations} \\
s\text{ equations}
\end{array}
\quad
\lambda\begin{bmatrix} \mathbf{p}_0 \\ \mathbf{q}_0 \end{bmatrix}
= \lambda
\underbrace{\begin{bmatrix} \mathbf{I} & \mathbf{O} \\ \mathbf{O} & \mathbf{I} \end{bmatrix}}_{\substack{m\text{ cols.}\quad s\text{ cols.}\\ \text{matrix }\mathbf{U}}}
\begin{bmatrix} \mathbf{p}_0 \\ \mathbf{q}_0 \end{bmatrix}
\tag{6.27}
$$

$$
= \underbrace{\begin{bmatrix} \mathbf{A} & \mathbf{O} \\ \mathbf{B} & \mathbf{I} \end{bmatrix}}_{\substack{n\text{ cols.}\quad s\text{ cols.}\\ \text{matrix }\mathbf{C}}}
\begin{bmatrix} \mathbf{r} \\ \mathbf{q} \end{bmatrix}
$$

we see that only $n$ out of the $m + n + 2s$ columns of $\mathbf{U}$ and $\mathbf{C}$ need be stored as full columns. The last $s$ columns of $\mathbf{U}$ and $\mathbf{C}$ are identical, and, since procedure A operates equally on the rows of $\mathbf{U}$ and $\mathbf{C}$, this identity holds throughout the analysis.

In the preliminary transformation of equation (6.23) into equation (6.24), the form of equation (6.27) implies that procedure A need only be applied to the first $m$ equations, since the remainder are already in the form required. Each application of procedure A in this phase reduces one of the full columns of $\mathbf{C}$ to a single 1 and generates a corresponding new full column of $\mathbf{U}$, so that the number of columns which require storage in full is always $n$. In the second phase of the analysis the application of procedure A reduces the column of the variable which becomes basic to a single 1, and generates a new full column for the variable which was basic in the previous stage but is now non-basic. Thus the total number of columns which require storage in full is still only $n$. This suggests the possibility of a system using only an array of $m + s$ rows and $n$ columns, with a book-keeping scheme to identify the variables associated with the columns.

A Fortran program using the method described in this section is given in Reference 27.

### 6.6 Generalization to the case of combined stress limits (by C. Gavarini— a discussion)

I would like to mention some results in the field of limit analysis using linear and non-linear programming, with particular reference to the duality question.

I am concerned with frames under single stress, i.e. there is a single stress active in plasticity (actually the bending moment). This assumption was also made in the paper on duality of Charnes, Lemke and Zienkiewicz[5]. The generalization to the case of structures under combined stresses is given in my paper published in 1966[28], concerned with one-dimensional structures, and in two later papers, published in 1968[29,30], concerned with two-and three-dimensional structures, dealing with a finite-difference approach.

To outline this development, let us consider an arch divided into elements by some sections; in every section we have a plastic domain in the space of the axial force-bending moment. The frontier of the domain is assumed to be linear, or is linearized. The plastic-collapse problem is therefore discrete, and can be presented, starting from the static theorem, as a linear-programming problem. The number of variables can be reduced to the minimum by considering the hyperstatic unknowns (three in the present case). By using these generalized variables we eliminate the equalities and we get a linear program with inequalities only.

It can be demonstrated that the dual program is a formulation of the kinematic method, and that the dual variables, corresponding to the sides of the plastic domains, govern the collapse mechanism. When, in the optimum solution, a dual variable is different from zero, the corresponding side is reached, i.e. the corresponding inequality is satisfied as an equality. In that manner the static and the kinematic formulations are the two dual aspects of a linear-programming problem, and well known techniques (e.g. the simplex method) give us the collapse multiplier and both the stress distribution at collapse and the collapse mechanism.

If the plastic domain frontier is not a polygon (or, in general, not a polyhedron) we go to non-linear programming, and the previous results can again be reached[31].

Duality is relevant also in the case of the shakedown problem. In another paper[32] published in 1969, Ceradini and myself present the shakedown problem as a linear-programming problem based on Bleich-Melan's theorem, and we show that the dual program is a formulation of the kinematic Koiter's theorem.

References 33–36 are also concerned with the topic of this chapter.

*Optimum Structural Design*

## APPENDIX 6A: DERIVATION OF THE EQUILIBRIUM EQUATIONS FOR A STRAIGHT MEMBER

The notation is shown in Figure 6A.1. We consider a member with a series of loads $W_i$, and consider the moment under a particular one of these $W_S$. The three stress resultants that we use to define the state of stress at all points in the member are $\tilde{p}$ (the mean axial tension), $\tilde{m}_1$ and $\tilde{m}_2$.

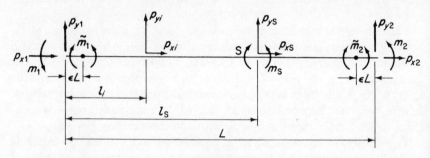

Figure 6A.1

### 6A.1 Expressions for the end-load vectors

Referring to Figure 6A.1, we have, from statics:

$$p_{x1} + p_{x2} + \sum W_{xi} = 0.$$

We therefore put

$$\left.\begin{array}{l} p_{x1} = -\tilde{p} - \tfrac{1}{2}\sum W_{xi} \\ p_{x2} = \tilde{p} - \tfrac{1}{2}\sum W_{xi}. \end{array}\right\} \tag{6A.1}$$

Also

$$\tilde{m}_1 + p_{y2}(1 - \varepsilon)L + m_2 = -\sum (l_i - \varepsilon L)W_{yi}$$

$$-\tilde{m}_2 = m_2 + p_{y2}\,\varepsilon\,L$$

$$\tilde{m}_1 = m_1 - p_{y1}\varepsilon L$$

$$p_{y2} + p_{y1} = -\sum W_{yi}.$$

where $\sum$ indicates summation.

Hence

$$p_{y2} = -\frac{\tilde{m}_1}{(1 - 2\varepsilon)L} + \frac{\tilde{m}_2}{(1 - 2\varepsilon)L} - \frac{\sum (l_i - \varepsilon L)W_{yi}}{(1 - 2\varepsilon)L}$$

$$p_{y1} = \frac{\tilde{m}_1}{(1 - 2\varepsilon)L} - \frac{\tilde{m}_2}{(1 - 2\varepsilon)L} - \frac{\sum (L - l_i - \varepsilon L)W_{yi}}{(1 - 2\varepsilon)L}$$

$$m_1 = \left(\frac{1 - \varepsilon}{1 - 2\varepsilon}\right)\tilde{m}_1 - \left(\frac{\varepsilon}{1 - 2\varepsilon}\right)\tilde{m}_1 - \frac{\varepsilon}{1 - 2\varepsilon}\sum (L - l_i - \varepsilon L)W_{yi}$$

$$m_2 = \left(\frac{\varepsilon}{1 - 2\varepsilon}\right)\tilde{m}_1 - \left(\frac{1 - \varepsilon}{1 - 2\varepsilon}\right)\tilde{m}_2 + \frac{\varepsilon}{1 - 2\varepsilon}\sum (l_i - \varepsilon L)W_{yi}.$$

$$(6A.2)$$

We now define

$$c_{1i} = L - l_i - \varepsilon L, \qquad c_{2i} = l_i - \varepsilon L,$$

$$E = \frac{1}{(1 + 2\varepsilon)L}, \qquad F_1 = \frac{1 - \varepsilon}{1 - 2\varepsilon}, \qquad F_2 = \frac{\varepsilon}{1 - 2\varepsilon}.$$

Equations (6A.2) may now be written

$$p_{y1} = E\tilde{m}_1 - E\tilde{m}_2 - E\sum C_{1i}W_{yi}$$
$$m_1 = F_1\tilde{m}_1 - F_2\tilde{m}_2 - F_2\sum C_{1i}W_{yi}$$
$$p_{y2} = -E\tilde{m}_1 + E\tilde{m}_2 - E\sum C_{2i}W_{yi}$$
$$m_2 = F_2\tilde{m}_1 - F_1\tilde{m}_2 + F_2\sum C_{2i}W_{yi}.$$

$$(6A.3)$$

Combining equations (6A.1) and (6A.3), we obtain

$$\begin{bmatrix} p_{x1} \\ p_{y1} \\ m_1 \end{bmatrix} = \begin{bmatrix} -1 & 0 & 0 \\ 0 & E & -E \\ 0 & F_1 & -F_2 \end{bmatrix} \begin{bmatrix} \tilde{p} \\ \tilde{m}_1 \\ \tilde{m}_2 \end{bmatrix} - \begin{bmatrix} \frac{1}{2}\sum W_{xi} \\ E\sum C_{1i}W_{yi} \\ F_2\sum C_{1i}W_{yi} \end{bmatrix}$$

$$\begin{bmatrix} p_{x2} \\ p_{y2} \\ m_2 \end{bmatrix} = \begin{bmatrix} 1 & 0 & 0 \\ 0 & -E & E \\ 0 & F_2 & -F_1 \end{bmatrix} \begin{bmatrix} \tilde{p} \\ \tilde{m}_1 \\ \tilde{m}_2 \end{bmatrix} - \begin{bmatrix} \frac{1}{2}\sum W_{xi} \\ E\sum C_{2i}W_{yi} \\ -F_2\sum C_{2i}W_{yi} \end{bmatrix}$$

or

$$p_1 = H_1 r - h_1$$
$$p_2 = H_2 r - h_2.$$

$$(6A.4)$$

## 6A.2 Transformation to system axes

Transforming into the system axes shown in Figure 6A.2, we have

$$
\begin{bmatrix} p'_x \\ p'_y \\ m' \end{bmatrix} = \begin{bmatrix} \cos\alpha & -\sin\alpha & 0 \\ \sin\alpha & \cos\alpha & 0 \\ 0 & 0 & 1 \end{bmatrix} \begin{bmatrix} p_x \\ p_y \\ m \end{bmatrix}.
$$

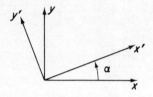

**Figure 6A.2**

Writing this as $p' = Tp$, equations (6A.4) become

$$p'_1 = TH_1 r - Th_1$$
$$p'_2 = TH_2 r - Th_2$$

or

$$
\left. \begin{array}{l} p'_1 = H'_1 r - h'_1 \\ p'_2 = H'_2 r - h'_2 . \end{array} \right\} \tag{6A.5}
$$

where

$$
H'_1 = TH_1 = \begin{bmatrix} -\cos\alpha & -E\sin\alpha & E\sin\alpha \\ -\sin\alpha & E\cos\alpha & -E\cos\alpha \\ 0 & F_1 & -F_2 \end{bmatrix}
$$

$$
H'_2 = TH_2 = \begin{bmatrix} \cos\alpha & E\sin\alpha & -E\sin\alpha \\ \sin\alpha & -E\cos\alpha & E\cos\alpha \\ 0 & F_2 & -F_1 \end{bmatrix}
$$

$$
h'_1 = Th_1 = \begin{bmatrix} \cos\alpha \sum W_{xi}/2 - E\sin\alpha \sum C_{1i}W_{yi} \\ \sin\alpha \sum W_{xi}/2 + E\cos\alpha \sum C_{1i}W_{yi} \\ F_2 \sum C_{1i}W_{yi} \end{bmatrix}
$$

$$h'_2 = Th_2 = \begin{bmatrix} \cos \alpha \sum W_{xi}/2 - E \sin \alpha \sum C_{2i}W_{yi} \\ \sin \alpha \sum W_{xi}/2 + E \cos \alpha \sum C_{2i}W_{yi} \\ -F_2 \sum C_{2i}W_{yi} \end{bmatrix}.$$

## 6A.3   Moments at intermediate points

Referring to Figure 6A.1, we consider the equilibrium of the member to the left of the point S. We have

$$p_{yi}l_S + m_S + \sum_{>} W_{yi}(l_S - l_i) - m_1 = 0$$

where $\sum_{>}$ indicates summation over only those loads for which $l_S > l_i$. Substituting for $p_{y1}$ and $m_1$ from equations (6A.3), we obtain

$$(E\tilde{m}_1 - E\tilde{m}_2 - E \sum C_{1i}W_{yi})l_S + m_S + \sum_{>} W_{yi}(l_S - l_i)$$

$$- F_1\tilde{m}_1 + F_2\tilde{m}_2 + F_2 \sum C_{1i}W_{yi} = 0$$

whence

$$-EC_{1S}\tilde{m}_1 - EC_{2S}\tilde{m}_2 + m_S = EC_{2S} \sum C_{1i}W_{yi} - \sum_{>} W_{yi}(l_S - l_i)$$

where, as before, $\sum$ indicates summation over *all* loads on the member. This may be written in the form

$$m_S = [EC_{2S} \sum C_{1i}W_{yi} - \sum_{>} W_{yi}(l_S - l_i)] - [0, -EC_{1S}, -EC_{2S}]r$$

or

$$m_S = m_{0S} - g^T r.$$

This corresponds to the form used in equation (6.3), $g^T$ being the row of $G$ corresponding to the element $m_S$ of $q$.

## APPENDIX 6B. PROOF OF THE COLLAPSE-MODE DERIVATION

The procedure described in Section 6.5 converts the set of equilibrium equations

$$\lambda p = Up = Cr \qquad (6B.1)$$

into a series of sets of equations of general form

$$\lambda p^* = U^*p = C^*r. \qquad (6B.2)$$

Equation (6B.2) is the result of a series of applications of procedure A to the *rows* of $U$ and $C$. The corresponding compatibility equations transform

*Optimum Structural Design*

into

$$\mathbf{d} = \mathbf{U}^T\mathbf{d}, \qquad \mathbf{e} = \mathbf{C}^T\mathbf{d} \qquad\qquad (6B.3)$$

$$\mathbf{d} = (\mathbf{U}^*)^T\mathbf{d}^*, \qquad \mathbf{e} = (\mathbf{C}^*)^T\mathbf{d}^*, \qquad\qquad (6B.4)$$

where $\mathbf{d}^*$ may be regarded as a vector of generalized displacements, which corresponds to $\mathbf{p}^*$ in a work sense.

If we look at the equation $\mathbf{e} = (\mathbf{C}^*)^T\mathbf{d}^*$ at the conclusion of the step described in Section 6.5.2., we find it has a form such as:

$$
\begin{array}{l}
\text{basic} \\
\text{non-basic} \\
\text{basic} \\
\text{non-basic} \\
\text{non-basic} \\
\text{basic} \\
\text{non-basic} \\
\text{non-basic} \\
\text{non-basic} \\
\boxed{\text{basic}} \\
\end{array}
\begin{bmatrix}
\cdot \\ \cdot \\ \cdot \\ \cdot \\ \cdot \\ \cdot \\ \cdot \\ \cdot \\ \cdot \\ e_{k'} \\ \cdot \\ \cdot \\ \cdot
\end{bmatrix}
=
\begin{bmatrix}
0 & 1 & 0 & 0 & 0 & 0. \\
\cdot & \cdot & \cdot & \cdot & \cdot & \cdot \\
0 & 0 & 1 & 0 & 0 & 0 \\
\cdot & \cdot & \cdot & \cdot & \cdot & \cdot \\
\cdot & \cdot & \cdot & \cdot & \cdot & \cdot \\
1 & 0 & 0 & 0 & 0 & 0 \\
\cdot & \cdot & \cdot & \cdot & \cdot & \cdot \\
\cdot & \cdot & \cdot & \cdot & \cdot & \cdot \\
\cdot & \cdot & \cdot & \cdot & \cdot & \cdot \\
0 & 0 & 0 & 1 & 0 & 0 \\
\cdot & \cdot & \cdot & \cdot & \cdot & \cdot \\
\cdot & \cdot & \cdot & \cdot & \cdot & \cdot \\
\cdot & \cdot & \cdot & \cdot & \cdot & \cdot
\end{bmatrix}
\begin{bmatrix}
\cdot \\ \cdot \\ \cdot \\ \cdot \\ d_k^* \\ \cdot \\ \cdot \\ \cdot
\end{bmatrix}
\qquad (6B.5)
$$

$\leftarrow$ row $k'$

$$\uparrow$$
$$\text{column } k$$

where column $k$ is identical to the pivotal row $k$ of $\mathbf{C}^*$, and $k'$ identifies the basic variable, which has reached one or other of its limits.

Since the basic stress resultant $r_{k'}$ has reached one of its limits, the corresponding deformation variable $e_{k'}$ may be non-zero. However, all the *other* basic stress-resultants $r_{i'}$ ($i' \neq k'$) are still strictly *within* their limits, so that the corresponding deformations must be zero. Since from equation (6B.5) each basic $e_{i'}$ is equal to its associated $d_i^*$, it follows that all the $d_i^*$ must be zero, except for $d_k^*$. Thus equation (6B.5) becomes

$$\mathbf{e} = \text{column } k \text{ of } (\mathbf{C}^*)^T \times d_k^*$$

$$= \text{row } k \text{ of } \mathbf{C}^* \times d_k^*$$

and equations (6B.4) give also

$$\mathbf{d} = \text{column } k \text{ of } (\mathbf{U}^*)^T \times d_k^*$$

$$= \text{row } k \text{ of } \mathbf{U}^* \times d_k^*.$$

This is the result we set out to prove, $d_k^*$ being the arbitrary multiplier determining the overall amount of deformation in the structure.

## References

1. T. C. Koopmans (Ed.), *Activity Analysis of Production and Allocation*, Wiley, New York, 1951.
2. G. B. Dantzig, *Linear Programming and Extensions*, Princeton University Press, 1963, p. 26.
3. B. G. Neal and P. S. Symonds, 'The calculation of the collapse loads for framed structures', *Jour. ICE*, **35**, 21 (1950).
4. A. Charnes and H. J. Greenberg, 'Plastic collapse and linear programming', Summer Meeting of the American Mathematical Society, 1951.
5. A. Charnes, C. E. Lemke and O. C. Zienkiewicz, 'Virtual work, linear programming and plastic limit analysis', *Proc. Roy. Soc. A.*, **251**, 110 (1959).
6. J. Foulkes, 'Minimum weight design and the theory of plastic collapse', *Quart. Appl. Math.*, **10**, 347 (1953).
7. J. Foulkes, 'Linear programming and structural design', in *Proc. 2nd Symp. on Linear Programming*, National Bureau of Standards, 1955, p. 177.
8. J. Heyman, 'Plastic design of beams and plane frames for minimum material consumption', *Quart. Appl. Math.*, **8**, 373 (1951).
9. J. Heyman, 'Plastic design of plane frames for minimum weight', *The Structural Engineer*, **31**, 125 (1953).
10. M. R. Horne, 'A moment distribution method for the analysis and design of structures by the plastic theory', *Proc. ICE,* **3**, Part III, 51 (1954).
11. W. Prager, 'Limit analysis and design', *Jour. Amer. Concrete Inst.*, **25**, No. 4, 297 (1953).
12. W. Prager, 'Minimum weight design of a portal frame', *Proc. ASCE*, **82**, EM4, paper 1073 (1956).
13. J. Heyman and W. Prager, 'Automatic minimum weight design of steel frames', *Jour. Franklin Inst.*, **266**, No. 5, 339 (1958).
14. R. K. Livesley, 'The automatic design of structural frames', *Quart. Jour. Mech. Appl. Math.*, **9**, 257 (1956).
15. J. Robinson, *Structural Matrix Analysis for the Engineer*, Wiley, 1966.
16. R. K. Livesley, 'The selection of redundant forces in structures', *Proc. Roy. Soc. A.*, **301**, 493 (1967).
17. J. P. Baty and J. Wright, 'Plastic analysis and design of framed structures', International symposium on the use of electronic digital computers in civil engineering, University of Newcastle, July, 1966, Paper 61.
18. G. B. Dantzig, *Linear Programming and Extensions*, Princeton University Press, 1963.
19. S. I. Gass, *Linear Programming*, 3rd ed., McGraw-Hill, 1969.
20. G. Hadley, *Linear Programming*, Addison-Wesley, 1962.
21. R. K. Livesley, *Matrix Methods of Structural Analysis*, Pergamon, 1964.
22. A. Charnes and C. E. Lemke, 'Computational theory of linear programming, I', ONR Res. Mem. No. 10, Graduate School of Industrial Administration, Carnegie Institute of Technology, 1954.
23. W. S. Dorn and H. J. Greenberg, 'Linear programming and plastic limit analysis of structures', *Quart. Appl. Math.*, **15**, 155 (1957).

24. A. R. Toakley, 'Some computational aspects of optimum rigid–plastic design', *Int. Jour. Mech. Sci.*, **10**, 531 (1968).
25. J. Heyman, *Plastic Design of Frames*, Vol. 2, Cambridge University Press, 1971, p. 132.
26. J. Heyman, 'Minimum weight of frames under shakedown loading', *Proc. ASCE*, **84**, EM4, paper 1790 (1958).
27. R. K. Livesley (to be published in *Int. Journ. Numerical Methods Engineering*).
28. C. Gavarini, 'Plastic analysis of structures and duality in linear programming', *Meccanica*, No. 3/4 (1966).
29. G. Ceradini and C. Gavarini, 'Calcolo a rottura e programmazione lineare. Continui bi e tri dimensionali. Nota I: fondamenti teorici', *Giornale del Genio Civile*, No. 2/3 (1968).
30. G. Ceradini and C. Gavarini, 'Calcolo a rottura e programmazione lineare. Continui bi e tri dimensionali. Nota II: applicazioni a piastre e volte di rivoluzione', *Giornale del Genio Civile*, No. 4 (1968).
31. C. Gavarini, 'Calcolo a rottura e programmazione non lineare', Istituto Lombardo di Scienze e Lettere, Rend. Sc., 1968.
32. G. Ceradini and C. Gavarini, 'Applicazione della programmazione lineare ai problemi di adattamento plastico statico e dinamico', *Giornale del Genio Civile*, No. 8 (1969).
33. G. Ceradini and C. Gavarini, 'Calcolo a rottura e programmazione lineare', *Giornale del Genio Civile*, No. 1/2 (1965).
34. C. Gavarini, 'The fundamental theorems of plastic analysis and duality in linear programming', *Ingegneria Civile*, No. 18 (1966).
35. G. Ceradini and C. Gavarini, 'Some applications of linear programming to plastic analysis of structures', International symposium on the use of electronic digital computers in civil engineering, University of Newcastle, July, 1966.
36. R. Casciaro and A. Di Carlo, 'Formulazione della analisi limite delle piastre come problema di minimax mediante una rappresentazione agli elementi finiti del campo delle tensioni', *Giornale del Genio Civile*, No. 2 (1970).

*Chapter 7*

# Shape Optimization and Sequential Linear Programming

*O. C. Zienkiewicz and J. S. Campbell*

## 7.1 Introduction

Structures such as dams, bridges and pressure vessels are encountered frequently in civil engineering. The functional constraints and those imposed by construction techniques frequently prevent anything approaching fully stressed design being reached in such structures. At the same time geometrical boundary freedom is often great, and the engineer can give the structure a variety of forms, as is borne out by, for example, the wide range of shapes occurring in dams.

The analysis of such structures has always presented serious difficulties owing to their 'continuum' nature, and it is only in the last decade that the finite-element method has allowed a properly constituted analysis to be carried out as a matter of course. It is, perhaps, not surprising therefore to find that often the engineer is satisfied if a *reasonable and feasible design is reached* and stops his efforts there. Nevertheless, if the possibilities offered by the advanced analysis processes are to be fully explored, some form of optimization is an obvious necessity, especially when the prime cost of such structures is considered.

In this chapter we shall explore the possibility of the optimization of such continuum structures with the assumptions that:

(a) Analysis is being carried out by the finite-element process.
(b) A reasonable constitutive behaviour of the structural material is properly described.
(c) Criteria governing the design are established and quantified.

The last two are, indeed, important assumptions, and the validity of any basis chosen here can often be questioned. Although the attention of the engineer must always be focused on improvement of the basis from which the design starts, the problem of achieving the best design *within a set of self-imposed rules* will now be considered.

In almost all formal optimization processes we are concerned with the determination of *sensitivities* (or gradients) of the *objective function W* and of the *constraints* $\mathbf{g}(\mathbf{x})$ for changes of the *design variables* $\mathbf{x}$.

109

In general only certain *explicit constraints* are simple linear functions of the design variables, and both the objective function and *implicit* constraints (usually concerned with stresses or deformations in the structure) are dependent in a non-linear manner on x. Further, the determination of the gradient of the implicit constraints in the problems discussed here is dependent on the *analysis* of the structure and often presents the most expensive part of the optimization process. We shall therefore devote the major part of this chapter to devising an economical method of determining the sensitivities of *stress/ deformation quantities*. We shall introduce here the concept of 'force derivatives', which is found to be an efficient process for the type of structure encountered.

Once the sensitivities for a particular design stage are known, the question of the optimization method to be used is open. Here obviously many possibilities could be used economically. The problems of optimization of shape are usually characterized as follows:

(1) A large number of design variables are involved (the shape of a concrete structure, for instance, may be varied widely and continuously without discontinuities of cost).

(2) Continuously defined criteria are expressed in the stress and displacement constraints.

Under such circumstances the process using a *sequence of linear-programming operations* is economical. We shall therefore describe here some details of this optimization method.

It should be borne in mind that the determination of the sensitivities is completely separable from the optimization process. Indeed, even if no formal optimization procedure is used (following, for instance, the interactive trial-and-error methods described in Chapter 17) sensitivities will obviously be useful as a guide to the best directions in which to implement changes, and the methods described here should be found to be of assistance.

## 7.2 A typical problem

To clarify the type of problem with which we are concerned, Figure 7.1 shows, schematically, two typical structures used in civil engineering for damming wide valleys. Within certain specified dimensions (such as the total height of the structure and/or the base width) a wide variety of possible shapes exist within each *type* of design shown.

The problem is to design the 'best' shape for either type of structure by varying certain design parameters. Obviously each type of structure would be considered separately.

*Design variables* will include parameters such as certain sizes or coefficients describing boundaries in a continuous manner. Material properties can also

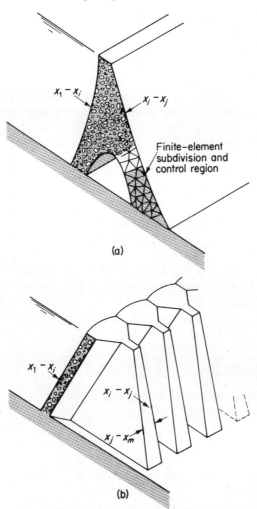

(a)

(b)

**Figure 7.1** Two types of dam for which shape optimization can be applied; design variables include the shape of the outside boundaries, the size of the hole or the thickness of the buttress

enter as 'design variables'. Although the limits on the concrete-deformation properties are small and not controlled by the designer, the properties of the foundation (which clearly constitutes here part of the structure) are only known within certain, often very wide, limits. It might be reasonable therefore to consider these in the design and to vary them within a prescribed range as to achieve the *worst* conditions. While this may be difficult in practice in this type of problem, we shall consider the derivatives with respect to material properties so as to achieve generality.

The *objective function* will obviously be determined by the volume of the concrete, the area of framework, etc. The costing of such features as, for example, the hole in the dam shown in Figure 7.1(a) will have to be taken into account at this stage. The important point is that, although the objective function is a non-linear function of $\mathbf{x}$, its derivatives can readily be determined by a direct variation of the design variables. We shall not, therefore, give further consideration to the determination of such gradients as $\nabla W = \partial W / \partial x_i$, etc.

*Constraints*, whether of an implicit or an explicit nature, have now to be formulated. Explicit constraints will limit directly the range of the design parameters and will be governed by *experience*, which forms an important part of the judgement concerned. *Implicit* constraints will, typically, concern stress magnitudes and, on occasions, deflections. For example, we shall usually try to limit the magnitude of the principal tensile stress and possibly impose a limit on some critical combination of stresses in the compressive region. A difficulty immediately arises owing to the continuous nature of the problem. Apparently an infinite number of constraints are necessary if the constraint is considered as a stress at any point. Even with a finite-element discrete representation the number of points at which the stress is given may be rather large. For practical purposes we shall restrict our attention to either:

(a) a few critical points at which stress is to be considered, or
(b) a maximum value of stress in a few critical subregions.

Thus, if we follow the second scheme and apply it to the example shown in Figure 7.1(a):

$$g_i(\mathbf{x}) = \sigma_T - \sigma_{Tm} \geqslant 0 \qquad (7.1)$$

is a constraint in which $\sigma_T$ is the permissible *tensile* stress and $\sigma_{Tm}$ stands for the value of maximum tensile stress in the shaded subregion composed of several elements.

To derive the sensitivity $\nabla g_i = \partial g_i / \partial x_j$, we should consider the worst stressed point of the region at a *particular design stage*. Once the design has been altered, however, we may well switch our attention to another point in the same region. Practical experience shows that often the process will coincide with consideration of one worst point, and discontinuities seldom arise.

Similar methods will, of course, be applied to deflections, strain or other implicit constraints.

*Loading* on the structure for which the design variables have to be optimized may be multiple. For instance, in a dam we shall have the situation of gravity loads acting either alone or combined with water-pressure loads. Thermal loads (strains) imposed on the structures may also have to be taken into account. *For each loading the constraints have to be applied and their gradients determined.*

A not dissimilar problem is one illustrated in Figure 7.2. Here the shape of a rotating turbine disk has to be optimized subject to a given load condition

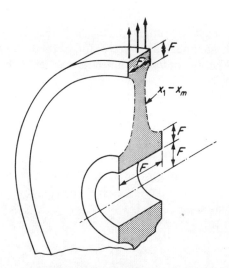

**Figure 7.2** A rotating turbine disk; dimensions marked $F$ may be prescribed, design variables $x_1$–$x_m$ define other dimensions. Forces prescribed and constraints imposed on second stress invariant

and a stress constraint. De Silva (see Chapter 8) considers this as a one-dimensional problem. In the present context a two-dimensional consideration of stress concentrations, etc., would be given.[1]

### 7.3 Relationship of continuum optimization to fully stressed design

In Chapter 3, attention was focused on the so-called 'fully stressed design' process as an approximation to the minimum-weight design. Various simple processes leading to the establishment of such fully stressed conditions were suggested (in particular, the stress-ratio method of scaling.) In the continuum problem the geometric constraints implied on the problem are such that, in general, fully stressed design is not possible and the procedures of scaling are not appropriate. In the dam illustrated in Figure 7.1(a) it is fairly obvious that the thicknesses of elements cannot be varied and, indeed, fully stressed conditions may only appear at the extremes of a section. Indeed, the objective function which now is used may be such as to cause an optimum in which the majority of the critical subregions is not fully stressed under any load combination.

### 7.4 Sensitivity—force-derivative concepts

In a linear elastic situation, when analysis is being carried out by the stiffness process with finite-element formulation implied[2], we have to solve a problem:

$$\mathbf{K\Delta} = \mathbf{P} \tag{7.2}$$

where $\mathbf{K}$ is the assembled stiffness matrix, $\mathbf{\Delta}$ is the vector of displacement parameters and $\mathbf{P}$ is a (nodal) force vector (or vectors). Usually both $\mathbf{K}$ and $\mathbf{P}$ will be dependent on the design parameters $\mathbf{x}$, which determine dimensions and material properties.

In a typical finite-element context, the major part of cost of analysis is in the solution of equation (7.2) for $\mathbf{\Delta}$, the computation of stresses, etc., being comparatively trivial and included in the normal output. We are therefore interested in the most economical way of obtaining the sensitivities $\partial \Delta_j / \partial x_i$.

While separate analyses for neighbouring values of $x_i$ and $x_i + \delta x_i$ are the obvious possibility, the cost of such an operation would require two analyses for each variable and be excessive. To avoid this we can differentiate equation (7.2), obtaining:

$$\mathbf{K} \frac{\partial \mathbf{\Delta}}{\partial x_i} + \frac{\partial \mathbf{K}}{\partial x_i} \mathbf{\Delta} = \frac{\partial \mathbf{P}}{\partial x_i} \tag{7.3}$$

and find the change of $\mathbf{\Delta}$ corresponding to $\delta x_i$ as

$$\delta \mathbf{\Delta} = \frac{\partial \mathbf{\Delta}}{\partial x_i} \delta x_i = \mathbf{K}^{-1} \left( \frac{\partial \mathbf{P}}{\partial x_i} \delta x_i - \frac{\partial \mathbf{K}}{\partial x_i} \cdot \mathbf{\Delta} \, \delta x_i \right). \tag{7.4}$$

This is equivalent to solving the same equation as equation (7.2) for a new set of forces, a process that can be carried out at a fraction of the cost required for a complete reanalysis. The quantity in parentheses in equation (7.4), reduced to a unit value of $\delta x_i$, is known as the *force derivative*, and this, if it is substituted directly into a standard finite-element program, will result immediately in the derivatives of the displacement and stress quantities (see also Chapters 8 and 11, p. 136 and p. 209).

The concept of the force derivative has been illustrated here for the case of *linear structures* where the solution is governed by equation (7.2). This limitation is, however, unnecessary and the process can be extended to structures in which material or geometric non-linearities are present. In such situations the analysis is often incremental[2], and equation (7.2) is valid for changes of force and displacement, i.e.:

$$\mathbf{K_T} \, \delta \mathbf{\Delta} = \delta \mathbf{P} \tag{7.5}$$

where $\mathbf{K_T}$ is the tangential stiffness matrix. The force derivatives can once again be determined from equation (7.4), but now using the appropriate substitutions for $\mathbf{K}$ and $\mathbf{P}$.

**7.5 Computation of force derivatives in the finite-element context**

To determine the 'force derivative'

$$\hat{\mathbf{P}}^i \equiv \frac{\partial \mathbf{P}}{\partial x_i} - \frac{\partial \mathbf{K}}{\partial x_i} \Delta \tag{7.6}$$

we can proceed in a variety of ways.

*Algorithm I*   The most obvious process is to vary the design parameter $x_i$ by a small amount $\delta x_i$ and compute, at a given design point (after the primary analysis has been carried out), the changes $\delta \mathbf{P}^l$ and $\delta \mathbf{K}^l$ for each element, carrying out the multiplication $\delta \mathbf{K}^l \Delta^l$ using the results of the analysis at the design points. Following the assembly rule of stiffness analysis, we have immediately for the $j$th component

$$\hat{\mathbf{P}}_j^i = \left( \sum_{l=1}^{m} \delta \mathbf{P}_j^l - \sum \delta \mathbf{K}^l \Delta^l \right) \Big/ \delta x_i. \tag{7.7}$$

In practice both element forces and stiffness matrices are frequently evaluated by the numerical integration of appropriate expressions. Two calculations of stiffness matrices will thus be required for a changed element shape or varied material properties if these are the design variable.

Although the details of the finite-element procedure are given elsewhere, it is convenient at this stage to summarize the 'displacement' method of approach. In this approach the displacements $\mathbf{f}$ through the body are determined approximately by prescribed shape functions $\mathbf{N}$ and the nodal displacement parameters throughout the volume $\Omega$. Thus

$$\mathbf{f} = \mathbf{N}_j \Delta_j = \mathbf{N}\Delta. \tag{7.8}$$

The strains at any point follow as

$$\varepsilon = \mathbf{B}\Delta \tag{7.9}$$

where once again $\mathbf{B}$ is a known function of position. By virtual work (or the minimization of potential energy), we have

$$\int_\Omega \delta \varepsilon^\mathrm{T} \sigma \, d\Omega - \int_\Omega \delta \mathbf{f} \, \mathbf{p} \, d\Omega - \delta \Delta^\mathrm{T} \mathbf{R}_j = 0$$

$$\int_\Omega \mathbf{B}^\mathrm{T} \sigma \, d\Omega - \int_\Omega \mathbf{N}^\mathrm{T} \mathbf{p} \, d\Omega - \mathbf{R}_j = 0 \tag{7.10}$$

in which $\mathbf{p}$ represents the body forces, $\mathbf{R}$ represents the *external* nodal forces and $\sigma$ represents the stresses. If an elastic constitutive relation is assumed, we have the standard form:

$$\sigma = \mathbf{D}\varepsilon$$

116     *Optimum Structural Design*

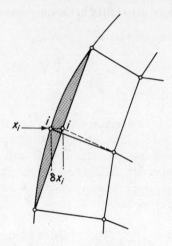

**Figure 7.3**  Effect of changing a design variable affecting the boundary of finite element

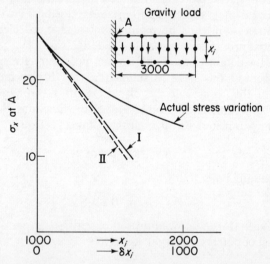

**Figure 7.4**  Stress variation at a point of a cantilever under gravity load with depth as a design variable, showing actual stresses and predictions by algorithms I or II

and

$$\mathbf{K}\boldsymbol{\Delta} = \int_{\Omega} \mathbf{B}^{\mathrm{T}}\boldsymbol{\sigma} \, d\Omega = \int_{\Omega} \mathbf{B}^{\mathrm{T}}\mathbf{D}\mathbf{B} \, d\Omega \, \boldsymbol{\Delta} \qquad (7.11)$$

$$\mathbf{P} = \mathbf{R} + \int_{\Omega} \mathbf{N}^{\mathrm{T}}\mathbf{p} \, d\Omega$$

where **D** is the matrix of elastic constants.

Determination of the force derivative now follows:

$$\delta \mathbf{K} = \int_{\Omega_1} \mathbf{B}^{\mathrm{T}} \mathbf{D} \mathbf{B} \, d\Omega - \int_{\Omega_2} \mathbf{B}^{\mathrm{T}} \mathbf{D} \mathbf{B} \, d\Omega \qquad (7.12a)$$

and

$$\delta \mathbf{P} = \int_{\Omega_1} \mathbf{N}^{\mathrm{T}} \mathbf{p} \, d\Omega - \int_{\Omega_2} \mathbf{N}^{\mathrm{T}} \mathbf{p} \, d\Omega \qquad (7.12b)$$

and $\Omega_1$ and $\Omega_2$ represent the domains before and after the variation of the design variable $x_i$.

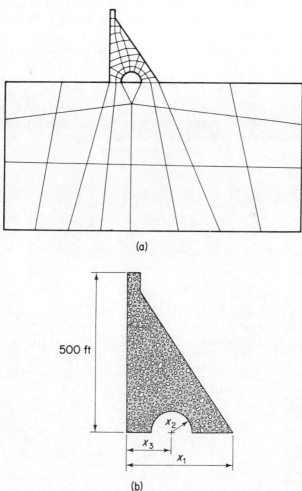

(a)

500 ft

(b)

**Figure 7.5** Mesh subdivision and design variables for the problem of a hollow gravity dam

The obvious process is thus to evaluate the force derivative using equations (7.12) and (7.7) for two 'neighbouring' element sizes. In Figure 7.3 we show, for instance, the element configurations for changes of a variable $x_i$ which here is simply the horizontal coordinate of a boundary point.

*Algorithm II*  The exact determination of the changes of $\delta K$ and $\delta P$ and the subsequent multiplication by $\Delta$ can be eliminated if we observe that the major change of these quantities is due to the changes of the integration region, notwithstanding the fact that the integrand itself is dependent on $x_i$.

Thus, for instance, the change $\delta P$ in equation (7.12b) could be found by retaining the previously calculated values of $N$ and evaluating the integral over the shaded region of Figure 7.3, which corresponds to the change of the domain of integration.

It is, however, more important to observe that, from equation (7.11),

$$\delta K \, \Delta = \int_{\Omega_1} B^T \sigma \, d\Omega - \int_{\Omega_1} B^T \sigma \, d\Omega \qquad (7.13)$$

and that one multiplication can be saved by simply integrating the quantity already computed during the analysis for the changes of the regions.

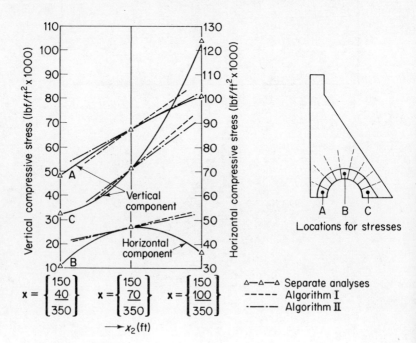

Figure 7.6  Stress changes and sensitivities predicted by algorithms I and II for the dam of Figure 7.5 subjected to gravity and water loading at $x_1 = 150, x_2 = 70, x_3 = 350$

Interpreted in this way, the force derivatives are not only simpler to calculate, but acquire a physical meaning. They are simply the nodal forces required to restore the equilibrium if the shaded portions of the structure are removed. This simple explanation of the force-derivative concept was provided first by Irons[3].

In both algorithms discussed here the obvious process of recomputing a specified quantity has been followed. This avoids the derivation of special expressions and the complication of available programs. It is clearly possible to derive, for specific elements, exact expressions for the derivatives of $K$ and $P$ arising in equation (7.4).

Comparison of the two algorithms with the actual variation of stress and displacement quantities for finite changes of geometric design parameters is shown in Figures 7.4 to 7.6. It will be observed that predictions for both algorithms are reasonable when compared with the actual non-linear variations, and the simpler one is therefore recommended.

### 7.6 Adaptation of finite-element programs for optimization

It is well known that solution of the basic stiffness equation [equation (7.2)] can be accomplished relatively cheaply for multiple-load vectors (providing, of course, that the program is not based on an iterative algorithm, but uses some form of elimination procedure). Here it is important to distinguish between two forms of process:

(1) *Re-solution*, in which the load vectors cannot be specified *a priori* and only one is done at a time.
(2) *Simultaneous* solution, in which a large set of load vectors is dealt with at the same time.

The first process is more costly, often requiring some 10–15 per cent of the original solution time for each additional vector, while the second will process a large set of vectors for a fraction of that cost.

Both types of calculation arise in the computation of sensitivities. First, analysis is carried out at the given design stage for all the load vectors specified. Secondly, using the information computed on stresses and displacements, the force derivatives are calculated. For maximum economy all force derivatives corresponding to each loading case and each design variable must be computed at this stage and the program re-entered for a *simultaneous* solution of multiple vectors.

At this second stage of sensitivity calculation, a very large number of vectors will arise [corresponding to the triple product of the number of load cases, the number of stress (displacement) constraints and the number of design variables].

This type of operation is envisaged in a computer system recently developed at the University College of Swansea, and provision is made for such a computation at minimal cost[4,5]. Figure 7.7 shows how a typical finite-element analysis sequence is modified to obtain sensitivities. It should be borne in mind that, unless the design variables are simply the nodal coordinates (which is seldom the case), a special 'preprogram' must be included to convert the design variables into the standard finite-element input. Generally such preprograms will be specially written for specific problems.

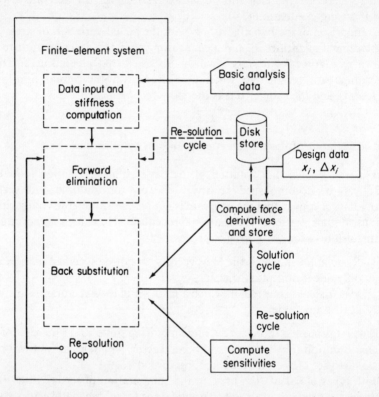

**Figure 7.7** Schematic representation of the force-derivative computational process

## 7.7 Sequential linear programs

Chapter 6 focused attention on problems for which linear programming provides a ready solution. Indeed, the background to linear programming and basic algorithms are discussed there in detail and show that many efficient programs exist for the solution of such problems. The speed with which optimization of problems with many variables can be accomplished by linear programming is such that its use in non-linear situations in an iterative manner becomes attractive. The first suggestions of such successive linearization were made by Cheney and Goldstein[6], Kellog[7] and Griffith and Stewart[8] in a mathematical context, and were followed by a series of applications in the structural field[9-16].

With the weight function $W$ and typical constraints $g_i \geqslant 0$ as non-linear functions of $x$, we can expand these in the neighbourhood of a design point $x^n$, neglecting higher-order terms, as

$$W = W^n + \left(\frac{\partial W}{\partial x_1}\right)^n \delta x_1 + \cdots + \left(\frac{\partial W}{\partial x_m}\right)^n \delta x_m$$

$$= W^n + \mathbf{V}_n^T W . \delta x \qquad (7.14a)$$

and similarly

$$g_i = g_i^n + \mathbf{V}_n^T g_i . \delta x \qquad (7.14b)$$

where $\mathbf{V}_n W$ and $\mathbf{V}_n g_i$ stand for the vectors of sensitivities with respect to all the design variables at design point $x^n$.

With such a linearization the non-linear-programming problem can be written as a typical linear-programming situation. Noting that

$$\delta x = x - x^n$$

and that with a suitable choice of origin all $x_i$ can be kept positive, we arrive at a standard linear-programming 'tableau':

$$
\begin{bmatrix}
\mathbf{V}^T W^n \\
\cdots \cdots \\
\mathbf{V}^T g_1 \\
\vdots \\
\mathbf{V}^T g_i \\
\cdots \cdots \\
g_{i+1} \\
\vdots \\
g_m \\
\cdots \cdots \\
\mathbf{I}
\end{bmatrix}
\mathbf{x} \geqslant
\begin{bmatrix}
\text{minimum} \\
\cdots \cdots \cdots \\
\mathbf{V}^T g_1 x^n - g_1^n \\
\vdots \\
\mathbf{V}^T g_i x^n - g_i^n \\
\cdots \cdots \cdots \\
0 \\
\vdots \\
0 \\
\cdots \cdots \cdots \\
\mathbf{O}
\end{bmatrix}
\left.\begin{matrix} \\ \\ \\ \\ \end{matrix}\right\} \text{explicit constraints}
\qquad (7.15)
$$

Solution for $x^{n+1}$ is achieved from this by procedures described in Chapter 6.

Figure 7.8 shows the process of linearization applied to a problem with three non-linear constraints. The linearization is carried out at a design stage $x^n$ which is in the 'non-feasible' region—a situation very frequently encountered. The optimum of the linearized problem is seen still to be

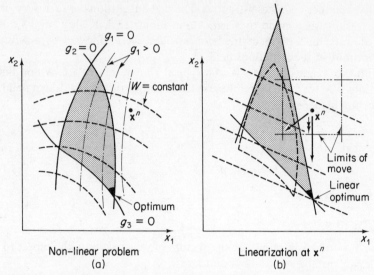

**Figure 7.8**    Linearization of optimization problem : (a) non-linear problem and (b) linearization at $\mathbf{x}^n$ for two design variables

non-feasible, but is not far from the true optimum. It is more than likely that, in this simple situation, one or two further linearizations would converge to the correct solution.

The situation is not, however, always so simple. First, in the full non-linear situation the true optimum may occur between two constraint intersections, as shown in Figure 7.9. A straightforward successive linearization will, in such a case, lead simply to an oscillation of the solution between widely separated values.

Secondly, even though the true optimum may be at a vertex, the starting design point may be so far from the true optimum that the solution still does not converge ( just as in the well known Newton method, where convergence cannot, in general, be guaranteed if the starting point is too distant from the true solution).

Various procedures can be used to overcome such difficulties.

*The cutting-plane* process[6,7,17] is one possibility. At each new iteration stage the previous constraints are retained in addition to the new ones created by the fresh linearization (see Figure 7.9). If all the linearized constraints fall *outside* the true feasible region, this process will obviously converge to the correct optimum. Unfortunately this proposition is only true for *convex* constraints, and *a priori* it is difficult to decide about their nature. If convexity does not exist, we may well exclude a considerable part of the feasible region from the minimization process and thus obtain an erroneous result. Further, the retention of constraints leads progressively

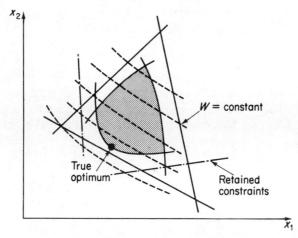

**Figure 7.9** The cutting-plane process

towards the solution of larger linear-programming problems at each stage and does not appear to be economical. Obviously variants of the process could be devised by discarding some constraints to improve its economy.

*Limited-move* methods avoid the difficulties of the previous process. Here a set of constraints is placed on the range of $\delta x$ admissible in any iteration. Figure 7.8 shows the effect of placing such constraints on the value of $\delta x_i$ at a particular linearized stage. It seems clear that economy and success of the solution will depend greatly on the choice of the move limits. If the move limits are made too small, the convergence will be very slow; if they are too large oscillation may occur. 'Adaptive' methods of various kinds in which the limit to be placed on the next move depends in some manner on the results of the previous step have been proposed and used[9,10]. This process of learning by 'experience' also allows the input of engineering 'know how' to be used to advantage and has much to commend it. However, much remains to be done in the development of such approaches.

A question remains open as to the best direction of the linearized move. In Figure 7.8(b) we have shown the difference between the simple linearized move and one in which a limit was placed on the range of $\delta x_1$ and $\delta x_2$, by adding the additional constraint explicitly to the linear-programming form. An alternative to this would appear to be the application of the move limit once the simple linear-programming problem is solved, retaining the same direction of the vector $\delta x$. This design-variable change is now indicated by the double arrow in Figure 7.8(b). Such a process not only reduces the linear-programming problem to be solved, but may result in an improved direction of the design changes. Indeed, the logic of this process can be made simple by restricting the *radius* within which the design variables are permitted to change. The radius $\delta R$ of this particular vector can, for this purpose, be defined simply by the scalar product

$$\delta R = \delta \mathbf{x}^T . \Delta \mathbf{x}. \qquad (7.16)$$

The consideration of a design-variable radius limitation does, to some extent, depend on the relative importance of the variables and would normally be used only if it is known that each of these has approximately the same influence. This necessitates some suitable scaling of the design variables.

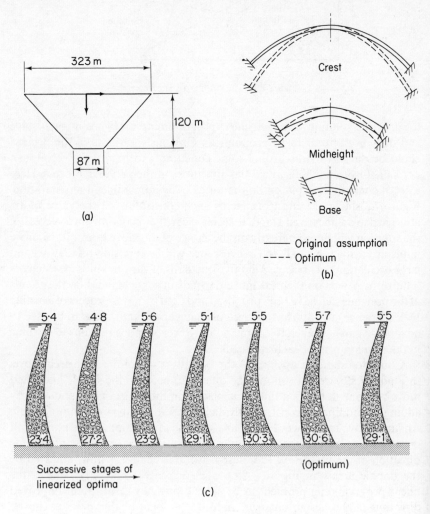

(a)

Crest

Midheight

Base

———— Original assumption
- - - - Optimum

(b)

5·4　　4·8　　5·6　　5·1　　5·5　　5·7　　5·5

23·4　27·2　23·9　29·1　30·3　30·6　29·1

Successive stages of
linearized optima

(Optimum)

(c)

**Figure 7.10**　Optimization of an arch dam (Sharpe[16]): (a) developed valley profile, (b) change of profile of valley arches and (c) successive changes of crown cross-section

## 7.8 Closure

We have indicated the possible methodology for the solution of shape-optimization problems in real engineering situations. Comparatively little work has so far been accomplished in this field, but the achievements so far made show that the processes will find an increasing application in engineering practice. For instance, Figure 7.10, adapted from the paper by Sharpe[16], shows how an automated procedure using successive linear-programming stages can be used to derive a shape for an arch dam that is acceptable from a designer's point of view.

Whether a fully automatic shape-design program can be economically used and developed for general purposes is, perhaps, questionable. Certainly this possibility should be explored for important and repetitive situations where accumulated experience can be used and special programs justify the effort. For less important one-off situations, in which more intuitive approaches may have to be used, the 'force-derivative' process of deriving sensitivities is in itself important. Here the generalization of computer programs is possible, and the 'sensitivities' will guide the designer, even in a 'trial-and-error' process, towards an economical solution in the 'feasible' region.

## References

1. L. Cavallaro, 'Stress analysis of rotating disks', *Nuclear Struct. Eng.*, **2**, 271–281 (1965).
2. O. C. Zienkiewicz, *The Finite Element Method in Engineering Science*, 2nd ed., McGraw-Hill, 1971.
3. B. M. Irons, International symposium on the use of electronic digital computers in civil engineering, University of Newcastle, July, 1966, Discussion Session 6, Paper 1.2.
4. J. S. Campbell, 'The FINESSE finite element computer system', Centre for Numerical Methods, in Engineering, Dept. of Civil Engineering, University of Wales, Swansea, 1971.
5. B. M. Irons, 'A frontal solution program for finite element analysis', *Int. J. Num. Methods Eng.*, **2**, 5–32 (1970).
6. E. W. Cheney and A. A. Goldstein, 'Newton's method for convex programming on Tchebycheff approximation', *Numerical Mathematics*, **1**, 253–268 (1959).
7. H. J. Kellog, 'The cutting plane method for solving complex programs', *SIAM J.*, **8**, 703–712 (1960).
8. R. E. Griffith and R. A. Stewart, 'A nonlinear programming technique for the optimization of continuous processing systems', *Management Science*, **7**, 379–392 (1961).
9. F. Moses, 'Optimum structural design using linear programming', *J. Struct. Div.*, *ASCE*, **90**, No. ST6, 89–104 (1964).
10. K. Reinschmidt, C. A. Cornell and J. F. Brotchie, 'Iterative design and structural optimization', *J. Struct. Div.*, *ASCE*, **92**, No. ST6, 281–318 (1966).
11. B. L. Karihaloo, P. R. Pathare and C. K. Ramash, 'The optimum design of space structures by linear programming using the stiffness method of analysis', in *Space Structures* (Ed. K. M. Davies), Blackwell, Oxford, 1967, pp. 278–290.

126 *Optimum Structural Design*

12. 'Some notes and ideas on mathematical programming methods for structural optimization', Norges Tekniske Høgskole, Institut for Shipbuilding, Trondheim, SKE 11/M8, Jan. 1967.
13. E. M. L. Beale, 'Numerical methods', in *Non-Linear Programming* (Ed. J. Abadie), North-Holland, 1967, pp. 97–131.
14. F. Moses and S. Onada, 'Minimum weight design of structures with application to elastic grillages', *Int. J. Num. Methods Eng.*, **1**, 311–331 (1969).
15. G. G. Pope, 'The application of linear programming techniques in the design of optimum structures', in *Proc. AGARD Symposium on Structural Optimization, Istanbul, Oct. 1969*, AGARD-CP-36-70.
16. R. Sharpe, 'The optimum design of arch dams', *Proc. Inst. Civ. Eng.*, Paper 7200S, Suppl. vol., 73–98 (1969).
17. R. L. Fox, *Optimization Methods for Engineering Design*, Addison-Wesley, 1971.

*Chapter 8*

# Feasible-direction Methods in Structural Optimization

*B. M. E. de Silva*

## 8.1 Introduction

This chapter describes procedures in the class of *feasible-direction methods* that have been applied to structural optimization problems. A feasible-direction method was, perhaps, the first of the non-linear-programming procedures to be employed in structural optimization by Schmit in 1960[1], and methods in this class enjoyed intensive development during the subsequent six years. They continue to be under development, but at a less rapid pace, and to be applied effectively to significant engineering problems, some of which are described in this chapter.

The basis of feasible-direction methods was outlined by Fletcher in Section 5.4. They are in the class of direct search algorithms and therefore address themselves to the determination of the distance $\alpha^k$ and direction $\mathbf{d}^k$ of travel from the $k$th to the $(k + 1)$th point in design space, i.e.

$$\mathbf{x}^{k+1} = \mathbf{x}^k + \alpha^k \mathbf{d}^k. \tag{8.1}$$

The direction $\mathbf{d}^k$ is *feasible* if a move in that direction does not cause constraint violation, i.e.

$$g_j(\mathbf{x}^k + \alpha^k \mathbf{d}^k) \leqslant 0, \qquad j = 1, 2, \ldots, m \tag{8.2}$$

for a system with $m$ constraints. This requires a negative dot product of the move direction and the gradient to each active constraint. Denote $\nabla g_l$ as the gradient of one of the $p$ active constraints $1, \ldots, l, \ldots, p$. Collecting these in an $n \times p$ matrix designated as $[\nabla g]$, where $n$ is the number of design variables, we have as the condition of feasibility

$$[\nabla g]^{\mathrm{T}} \mathbf{d}^k \leqslant 0. \tag{8.3}$$

A desirable condition on the direction of move is that it also results in a reduction of the merit function, i.e. it is *usable*. In this case, the mathematical condition is stated as

$$(\nabla W)^{\mathrm{T}} \mathbf{d}^k \leqslant 0. \tag{8.4}$$

Furthermore, note should be taken of the *side constraints*, which define upper $(U_j)$ and lower $(L_j)$ bounds on each design variable $x_j$, i.e.

$$L_j \leqslant x_j \leqslant U_j, \qquad j = 1, \ldots, n. \tag{8.5}$$

Nearly all the applications to be described here employ an accelerated steepest-descent mode (see Section 5.2) to travel from an initial feasible design point to a constraint. (A steepest-ascent mode is used if the initial point is infeasible.) When the constraint is reached, it is impossible to move in the steep-descent direction without violating the constraint. An alternate redesign procedure that ensures continuation of the optimum design process is therefore required. The development of efficient directions and distances of search from the boundary of the constraint set constitutes a central phase of the feasible-direction procedure; it is studied in this chapter under the following categories:

(1) Constant merit redesign (Section 8.2).
(2) Travel on the constraint surface, with the direction of travel $\mathbf{d}^q$ being a projection of the merit-function gradient on the constraint boundary (Section 8.3).
(3) Travel in a direction between the limits defined in (1) and (2), with the direction chosen 'optimally' using a linear-programming algorithm (Section 8.4).

These alternative procedures will now be discussed in turn.

## 8.2 Constant merit redesign

### 8.2.1 Method of alternate base planes

Among the first successful attempts at the boundary-redesign problem was the method of *alternate base planes* used by Schmit et al.[1-3] for the minimum-weight design of trusses and waffle plates. This is a quasirandom method, which seeks a feasible design on the constant-weight contour through a main constraint. The problems were characterized by linear side constraints, which were handled separately to ensure designs, most of which lie within the lower $(L_j)$ and upper $(U_j)$ bounds on the design variables. The basic steps of the algorithm are as follows (Figure 8.1):

(i) The program begins by generating random search directions $\mathbf{d}^i$ in planes normal to the coordinate lines $Ox_1, Ox_2, \ldots, Ox_n$ in turn. This scanning is controlled by a counter $i$ which is initially set to unity.

(ii) The direction cosines of the straight line of travel are generated:

$$d_j^i = R_j \bigg/ \left( \sum_{j \neq i}^{n} R_j^2 \right)^{\frac{1}{2}}, \qquad j = 1, 2, \ldots, n, \qquad j \neq i$$

$$= 0, \qquad j = i$$

where $R_j$ are random numbers.

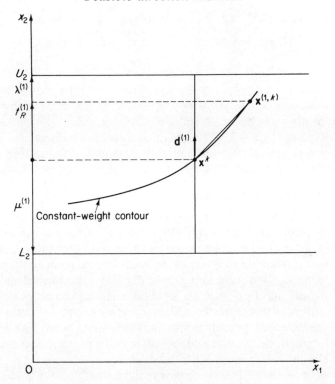

**Figure 8.1** Method of alternate base planes

(iii) The distance to the side constraints is calculated:

$$\alpha_j^i = (L_j - x_j^k)/d_j^i, \qquad j = 1, 2, \ldots, n, \qquad j \neq i$$

$$\alpha_{n+j}^i = (U_j - x_j^k)/d_j^i, \qquad j = 1, 2, \ldots, n, \qquad j \neq i.$$

This ensures that the proposed designs satisfy most of the side constraints. From this set of values $\alpha^i$ the smallest positive value is selected and designated $\lambda^i$, and the negative value having the smallest absolute value is designated $\mu^i$.

(iv) Six random numbers $R_q$ ($q = 1, 2, \ldots, 6$) between 0 and 1 are generated in two sets of three and are multiplied by $\lambda^i$ and $\mu^i$ to give the distance of travel in the base plane, designated by $\alpha_q^i$, i.e.

$$\alpha_q^i = R_q \lambda^i, \qquad q = 1, 2, 3$$

$$= R_q \mu^i, \qquad q = 4, 5, 6.$$

(v) The proposed new designs are given by

$$\mathbf{x}_q^i = \mathbf{x}^k + \alpha_q^i \mathbf{d}^i$$

130     *Optimum Structural Design*

where $x^i_{q_i}$ is calculated from the constant-weight condition

$$W(\mathbf{x}^k) = W(x^k_1 + \alpha^i_q d^i_1, \ldots, x^k_{i-1} + \alpha^i_q d^i_{i-1}, x^i_{q_i},$$
$$x^k_{i+1} + \alpha^i_q d^i_{i+r}, \ldots, x^k_n + \alpha^i_q d^i_n).$$

(vi) Check the six proposed designs against the behavioural constraints in the order $q = 1, 2, \ldots, 6$. If any one of $x^i_q$ is feasible, steepest-descent motion continues as before. Otherwise go to step (vii).

(vii) Replace $i$ by $i + 1$, go to step (ii) and repeat iterations.

Step (vii) is equivalent to changing the base plane. If a feasible design is still not forthcoming, the boundary point is taken as the optimum.

### 8.2.2 A hill-climbing procedure

The method described in the previous section was applied by de Silva[4,5] to the minimum-weight design of disks subject to stress and vibration constraints. The method consumed considerable computer time in searching through the random directions to determine a feasible point on the constant-weight contour, and deteriorated rapidly for high-dimensional design spaces. Schmit and Fox[6] used a simple hill-climbing technique based on a zigzag concept to determine the optimal response of a spring–mass–damper system characterized by merit contours with a sharp ridge. This is a more rational method, based on an understanding of the problem, and a modification of this procedure by de Silva[5] is as follows.

$\mathbf{x}^{k-2}, \mathbf{x}^{k-1}$ and $\mathbf{x}^k$ are three successive designs generated by a gradient mode of travel with $\mathbf{x}^k$ a boundary point on a behavioural constraint

$$W(\mathbf{x}^k) < W(\mathbf{x}^{k-1}) < W(\mathbf{x}^{k-2}) \tag{8.6}$$

where

$$g_j(\mathbf{x}^k), g_j(\mathbf{x}^{k-1}), g_j(\mathbf{x}^{k-2}) \leqslant 0, \qquad j = 1, \ldots, m$$

and

$$g_l(\mathbf{x}^k) = 0 \qquad \text{for at least one } l \text{ in } 1 \leqslant l \leqslant m.$$

For the disk problem that will be discussed in further detail later, the $g_l$ correspond to the vibration constraints in which the fundamental frequencies are required to exceed specified lower bounds. Let $\mathbf{x}$ be the foot of the perpendicular from $\mathbf{x}^k$ onto the gradient-mode vector $\mathbf{d}^{k-2}$ from $\mathbf{x}^{k-2}$ (Figure 8.2). Therefore

$$\mathbf{x} = \left(1 + \frac{\alpha^{k-1}}{\alpha^{k-2}} \cos \theta\right)\mathbf{x}^{k-1} - \frac{\alpha^{k-1}}{\alpha^{k-2}} \cos \theta \, \mathbf{x}^{k-2} \tag{8.8}$$

where

$$\cos \theta = \mathbf{d}^{k-2} \cdot \mathbf{d}^{k-1}.$$

This is the scalar product of the (normalized) steepest-descent vectors $\mathbf{d}^{k-2}$ and $\mathbf{d}^{k-1}$ with associated step lengths $\alpha^{k-2}$ and $\alpha^{k-1}$. The angle $\theta$ measures the amount of zigzag. In the absence of a sharp ridge on the merit contours, $\theta$ is small, $\cos\theta > 0$ and the point $\mathbf{x}$ will be close to, but seldom on, the behavioural constraint $g_l$, which is, essentially, a numerical or non-analytic constraint.

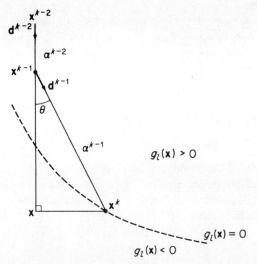

**Figure 8.2** Estimate for direction of bounce given three successive steepest-descent-mode designs

Consider a direction with direction ratios defined by

$$\mathbf{d}_l^k = \mathbf{x} - \mathbf{x}^k \quad \text{if } g_l(\mathbf{x}) < 0$$
$$= \mathbf{x}^k - \mathbf{x} \quad \text{otherwise.} \quad (8.9)$$

Under suitable conditions, $\mathbf{d}^k$ approximates a tangent move towards the interior of the feasible set. In the disk problem the weight is a quadratic in $x_n$ but linear in the remaining variables $x_1, x_2, \ldots, x_{n-1}$. The proposed direction of search is obtained by projecting the normalized direction [equation (8.9)] onto the constant-weight hyperplane:

$$W(x_1, \ldots, x_{n-1}, x_n^k) = W(x_1^k, \ldots, x_{n-1}^k, x_n^k). \quad (8.10)$$

The distance of travel yields an alternate step within the design-variable bounds [equation (8.5)]:

$$\alpha^k = \min_{1 \leqslant j \leqslant n} \{(x_j^k - L_j), (U_j - x_j^k)\}. \quad (8.11)$$

For a design violating a main constraint, the step length [equation (8.11)] is progressively reduced. For multiple constraints ($p$ in number) the direction given by equation (8.9) is replaced by the weighted sum

$$\mathbf{d}^k = \sum_{l=1}^{p} c_l \mathbf{d}_l^k \tag{8.12}$$

where $c_l$ are non-negative weighting factors determined using Zoutendijk-type procedures[7].

A different alternate-step mode uses the distance of travel [equation (8.11)] to generate the direction of bounce into the feasible regions. The direction cosines $d_i^k$, $i = 1, 2, \ldots, n$, are constrained by the condition that the objective function remains constant:

$$W(\mathbf{x}^k) = W(\mathbf{x}^k + \alpha^k \mathbf{d}^k) \tag{8.13a}$$

and the condition that $\mathbf{d}^k$ be normalized:

$$(\mathbf{d}^k)^\mathrm{T} \mathbf{d}^k = 1 \tag{8.13b}$$

where $\alpha^k$ is the step length defined by equation (8.11). The system [equations (8.13)] is indeterminate for $n > 2$. Complete solutions are obtained by using the physics of the problem to specify $n - 2$ components of $\mathbf{d}^k$ and calculating the rest from equations (8.13).

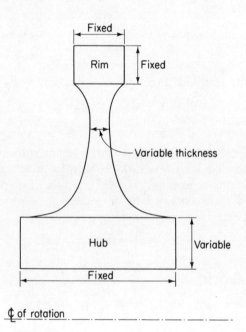

**Figure 8.3**  Cross-section of typical turbine disk

The method was applied to the minimal-weight design of disks[4] in which the stresses were constrained to lie below the yield stress for the material. One such turbine disk to be optimized is shown in Figure 8.3. The problem is made discrete by using a piecewise linear representation for the disk profile (Figure 8.4). The shape function $h(r)$ (where $r$ is the radial distance from the

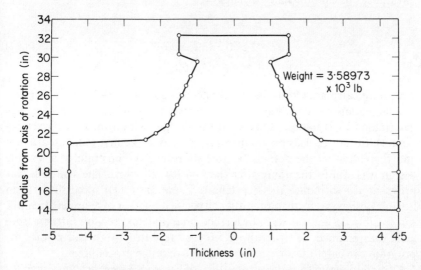

**Figure 8.4** Numerical example: initial design

axis of rotation) is therefore approximated by a sequence of discrete thickness variables $b_j, j \in J$, at specified radial distance $a_j, j \in J$. Engineering design considerations specify the width of the hub and the rim shape, while the hub radius $a_2$ is variable. Thus the design variables are $b_j, j \in J$, and $a_2$. The weight is linear in $b_j$ but quadratic in $a_2$.

The stress computations were based on Donath's method (see Reference 4), and cannot be expressed as analytically defined functions of the design variables. The behaviour variables are functions only in the sense that they are computer-oriented rules for determining the behaviour associated with a given design and are not given in a closed analytical form in terms of the design variables. The behaviour variables may be regarded as a 'black box' into which are put the design variables representing a given design and out of which come the corresponding behavioural variables for that design. These are then checked against the behavioural constraints. The side constraints are essentially linear and are of the form $b_j \leqslant \varepsilon; j \in J; L \leqslant a_2 \leqslant U$ where $\varepsilon$, $L$ and $U$ are specified positive tolerances.

The computer program starts from an initial feasible design and enters an accelerated steepest-descent mode of travel, continuing in this mode until a

134 *Optimum Structural Design*

constraint is encountered. It is then no longer possible to move in this mode without violating the constraint. In this problem, this situation occurs when a section $b_l$, $l \in J$, of the disk is at the yield stress. A feasible design is sought by thickening this section and thinning the section $b_s$, $s \in J$, furthest from the yield stress in such a manner as to leave the weight unchanged. All the other thicknesses remain unchanged. Thus:

$$d_i = 0, \quad i \neq l, s$$

$$> 0, \quad i = l \qquad (8.14)$$

$$< 0, \quad i = s.$$

$d_l, d_s$ were calculated from the simultaneous equations (8.13) and gave polynomial equations consistent with expressions (8.14). The step size was determined by equation (8.11) to ensure designs within the design-variable bounds. Initially, a feasible point was obtained at the first redesign attempt and thereafter, as the designs became more highly constrained, a feasible design was still forthcoming after the first few attempts. The synthesis was programmed to reduce the step length [equation (8.11)] successively if no feasible design was forthcoming after a specified number of redesign attempts. If a feasible design was still not forthcoming, the next section furthest from yield was thinned and the above process repeated. As a last resort, the program can enter the alternate-base-plane redesign procedure.

Optimum designs for the turbine disk of Figure 8.3 were accomplished for design spaces which ranged from four to eleven design variables. The initial design for an eleven-dimensional representation is illustrated in Figure 8.4.

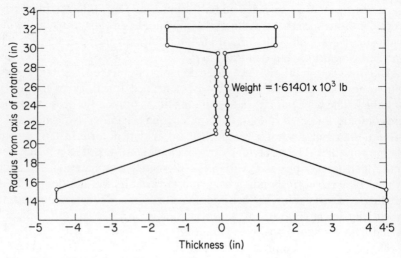

**Figure 8.5** Numerical example: final design at 186 cycles via selective search procedure

*Feasible-direction Methods* 135

Results are shown for the selective and random search procedures, respectively, in Figures 8.5 and 8.6. Figure 8.7 shows the variation of the design weight as a function of the number of redesign attempts. Other results for this problem and complete details of the method are presented in Reference 4.

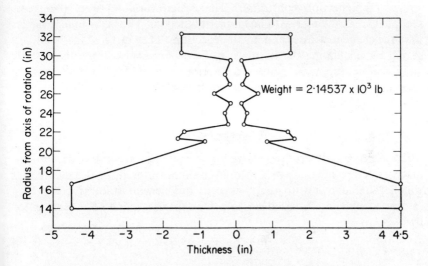

**Figure 8.6** Numerical example: final design via random search procedure

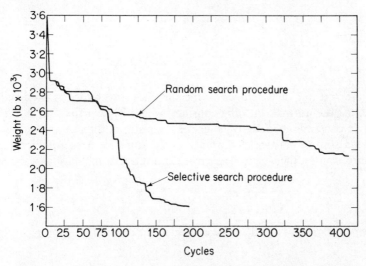

**Figure 8.7** Numerical example: selective versus random search procedures

*8.2.3  Structural-analysis-influenced travel*

The alternate-step modes studied hitherto do not utilize the mechanisms inside the structural-analysis packages to influence the design optimizations from a main constraint. Gellatly and Gallagher[8] have used constant-merit redesign techniques for the minimum-weight design of trusses subject to stress and deflection constraints. The design variables were the cross-sectional areas, giving rise to linear merit functions. The behaviour variables were the element stresses and nodal deflections. They directed the boundary search by calculating the normals to the behavioural constraints in static and dynamic response regimes. To describe the associated formulations, we designate the relevant equations of matrix displacement analysis as

$$\mathbf{P} = \mathbf{K}\boldsymbol{\Delta}$$
$$\boldsymbol{\sigma} = \mathbf{S}\boldsymbol{\Delta} \tag{8.15}$$

where $\mathbf{K}$, $\mathbf{S}$ and $\mathbf{P}$ are, respectively, the stiffness, stress and design-load matrices. The stiffness matrix at a given point in the design sequence $(\mathbf{K}_0)$ will be altered owing to the change in the element stiffness matrix $(\mathbf{K}_i)$ associated with the $i$th design variable. Thus the new stiffness $\mathbf{K}$ is represented by

$$\mathbf{K} = \mathbf{K}_0 + \sum_i \delta x_i \mathbf{K}_i \tag{8.16}$$

where $\delta x_i$ is the change in the associated design variable. Reference 8 demonstrates that a local approximation to the normals to the behavioural constraints is then given by [see also equation (7.3)]

$$\frac{\partial \boldsymbol{\Delta}}{\partial x_i} = -\mathbf{K}_0^{-1}\mathbf{K}_i\boldsymbol{\Delta}$$
$$\frac{\partial \boldsymbol{\sigma}}{\partial x_i} = S\frac{\partial \boldsymbol{\Delta}}{\partial x_i} = -\mathbf{S}^k\mathbf{K}_0^{-1}\mathbf{K}_i\boldsymbol{\Delta}. \tag{8.17}$$

In the method of Reference 8, the direction of bounce is obtained by projecting the normal onto the constant-weight hyperplane. For points on multiple constraints, Gellatly[9] suggests a constraint direction based on the weighted sum of constraint normals, of the same form as equation (8.12). The direction on the weight hyperplane is a linear combination of the form

$$\mathbf{d}_D^k = c\mathbf{d}_D^k + \sum_{l=1}^p c_l \mathbf{d}_l \tag{8.18}$$

where $c$ is a constant and $\mathbf{d}_D^k$ is the normal to the constant-weight hyperplane, i.e.

$$\mathbf{d}_W^k \cdot \mathbf{d}_D^k = 0. \tag{8.19}$$

To bounce back into the feasible regions, the direction $\mathbf{d}_W^k$ must make acute angles with all constraint normals. This condition is expressible in the form

$$\mathbf{d}_l^k \cdot \mathbf{d}_W^k = \varepsilon_l > 0 \tag{8.20}$$

where the $\varepsilon_l$ are specified tolerances, usually selected to be unity. From equation (8.18)

$$c\mathbf{d}_W^k \cdot \mathbf{d}_D^k + \sum_{l=1}^{p} c_l \mathbf{d}_l^k \cdot \mathbf{d}_D^k = 0$$

$$\tag{8.21}$$

$$c\mathbf{d}_W^k \cdot \mathbf{d}_m^k + \sum_{l=1}^{p} c_l \mathbf{d}_l^k \cdot \mathbf{d}_m^k = \varepsilon_m \qquad \text{for all } m.$$

These equations form a determinate system for $c_l$. The matrix of coefficients tends to be ill conditioned in the neighbourhood of an optimum.

## 8.3 Constrained boundary motion

### 8.3.1 Best's method[10,11]

One of the earliest applications of travel along the constraints in the context of the structural design problem is due to Best[10,11]. His method starts from a trial design in the feasible region, and steeply descends to the nearest (main) constraint. From a boundary point the method moves on the constraint surface in a direction in which the merit decreases most rapidly.

Suppose the point lies at the intersection of $p$ constraint hypersurfaces. The normals are determined using techniques similar to equation (8.17) and are collected in the matrix $[\nabla g]$. The direction of travel $(\mathbf{d}^k)$ is orthogonal to $[\nabla g]$.

$$[\nabla g]^{\mathrm{T}} \mathbf{d}^k = 0 \tag{8.22}$$

and is assumed to be normalized, i.e. equation (8.13b) applies. The rate of decrease of the weight in the direction $\mathbf{d}^k$ is determined by

$$-\frac{\mathrm{d}}{\mathrm{d}\alpha^k} W[\mathbf{x}^k + \alpha^k \mathbf{d}^k] = -\sum_{i=1}^{n} \frac{\partial W}{\partial x_i} \mathbf{d}_i^k = -[\mathbf{d}^k]^{\mathrm{T}} \nabla W. \tag{8.23}$$

The problem consists in maximizing equation (8.23) subject to the constraint conditions (8.13b) and (8.22) so as to give the optimal direction of travel $\mathbf{d}^k$. We can accomplish this by the Lagrange-multiplier technique. Introduce the Lagrange multipliers $\lambda_0, \lambda_1, \ldots, \lambda_p$. Then

$$-\nabla W + [\nabla g]\lambda + 2\lambda_0 \mathbf{d}^k = \mathbf{O} \tag{8.24}$$

where $\lambda = \{\lambda_1, \ldots, \lambda_p\}$.

From conditions (8.13b), (8.22) and (8.24), the direction of travel is given by

$$\mathbf{d}^k = \frac{\mathbf{H}\nabla W}{2\lambda_0} \tag{8.25}$$

where

$$\mathbf{H} = \mathbf{I} - [\nabla g]\{[\nabla g]^{\mathrm{T}}[\nabla g]\}^{-1}[\nabla g]^{\mathrm{T}}$$
$$\lambda_0 = -\tfrac{1}{2}[(\mathbf{H}\nabla W)^{\mathrm{T}}(\mathbf{H}\nabla W)]^{\frac{1}{2}}. \tag{8.26}$$

The operator $\mathbf{H}$ plays a central role in the gradient-projection method, as will be shown in Section 8.3.2.

The distance of travel is estimated to the nearest constraint, so that, to first order

$$g_j(\mathbf{x}^k) + \alpha^k(\mathbf{d}^k)^{\mathrm{T}}\nabla g_j(\mathbf{x}^k) = 0. \tag{8.27}$$

The required step length is then

$$\alpha^k = \min_j \left\{ \frac{-g_j[\mathbf{x}^k]}{[\mathbf{d}^k]^{\mathrm{T}}\nabla g_j[\mathbf{x}^k]} \right\}. \tag{8.28}$$

The method was applied by Best[10,11] to the minimum-weight design of cantilever box structures in the presence of stress and deflection constraints. The method is primarily applicable to problems with very 'flat' constraints in which movement in the direction $\mathbf{d}^k$ does not give rise to significant constraint violation. This condition is usually not satisfied by behavioural constraints in structural mechanics. A modification has been proposed by Schmit[12], in which condition (8.22) is replaced by condition (8.3). This reduces the problem to one with inequality constraints, with a corresponding increase in complexity.

Constrained boundary motion in conjunction with a dynamic constraint was used by Zarghamee[13] to maximize frequency subject to a linear weight constraint. The frequency is calculated from the eigenvalue equation

$$[\mathbf{k} - \omega^j \mathbf{M}]\Delta^j = \mathbf{O} \tag{8.29}$$

where $\mathbf{K}$, $\mathbf{M}$ are the stiffness and mass matrices, respectively, and $\Delta^j$ is the modal shape corresponding to the eigenfrequency $\omega^j$. The modified stiffness matrix is given by equation (8.16) and, for the modified mass matrix,

$$\mathbf{M} = \mathbf{M}_0 + \sum_i \delta x_i \mathbf{M}_i. \tag{8.30}$$

Differentiating equation (8.29) partially with respect to $x_i$, and using equations (8.16) and (8.30):

$$[\mathbf{k}_i - \omega^j \mathbf{M}_i]\Delta^j - \frac{\partial \omega^j}{\partial x_i}\mathbf{M}\Delta^j + [\mathbf{k} - \omega^j \mathbf{M}]\frac{\partial \Delta^j}{\partial x_i} = 0. \tag{8.31}$$

We assume that the eigenvectors $\mathbf{\Delta}^j$ form a complete set so that we can express their gradients as

$$\frac{\partial \mathbf{\Delta}^j}{\partial x_i} = \sum_k \beta_{k_i,j} \mathbf{\Delta}^k \tag{8.32}$$

where the $\beta_{k_i,j}$ are constants.

Also, we take note of the orthogonality property of the eigenvalues with respect to $\mathbf{M}$ as a weighting matrix:

$$(\mathbf{\Delta}^i)^{\mathrm{T}} \mathbf{M} \mathbf{\Delta}^j = \delta_{ij} \tag{8.33}$$

where $\delta_{ij}$ is the Kronecker delta ($\delta_{ij} = 0$ if $i \neq j$ and $\delta_{ij} = 1$ if $i = j$).
From equations (8.31)–(8.33)

$$\frac{\partial \omega^j}{\partial x_i} = [\mathbf{\Delta}^j]^{\mathrm{T}} [\mathbf{k}_i - \omega^j \mathbf{M}_i] \mathbf{\Delta}^j. \tag{8.34}$$

This measures the rate of change of the frequency in terms of the corresponding eigenvector. The constraint on the total weight is of the form

$$W(\mathbf{x}) = W_0 + \sum_{i=1}^{n} W_i x_i \tag{8.35}$$

where $W(\mathbf{x}) \leqslant W_0$. Hence we have the linear constraint

$$\sum_{i=1}^{n} W_i x_i \leqslant 0. \tag{8.36}$$

The problem therefore consists in maximizing the frequency $\omega^j(\mathbf{x})$ subject to the linear constraint. The solution was based on the gradient-projection method for linear constraints, in which the gradient direction [equation (8.34)] is projected onto the linear constraint using the projection operator $\mathbf{H}$. This gradient-projection method is taken up in Section 8.3.2. Generalizations of the analysis to more complex structures are given by Turner[14,15].

### 8.3.2 *Gradient-projection method*

The gradient-projection method due to Rosen[16] has proved to be of value in structural optimization, and applications to various structural systems are given by Brown and Ang[17].

The method offers considerable flexibility and scope for non-linear constraints, and consists in orthogonal projection of the gradient into the linear manifold of the supporting hyperplanes to the active constraints. The basic steps of the algorithm are summarized as follows.

Suppose $\mathbf{x}^k$ lies on $p$ constraint surfaces. Using prior symbolism, the $n \times p$ matrix of normals is designated as $[\mathbf{Vg}]$, where each column is assumed to be linearly independent of the rest. The projection operator $\mathbf{H}$ for the

140     *Optimum Structural Design*

linear manifold spanned by the supporting hyperplanes is given by equation (8.26). The normalized direction of travel ($\mathbf{d}^k$) is therefore defined by

$$\mathbf{d}^k = \frac{\mathbf{H}\nabla W}{|\mathbf{H}\nabla W|}. \tag{8.37}$$

The gradient vector $\nabla W(\mathbf{x}^k)$ can be written as a linear combination of the projected gradient and the normals $\Gamma g_j(\mathbf{x}^k)$ to the active constraints:

$$-\nabla W(\mathbf{x}^k) = -\mathbf{H}\nabla W(\mathbf{x}^k) + \sum_{i=1}^{p} r_i \nabla g_i(\mathbf{x}^k) \tag{8.38}$$

where the $r_i$ are constants. It can be shown that, if $-\mathbf{H}\nabla W = \mathbf{O}$ and $r \leqslant 0$, then $\mathbf{x}^k$ is a local optimum. Whenever $|\mathbf{H}\nabla W| > 0$, a small step length $\alpha^k$ is taken in the projected direction [equation (8.37)] to a point of improved merit. Because of the curvature of the boundary, this will be a non-feasible point and use is made of an interpolation procedure detailed in Reference 17.

When $-\mathbf{H}\nabla W = 0$ and $r_i > 0$ for some $i$ ($i = 1, \ldots, p$), the constraints for which $r_i > 0$ are removed and the analysis is performed on the intersection of the remaining constraints. This is represented by sets of recursion relations on $\mathbf{H}$ and $\mathbf{r}$, and are given in in Reference 16.

### 8.4   Linear-programming-type methods

Another method of boundary redesign is Zoutendijk's method of feasible directions[7], which has been applied by Pope[18] to static problems, and by Fox and Kapoor[19,20] to minimum-weight design problems that include inequality constraints on the natural frequencies. The method consists in reducing the problem to a series of linear programs. We describe the method with reference to the problem treated by Fox and Kapoor.

The method first requires calculation of gradients to the active constraints. Equation (8.34) can be adapted to the calculation at the normal to the frequency constraint. The normals to the deflection constraints are given by the derivatives to the eigenvectors, as follows.

By differentiation of equation (8.30) with respect to $x_i$ and using equation (8.33), we have

$$[\mathbf{k} - \omega^j \mathbf{M}] \sum_k \beta_{k_{i,j}} \Delta^k + \left[ \frac{\partial \mathbf{k}}{\partial x_i} - \omega^j \frac{\partial \mathbf{M}}{\partial x_i} - \frac{\partial \omega^j}{\partial x_i} \mathbf{M} \right] \Delta^j = \mathbf{O}. \tag{8.39}$$

Premultiplying by $(\Delta^k)^T$ and using the orthogonality condition [equation (8.33)], we have, for $k \neq j$,

$$(\Delta^k)^T [\mathbf{k} - \omega^j \mathbf{M}] \sum_k \beta_{k_{i,j}} \Delta^k + (\Delta^k)^T \left[ \frac{\partial \mathbf{k}}{\partial x_i} - \omega^j \frac{\partial \mathbf{M}}{\partial x_i} \right] \Delta^j = 0 \tag{8.40}$$

from which

$$\beta_{k_{i,j}} = (\Delta^k)^\mathrm{T}\left[\frac{\partial \mathbf{k}}{\partial x_i} - \omega^j\frac{\partial \mathbf{M}}{\partial x_i}\right]\Delta^j/(\omega^j - \omega^k). \qquad (8.41)$$

Also, for $k = j$, we have, by differentiation of equation (8.33) with respect to $x_i$ and other operations,

$$\beta_{k_{i,j}} = -\frac{1}{2}(\Delta^j)^\mathrm{T}\frac{\partial \mathbf{M}}{\partial x_i}\Delta^j. \qquad (8.42)$$

Equations (8.34), (8.41) and (8.42) determine the normals to the behavioural constraints. The linear program for the problem is now formulated as the determination of a direction $\mathbf{d}^k$ which minimizes the linear function $(\mathbf{d}^k)^\mathrm{T}\nabla W$ subject to the constraints represented by equations (8.3) and (8.4), except that $\alpha^k$ is determined by equation (8.11) for linear side constraints.

## 8.5 Closure

This chapter has described some of the more commonly used boundary-redesign techniques for structural problems. Many of these have structural-analysis packages which, although relatively simple from a mathematical standpoint, involve extremely long and complex programming routines that consume considerable computer space and time. This limits a fuller utilization of classical non-linear-programming algorithms. The objective of structural optimization is not the determination of the numerical optimum to the constrained problem, but rather improving the efficiency of existing structural systems. As a result of these considerations, there is a growing tendency to utilize the structural-analysis procedures to solve the boundary-redesign problem. Analysis procedures based on finite-element procedures enable a more automatic coupling of the analysis and synthesis phases of the design process.

## References

1. L. A. Schmit, 'Structural design by systematic synthesis', in *Proc. 2nd Conf. on Elect. Comp., ASCE, Pittsburgh, 1960.*
2. L. A. Schmit and W. M. Morrow, 'Structural synthesis with buckling constraints', *J. Struct. Div. ASCE*, **89**, 107–126 (1963).
3. L. A. Schmit, T. P. Kicher and W. M. Morrow, 'Structural synthesis capability for integrally stiffened waffle plates', *AIAA J.*, **1**, 2820–2836 (1963).
4. B. M. E. de Silva, 'Application of nonlinear programming to the automated minimum weight design of rotating discs', in *Optimization* (Ed. R. Fletcher), Academic Press, 1969, pp. 115–150.
5. B. M. E. de Silva, 'Minimum weight design of discs, using a frequency constraint', *Trans. ASME, J. Eng. Ind.*, **91**, 115–150 (1969).

6. L. A. Schmit and R. L. Fox, 'Synthesis for a simple shock isolator', NASA CR-55, 1964.
7. G. Zoutendijk, *Methods of Feasible Directions*, Elsevier, 1960.
8. R. A. Gellatly and R. H. Gallagher, 'A procedure for automated minimum weight structural design; Part I: Theoretical basis,' *Aeron. Quart.*, **17**, 332–342 (1966).
9. R. A. Gellatly, 'Development of procedures for large scale automated minimum weight structural design', AFFDL-TR-66-180, December 1966.
10. G. Best, 'A method of structural weight minimization suitable for high speed digital computers', *J. Aircraft*, **1**, 129–133 (1964).
11. G. Best, 'Completely automatic weight minimization method for high speed digital computers', *J. Aircraft*, **1**, 129–133 (1964).
12. L. A. Schmit, 'Comments on "Completely automatic weight minimization method for high speed digital computers" ', *J. Aircraft*, **1**, 375–376 (1964).
13. M. S. Zarghamee, 'Optimum frequency of structures', *AIAA J.*, **6**, 749–750 (1968).
14. M. J. Turner, 'Design of minimum mass structures with specific natural frequencies', *AIAA J.*, **5**, 406–412 (1967).
15. M. J. Turner, 'Optimization of structures to satisfy flutter requirements', *AIAA J.*, **7**, 945–951 (1969).
16. J. B. Rosen, 'The gradient projection method for nonlinear programming; Part I: Linear constraints', *SIAM J.*, **8**, 181–217 (1960); 'Part II: Nonlinear constraints', *SIAM J.*, **9**, 514–532 (1961).
17. D. M. Brown and A. H. S. Ang, 'Structural optimization by nonlinear programming', *J. Struct. Div.*, *ASCE*, **92**, 319–340 (1966).
18. G. G. Pope, 'The design of optimum structures of specified basic configuration', *Int. J. Mech. Sci.*, **10**, 251–263 (1968).
19. R. L. Fox and M. P. Kapoor, 'Structural optimization in the dynamic response regime: a computational approach', *AIAA J.*, **8**, 1798–1804 (1970).
20. R. L. Fox and M. P. Kapoor, 'Rates of change of eigenvalues and eigenvectors', *AIAA J.*, **6**, 2426–2429 (1968).

# Chapter 9

# Penalty-function Methods

*Johannes Moe*

## 9.1 Introduction

Penalty-function methods, whose origin and role in mathematical programming was discussed by Fletcher in Chapter 5, have proved to be highly advantageous in practical structural-design problems. The solutions to such problems are invariably located on the border of the feasible region, i.e. one or several of the imposed stress and deflection constraints, etc., will govern the design. This characteristic of the optimum design is utilized in many optimization algorithms that involve systematic searches for the best solution at, or in the close vicinity of, the boundary between the feasible and the infeasible region.

While this approach is ideally suited for linear-programming problems, considerable difficulties arise when the constraint surfaces are highly nonlinear, as in most cases of structural design. In such cases it has become increasingly popular to apply methods by which the constrained problem is transformed into a sequence of unconstrained non-linear-programming problems.

Once this has been accomplished there exist, as seen in Chapter 5, a host of rather efficient and reliable algorithms designed to minimize unconstrained non-linear multivariate functions.

The idea behind penalty-function methods is extremely simple. Rather than trying to solve the optimum-structural-design problem posed in Chapter 2, it is transformed as follows:

$$\min_{x} \phi(\mathbf{x}, r_k) = W(\mathbf{x}) + r_k \sum_{m=1}^{M} G[g_m(\mathbf{x})] \qquad (9.1)$$

where $M$ is the number of constraints.

The function $G$ is selected so that, if this minimization is performed for a sequence of values of $r_k$, the solution may be forced to converge to that of the constrained problem. The second term on the right-hand side of equation (9.1) may be interpreted as a penalty term that somehow takes care of the constraints. The factor $r_k$ performs the weighting between the objective-function value and the penalty term. The factor $r_k$ is often called the *response factor*, and the $\phi(\mathbf{x}, r_k)$ surfaces are correspondingly termed response surfaces.

The first person to suggest the use of a type of penalty-function method seems to have been Courant[1] in 1943. In 1955 Frisch[2] presented a 'logarithmic potential method' which involved a penalty-function representation. The penalty function introduced by Carroll[3] furnished motivation for much of the current work in this area. Fiacco and McCormick[4] amplified this approach and have contributed much to the development of penalty-function methods. While Carroll called his approach a 'created response surface technique', Fiacco and McCormick introduced the term 'sequential unconstrained minimization technique' or Sumt, which is now widely used.

## 9.2  Elementary examples

In this section we formulate and solve a number of simple examples as we describe the penalty-function approach. First, however, some basic definitions in this area, given previously in Chapter 5, will be restated.

One of the much used formulations of the transformed function is as follows:

$$\phi(\mathbf{x}, r_k) = W(\mathbf{x}) + r_k \sum_{m=1}^{M} \frac{1}{g_m(\mathbf{x})}. \tag{9.2}$$

This function is minimized for a sequence of decreasing values of $r_k$ such that $r_k \to 0$ when $k$ increases. It is readily seen that if any one of the constraint functions $g_m(\mathbf{x})$ approaches zero, the penalty term in equation (9.2) increases very rapidly. If the original problem is convex, this is also true for $\phi(\mathbf{x}, r_k)$, and the minimum of $\phi(\mathbf{x}, r_k)$ must be inside (i.e. interior to) the feasible region. Hence an unconstrained minimization algorithm can be applied in the search for the minimum point of $\phi(\mathbf{x}, r_k)$. One should observe, however, that to find this optimum it is necessary to start the search from an interior point. The penalty function given by equation (9.2) is thus called an *interior* penalty function and has the advantage that all intermediate designs during the search for the optimum solution are feasible. One may therefore stop the search at any time and end up with a feasible and, hopefully, usable design.

We now examine examples which illustrate these ideas in more detail.

**Example 9.1.** *The appearance of the interior penalty function of equation (9.2) is demonstrated in Figure 9.1 for the following very simple example:*

$$\min_x ax \tag{9.3}$$

*subject to $g_1(x) = x - b \geqslant 0$.*
*This yields the following transformed problem:*

$$\min_x \phi(x, r_k) = ax + r_k/(x - b). \tag{9.4}$$

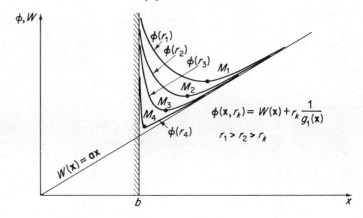

**Figure 9.1**  Objective function and penalty functions

*It is easy to obtain the following analytical expression for the minimum point:*

$$x^* = x_{\min} = b + \sqrt{r_k/a}$$
$$\phi(x^*, r_k) = ab + 2\sqrt{(ar_k)}.$$

$$(9.5)$$

*The limiting values for $r_k \to 0$ are*

$$x^* = b \quad and \quad \phi(b) = W(b) = ab. \tag{9.6}$$

*Thus it is verified that the solution to the unconstrained problem tends towards the constrained solution as $r_k$ tends towards zero. A formal proof of this property for multivariate problems is presented in Reference 4.*

Consider now an exterior-penalty-function representation of this problem. We use exterior penalty functions of the types suggested by Ablow and Brigham[5]:

$$\phi(\mathbf{x}) = W(\mathbf{x}) - r_k \sum_{m=1}^{M} \min \{0, g_m(\mathbf{x})\} \tag{9.7}$$

and

$$\phi(\mathbf{x}) = W(\mathbf{x}) + r_k \sum_{m=1}^{M} [\min \{0, g_m(\mathbf{x})\}]^2. \tag{9.8}$$

These formulae are shown in Figures 9.2(a) and (b), respectively, for the above problem and a sequence of $r_k$. Note that in these formulations $r_1 < r_2 < \cdots < r_k$.

Exterior penalty functions are also of special interest in connexion with equality constraints. Assume that the above problem is supplemented with

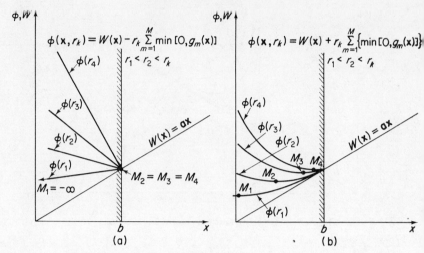

**Figure 9.2** Exterior penalty functions

the following additional requirements:

$$h_j(\mathbf{x}) = 0, \qquad j = 1, \ldots, J. \tag{9.9}$$

There is no interior region with respect to the equality constraints, which are therefore usually included in the penalty function by an exterior-type formulation, such as:

$$\phi(\mathbf{x}, r_k, s_k) = W(\mathbf{x}) + r_k \sum_{m=1}^{M} 1/g_m(\mathbf{x}) + s_k \sum_{j=1}^{J} [h_j(\mathbf{x})]^2. \tag{9.10}$$

One would expect difficulties in seeking to obtain a proper scaling between the two penalty terms. Fiacco and McCormick[4] report on some success with $s_k = r_k^{-\frac{1}{2}}$, but the inclusion of equality constraints certainly complicates the problem considerably.

**Example 9.2.** *Further insight into the method may be obtained by an examination of the following example with two variables. A hatch opening is covered by a box beam of span length and width as shown in Figures 9.3(a) and (b), respectively. The only free design variables are the flange thickness ($x_1$) and the beam height ($x_2$). The hatch cover, which is subjected to a maximum load of 6000 N/m, has to be designed in aluminium with modulus of elasticity $E = 70,000 \ N/mm^2$ to satisfy the following requirements:*

*(1, 2) Positive dimensions ($g_1 : x_1 \geqslant 0, g_2 : x_2 \geqslant 0$).*
  *(3) Maximum shearing stress $\bar{\tau} = 45 \ N/mm^2$ ($g_3 : \bar{\tau}/\tau - 1 \geqslant 0$).*
  *(4) Maximum bending stress: $\bar{\sigma} = 70 \ N/mm^2$ ($g_4 : \bar{\sigma}/\sigma_b - 1 \geqslant 0$).*

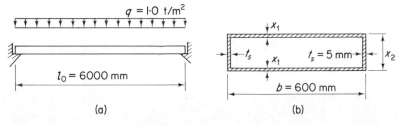

**Figure 9.3** Hatch-cover example

(5) *Buckling of the flange must not occur* $(g_5 : \sigma_k/\sigma_b - 1 \geqslant 0)$.
(6) *Maximum deflection* $\delta = l/400\ (g_6 : 1 - 400\ \delta/l \geqslant 0)$ *where*

$$\tau = \frac{Q}{2t_s x_2} = \frac{180}{x_2}$$

$$\sigma_b = \frac{M}{bx_1 x_2} = \frac{450}{x_1 x_2}$$

$$\sigma_k = \frac{\pi^2 E}{3(1 - v^2)}\left(\frac{x_1}{b}\right)^2 = 70x_1^2$$

$$\delta/l = \frac{5}{384}\frac{ql^3}{EI} = \frac{0\cdot8}{x_1 x_2}.$$

*Assume that the weight of the cover is to be minimized. For simplicity, the weights of transverse stiffener arrangements, etc., are disregarded, and the following objective function is selected:*

$$W(x_1, x_2) = (W_t)/\rho l = 120\ x_1 + x_2$$

*where* $(W_t)$ *is the total weight and* $\rho$ *is the density of aluminium.*

The design plane and the restrictions for this problem are shown in Figure 9.4, while Figures 9.5(a)–(c) show the isomerit curves of the $\phi(\mathbf{x}, r_k)$ functions for three consecutive values of $r_k$ $(r_1 > r_2 > r_3)$. The search patterns obtained using Powell's method of direct search[6] are also depicted in the figures.

As we have emphasized, the interior penalty functions have the disadvantage that they require a feasible starting point. In some of the most powerful search schemes there is also, as we shall see later, a risk that one may drop out of the feasible region during the search. Even if algorithms for automatic generation or regeneration of feasible starting points are a desirable feature of any program of this type, a promising alternative approach is to use the following type of extended penalty function proposed

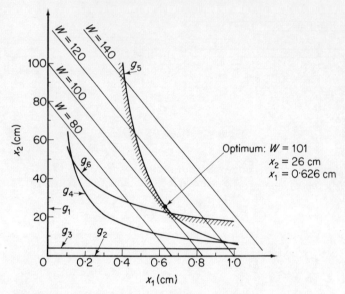

**Figure 9.4**   Design plane and constraints for the hatch-cover example

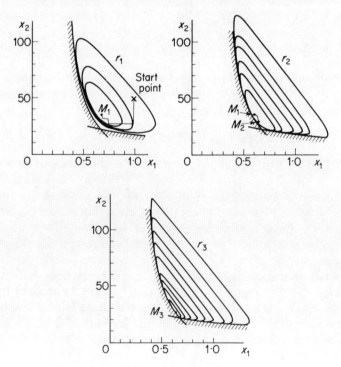

**Figure 9.5**   Isomerit curves for the hatch-cover problem

by Kavlie[7]:

$$G[g_m(\mathbf{x})] = \begin{cases} 1/g_m(\mathbf{x}) & \text{for } g_m(\mathbf{x}) \geqslant \varepsilon \\ [2\varepsilon - g_m(\mathbf{x})]/\varepsilon^2 & \text{for } g_m(\mathbf{x}) < \varepsilon \end{cases} \qquad (9.11)$$

where

$\varepsilon = r_k/\delta$

$\delta =$ a constant that defines the transition between the two types of penalty terms (see Figure 9.6).

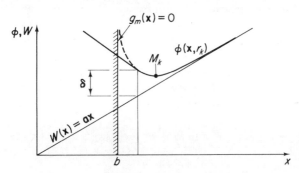

**Figure 9.6**   Extended penalty function [equation (9.11)]

**Example 9.3.** *In a structural-design problem one may consider two alternative formulations, one with and one without equality constraints. We must then introduce the concept of behaviour variables*

$$\boldsymbol{\sigma} = \{\sigma_1, \sigma_2, \ldots, \sigma_T\}.$$

*These variables characterize the behaviour or response of the structure under the relevant loading conditions in terms of stresses, displacements, etc. For any well formulated problem there is always a unique vector* $\boldsymbol{\sigma}$ *corresponding to a given vector* $\mathbf{x}$. *The relationship between the two vectors is governed by equations of equilibrium and compatibility, etc. These equations may be written as follows:*

$$h_t(\mathbf{x}, \boldsymbol{\sigma}) = 0, \qquad t = 1, 2, \ldots, T. \qquad (9.12)$$

*Alternative 1:*

*Select* $\mathbf{x}$ *as the design variables, solve equations (9.12) to obtain*

$$\sigma_t = c_t(\mathbf{x}), \qquad t = 1, 2, \ldots, T$$

*and introduce these relationships into the inequality constraints to obtain the following form:*

$$g_m(\mathbf{x}) \geqslant 0, \qquad m = 1, 2, \ldots, M.$$

*This is the approach described in Example 9.2.*

*Alternative 2:*

*Select* $\mathbf{y} = \begin{bmatrix} \mathbf{x} \\ \boldsymbol{\sigma} \end{bmatrix}$ *as the vector of free variables and introduce simultaneously the following constraints:*

$$h_t(\mathbf{y}) = 0, \qquad t = 1, 2, \ldots, T.$$

$$g_m(\mathbf{y}) \geqslant 0, \qquad m = 1, 2, \ldots, M.$$

*Alternative 1 is generally preferable, because of the fewer variables, and hence the lower dimensionality of the problem. Some exceptions exist, however.*

In cases with non-linear equality constraints, it may not be feasible to solve the equations as proposed in Alternative 1 of the preceding example. Fox[8] suggests a different approach to the problem, in which the sum of the squares of the residuals is selected as the function to be minimized. The problem is then posed as follows:

$$\min_{\mathbf{x}} R(\mathbf{x}) = \sum_{j=1}^{J} [h_j(\mathbf{x})]^2$$

subject to

$$g_m(\mathbf{x}) \geqslant 0, \qquad m = 1, \ldots, M \qquad (9.13)$$

$$M_0 - f(\mathbf{x}) \geqslant 0.$$

Here $M_0$ is a selected goal for the object function. This problem must be minimized sequentially for a series of decreasing values of $M_0$. If an $\mathbf{x}$ is found for which $R(\mathbf{x}) = 0$, we have an acceptable design for which $f(\mathbf{x})$ is less than the current $M_0$ value. Eventually a value of $M_0$ will be introduced for which $\min R(\mathbf{x}) > 0$. The optimum design then lies between the two last values of $M_0$. Problem (9.13) may be solved by the penalty-function method as well as by other methods of constrained minimization.

Parametric inequality constraints of the type

$$g_m(\mathbf{x}, z) \geqslant 0 \qquad \text{for } z_1 \geqslant z \geqslant z_2, \qquad m = 1, 2, \ldots, M \qquad (9.14)$$

may also be treated by means of the penalty-function method, by the introduction of the following penalty function[9]:

$$\phi(\mathbf{x}, r_k) = W(\mathbf{x}) + r_k \sum_{m=1}^{M} \frac{1}{z_2 - z_1} \int_{z_1}^{z_2} \left| \frac{dz}{g_m(\mathbf{x}, z)} \right|. \qquad (9.15)$$

This may, for instance, be of interest in connexion with the optimum design of structures subject to moving loads (see Figure 9.7). A word of caution may be in order here, since, theoretically, the integral may have a finite value even if a constraint value is zero at a certain $z$ value.

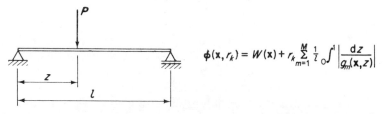

$$\phi(\mathbf{x}, r_k) = W(\mathbf{x}) + r_k \sum_{m=1}^{M} \frac{1}{l} \int_0^l \left| \frac{\mathrm{d}z}{g_m(\mathbf{x}, z)} \right|$$

**Figure 9.7**  Penalty function for moving-load situation

### 9.3  Unconstrained minimization of multivariate functions

The selection of an efficient method for unconstrained minimization becomes extremely important in penalty-function methods, since a sequence of such minimizations has to be performed. There is a wide choice of methods of search. In directed methods the minimization is performed by successive unidirectional searches in a number of different directions with the same search formula as in prior chapters:

$$\mathbf{x}^{k+1} = \mathbf{x}^k + \alpha^k \mathbf{d}^k \tag{9.16}$$

where now $\mathbf{x}^{k+1}$ is the design vector corresponding to the minimum of the unconstrained function to be minimized along the current direction $\mathbf{d}^k$.

The available methods differ mainly in the manner in which the search directions $\mathbf{d}^k$ are generated. One may distinguish between methods of 'direct search' which are making use only of function values $\phi(\mathbf{x})$, and 'gradient' methods which also require the computation of gradients of $\phi(\mathbf{x})$:

$$\mathbf{\nabla}\phi(\mathbf{x}) = \left[ \frac{\partial \phi(\mathbf{x})}{\partial x_1} \cdots \frac{\partial \phi(\mathbf{x})}{\partial x_N} \right]. \tag{9.17}$$

Fletcher presents a review of the more important methods available and their relative merits in Chapter 5. It appears that Powell's method of direct search[6] and the variable metric method of directed search[10] are among the best within the respective classes. Kowalik[11] has described in detail these two methods as applied in connexion with Sumt.

They have been applied to a wide range of problems in structural optimization (see Section 9.7) and the following estimates can be given for required total computer times:

$$T_{\mathrm{DS}} = k_{\mathrm{DS}} N(N + 1) K n_{\mathrm{LS}} T_{\mathrm{F}} \tag{9.18}$$

$$T_{\mathrm{VM}} = k_{\mathrm{VM}} N K(n_{\mathrm{LS}} T_{\mathrm{F}} + T_{\mathrm{G}}) \tag{9.19}$$

where

$T_{DS}$ = total time required when Powell's direct search method is used
$T_{VM}$ = total time required when the variable metric method is used
$N$ = number of free variables
$K$ = number of response surfaces
$n_{LS}$ = average number of function evaluations in the unidirectional searches
$T_F$ = time required for one function evaluation
$T_G$ = time required for one gradient vector evaluation
$k_{DS}, k_{VM}$ = empirical coefficients depending on choice of convergence criteria, etc.

Some typical numerical values of the coefficients may be: $n_{LS} = 6$–$12$ if the Golden Section method is used in the unidirectional search, $n_{LS} = 3$ if an interpolation procedure of the type described in Section 9.4 is used, $K = 3$–$5$, $k_{DS} = 1·2$, and $k_{VM} = 1·5$. The latter two values clearly depend on the selected convergence criteria (see Section 9.5.5).

The choice of method should depend on the type of problem dealt with. The expressions given above indicate that, if the computational effort to determine the gradient corresponds to $N$ function evaluations, the variable metric method by Fletcher and Powell is generally slightly superior to Powell's method of direct search.

One other important practical consideration is the additional effort required if analytical expressions for the gradients need to be programmed. Gradient directions may alternatively be derived by a finite-difference scheme[12].

Fast convergence of gradient methods can only be expected for functions that are well approximated by a quadratic function in the region considered. In the design of statically indeterminate structures, experiences seem to indicate that the $\phi$ functions are quite well conditioned.

Computer time is a critical factor in any optimization effort of reasonable complexity. It is therefore mandatory to scrutinize all possible avenues toward improved efficiency. Formulae (9.18) and (9.19) indicate the following possibilities for improvements, which are discussed further in the sections indicated:

(a) Reductions in $n_{LS}$ through efficient methods of unidirectional search (Section 9.4).
(b) Reductions in the number of response surfaces, for instance through extrapolation techniques (Section 9.5.4).
(c) Reductions in $T_F$ and $T_G$ through efficient methods of analysis (Section 9.6).
(d) Reductions in $k_{DS}$ and $k_{VM}$ through judicious selection of convergence criteria and other program parameters (Sections 9.5.3 and 9.5.5).

## 9.4 Unidirectional search

The search along any particular direction $\mathbf{d}^k$ described in the preceding section is one dimensional. To locate the minimum point along such a vector does not pose any theoretical problem. It is, however, important that the step length $\alpha^k$ [equation (9.16)] be found with as few trial designs as possible.

A typical section through the $\phi$, $W$ and $g$ functions in some arbitrary direction $\mathbf{d}^k$ is shown in Figure 9.8. In this direction the functions are dependent on one variable ($\alpha$) only. The $k$th constraint is assumed to be zero for $\alpha = \alpha_l$, and the function plane is thus divided into one acceptable and one unacceptable region. The $\phi$ function [equation (9.2)] approaches infinity as $\alpha$ approaches $\alpha_l$.

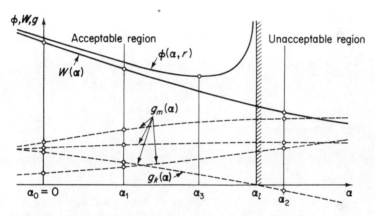

**Figure 9.8** Section through penalty function, objective function and constraints

With decreasing $r$ values, the minimum point of $\phi(\alpha, r)$ will approach $\alpha_l$, which is the true minimum in this local direction. The Golden Section method of search[11], which is widely used, may, typically, require six to twelve evaluations of the $\phi$ function in order to find the minimum along the vector $\mathbf{d}^k$, i.e. $n_{LS} = 6-12$. In this method a considerable amount of information about the $\phi$ function is generated during the search, but the greater part of this information is not used efficiently. Hence the rather large number of function evaluations.

In the search for more efficient methods, a direct application of quadratic or cubic polynomial approximations to the $\phi$ function naturally suggests itself[8], but is usually not successful, since the function tends to exhibit rapid changes in its derivatives near minimum. Lund[13] has found that the use of polynomial approximations to the object and the constraint functions yield much better results. These functions are usually rather smooth and may be well approximated by polynomials in any direction. The $\phi$ function is then expressed in terms of the approximate object and constraint functions, and the minimum along the vector $\mathbf{d}^k$ is found by means of a search on this approximate $\phi$ function.

In cases where accurate calculations of the $W$ and $g$ functions are time consuming, this approach is very efficient.

With reference to Figure 9.8, the process is as follows:

(1) From the starting point $\alpha_0 = 0$, where $W(\alpha_0')$ and $g_m(\alpha_0)$ are known, a step length $\alpha_1$ is taken, and $W(\alpha_1)$ and $g_m(\alpha_1)$ are calculated.

(2) The $W$ and $g$ functions are approximated by straight lines through the function values in $\alpha_0$ and $\alpha_1$, by generating the coefficients $a$, $b$ and $d_m, e_m, m = 1, \ldots, M$:

$$W_1(\alpha) = a + b\alpha$$
$$g_{1m}(\alpha) = d_m + e_m\alpha. \tag{9.20}$$

An approximate expression of the $\phi$ function is then

$$\phi_1(\alpha, r) = a + b\alpha + r \sum_{m=1}^{M} \frac{1}{d_m + e_m\alpha}. \tag{9.21}$$

The derivative is:

$$\phi_1'(\alpha, r) = b - r \sum_{m=1}^{M} \frac{e_m}{(d_m + e_m\alpha)^2}. \tag{9.22}$$

(3) A new step $\alpha_2$ is taken. Depending on the $\phi_1$ and $\phi_1'$ values in $\alpha_0$ and $\alpha_1$, $\alpha_2$ will be to the right of $\alpha_1$, in between $\alpha_0$ and $\alpha_1$, or to the left of $\alpha_0$. $W(\alpha_2)$ and $g_m(\alpha_2)$ are calculated.

(4) The $W$ and $g$ functions are now approximated by parabolas through the function values in $\alpha_0$, $\alpha_1$ and $\alpha_2$ by generating the coefficients $a, b, c$ and $d_m, e_m, f_m, m = 1, \ldots, M$:

$$W_2(\alpha) = a + b\alpha + c\alpha^2$$
$$g_{2m}(\alpha) = d_m + e_m\alpha + f_m\alpha^2. \tag{9.23}$$

A new approximate expression for the $\phi$ function is:

$$\phi_2(\alpha, r) = a + b\alpha + c\alpha^2 + r \sum_{m=1}^{M} \frac{1}{d_m + e_m\alpha + f_m\alpha^2} \tag{9.24}$$

and the derivative is:

$$\phi_2'(\alpha, r) = b + 2c\alpha - r \sum_{m=1}^{M} \frac{e_m + 2f_m\alpha}{(d_m + e_m\alpha + f_m\alpha^2)^2}. \tag{9.25}$$

(5) From here on the search for a minimum is performed entirely by means of equations (9.24) and (9.25). If the interval for a minimum is not covered by the smallest and largest value of $\alpha_0$, $\alpha_1$ and $\alpha_2$, such an interval is found by trials with increasing or decreasing $\alpha$ values. When the interval is found, a search for the point of zero slope $\phi_2'$ is performed. The interval is divided into two equal parts, and the derivative in the middle point is checked. The sign of the derivative indicates where the next division should take place, and the process is repeated until the convergence criterion, usually based on the absolute value of the derivative, is satisfied.

The $W$ and $g_m$ values are calculated for this minimum point $\alpha_3$, and a new approximate $\phi_2$ function may, if necessary, be generated using these values, and the values for the closest pair of $\alpha_0$, $\alpha_1$ and $\alpha_2$. The minimum point $\alpha^i = \alpha_3$ (see Figure 9.8) obtained by the first approximate $\phi_2$ function is, however, usually very accurate, and is accepted as minimum for the $\phi$ function.

Thus only three evaluations of the $W$ and $g_m$ functions are necessary for each minimization along the line. In addition, the coefficients of the polynomial approximations have to be calculated, and five to eight evaluations of the derivative $\phi_2'$ usually have to be performed.

**9.5  Search strategy**

*9.5.1  Starting point*

All of the directed methods of unconstrained minimization need to be supplied with a starting point—an initial design—from which the search may depart. If an interior penalty formulation, such as that presented in equation (9.2), is used, this starting point needs to be feasible. In complex problems it may be rather time consuming to find a feasible point by guessing or simple estimation. It is, however, simple to supply the optimization program with a special routine that generates a feasible starting point. If the program user has supplied a starting point $(\mathbf{x}^0)$ that violates, say, $p$ constraints, those may be arranged as the first $p$ constraints such that

$$g_p \leqslant g_{p-1} \leqslant \cdots \leqslant g_1 \leqslant 0.$$

The one that is most negative is selected as the objective function in the following formulation:

$$\min_{\mathbf{x}} \left[ -g_p(\mathbf{x}) \right] \tag{9.26}$$

subject to

$$g_m(\mathbf{x}) - g_m(\mathbf{x}^0) \geqslant 0, \qquad m = 1, 2, \ldots, p - 1$$

$$g_m(\mathbf{x}) \geqslant 0, \qquad m = p + 1, \ldots, M.$$

This problem is solved by means of the penalty-function method. The search is terminated as soon as the objective function becomes negative, i.e.

$$g_p(\mathbf{x}) > 0$$

and a new test for feasibility is performed. The process is repeated until all of the constraints are satisfied.

Another approach is to define the first response surface also in the infeasible region, for instance by means of the extended penalty function presented by equations (9.11). Then infeasible starting points are immediately acceptable to the minimization routine.

### 9.5.2 Scaling of the variables and the functions

In general-purpose unconstrained minimization routines, it is advantageous to work with scaled variables

$$x'_n = c_n x_n, \qquad n = 1, \ldots, N \tag{9.27}$$

where $c_n$, $n = 1, \ldots, N$, are the scaling factors. There is no easy way of finding optimum scaling factors. The theory of scaling employs the matrix of second derivatives, which is not generally available. In terms of the two-dimensional case, one would like to transform the contours of the function to be minimized as nearly as possible into circles, since gradient directions lead directly to the minimum in such a case. In practice the variables may be scaled so that they are all of the same order of magnitude. It has been found to be satisfactory to select

$$c_n = 1/x_n^0 \tag{9.28}$$

where $x_n^0$ is the selected initial value of $x_n$. If the initial value is reasonably close to the optimum value, the scaled variable will vary in the neighbourhood of one.

It is also advisable to scale the constraint functions so that they contribute similarly to the penalty term. In Example 9.2 this is done by reformulating requirements such as

$$\bar{\sigma} - \sigma_b \geqslant 0$$

to

$$\bar{\sigma}/\sigma_b - 1 \geqslant 0.$$

### 9.5.3 Response factors and number of response surfaces

The proper choice of the initial value of the response factor $(r_1)$ is of considerable significance for the efficiency of the Sumt method. If the initial response factor is chosen too small, the first response surface becomes quite ill behaved, with narrow valleys along some of the boundaries of the feasible region. It will then, at best, take much effort to find the minimum point on this surface if the starting point is at some distance from the minimum. If too large an initial response factor is selected, the minimum of the first response surface will be found far from the boundaries, and one may discover that the first sequence has carried the solution further away from the true optimum than was the starting point. Fox[8] proposes that one could avoid this annoying experience by adding temporarily a new constraint in the form of

$$g_{M+1} = W(\mathbf{x}^0) - W(\mathbf{x}) + \varepsilon \geqslant 0 \tag{9.29}$$

where $\varepsilon$ is some small increase in $W(\mathbf{x})$ that must be permitted in order to have a starting point in the interior of the feasible region. In this manner one may start with a relatively large initial response factor. Constraint (9.29) will prohibit a possible move away from the optimum. This may not, however, always be desirable, as indicated by the following example which deals with a non-convex problem.

**Example 9.4.** *Figure 9.10 illustrates the appearance of several $\phi$ functions [equation (9.2)] for the grillage design problem shown in Figure 9.9. In this non-convex problem there are three distinct relative minima. The first response*

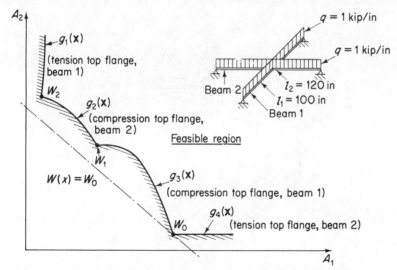

Reproduced by permission from D. Kavlie and J. Moe, 'Automated design of frame structures', *J. Struct. Div.*, *ASCE*, **97**, No. ST1, 33–62 (January 1971)

**Figure 9.9**  Design plane for one-by-one grillage with cross-sectional areas of beams as variables (Reference 14)

*surface shown in Figure 9.10(a) has only one minimum, which is located in the interior of the feasible region. This minimum $\bar{\mathbf{x}}^1$ is, in fact, fairly close to the global optimum. Figure 9.10(b) shows that the second response surface, obtained*

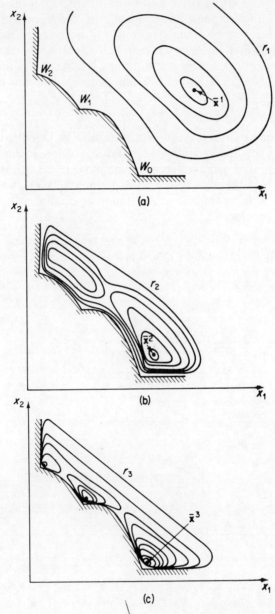

(a)

(b)

(c)

Reproduced by permission from D. Kavlie and J. Moe, 'Automated design of frame structures', *J. Struct. Div.*,
ASCE, **97**, No. ST1, 33–62 (January 1971)

**Figure 9.10**   Response surface for one-by-one grillage of Figure 9.9

with $r_2 = r_1/50$, has two relative minima. The global minimum $\bar{\mathbf{x}}^2$ was reached in this case. Finally Figure 9.10(c) shows that the function $\phi(\mathbf{x}, r_3)$ has three local minima, one corresponding to each of the relative minima of the design problem. Starting from $\bar{\mathbf{x}}^2$ leads to $\bar{\mathbf{x}}^3$ which is close to the global minimum $W_0$. The initial response factor $r_1$ was in this case selected so that the value of the penalty term at point $\bar{\mathbf{x}}^1$ was approximately half the magnitude of the objective function. It is, however, clear that if the initial response factor had been selected smaller, say equal to $r_1/50$, there would have been a considerable risk of ending up at either of the relative minima.

Fiacco and McCormick[4] suggest selecting the magnitude of the initial response factor $r_1$ so that it minimizes the norm of the gradient of the $\phi$ function at the starting point, i.e.

$$|\nabla\phi(\mathbf{x}^0, r_1)| = \min_r |\nabla\phi(\mathbf{x}_0, r)| \qquad (9.30)$$

subject to the condition that $r > 0$.

Using the penalty function given in equation (9.1), condition (9.30) yields

$$r_1 = -\frac{\nabla W(\mathbf{x}^0)^\mathsf{T} \nabla G_1(\mathbf{x}^0)}{|\nabla G_1(\mathbf{x}^0)|^2} \qquad (9.31)$$

where

$$G_1(\mathbf{x}) = \sum_{m=1}^{M} G[g_m(\mathbf{x})]. \qquad (9.32)$$

Although Fiacco and McCormick and others report on successful use of equations (9.31) and (9.32), Example 9.4 above indicates that their method may not always lead to the best choice of $r_1$ for practical problems for which it may be difficult to judge whether the feasible region is convex or not. A simpler and much used approach is to select $r_1$ so that the penalty term adds a certain percentage to the objective function at the starting point, i.e.

$$r_1 \sum_{m=1}^{M} G[g_m(\mathbf{x}^0)] = \frac{p}{100} W(\mathbf{x}^0) \qquad (9.33)$$

where $p$ is the selected percentage. In Example 9.4 $p = 50$ was used. Typical values for well behaved problems with little risk of meeting non-convex boundaries seems to be $p = 1$–$50$ (References 15, 16 and 17). A recommended initial guess may be $p = 10$. Use of equation (9.33) in this simple form is, however, not quite reliable. If a starting point close to one or more of the constraints is selected, the initial response factor, calculated according to equation (9.33), may become too small.

The later response factors may be calculated by the formula

$$r_k = r_{k-1}/t_k, \qquad k = 2, \ldots, K \qquad (9.34)$$

where

$$t_k > 1 \cdot 0.$$

It seems to be the general opinion that the selection of the sequence of $t_k$ values is not too critical. If large values of $t_k$ are selected, one should be able to reach a certain accuracy with few response surfaces. The search for a minimum on each response surface is, however, likely to require relatively many steps. The total number of required steps may prove to be rather similar if smaller $t_k$ values and a correspondingly larger number of response surfaces are employed. Typical numbers for $t_k$ and $K$ are $t_k = 10\text{–}50$ and $K = 6\text{–}3$.

**Example 9.5.** *Weight minimization of the plates of the corrugated bulkhead shown in Figure 9.11 has been carried out by means of the Sumt method, using a general optimization program described by Lund[16]. The program is an*

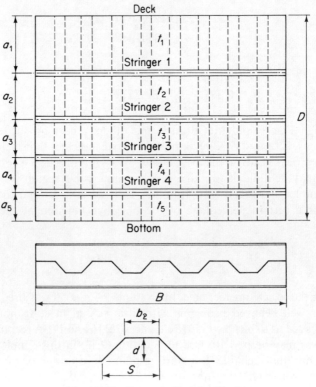

**Figure 9.11**   Corrugated bulkhead in ship

extended version of that discussed in Reference 18, called Koropt. Powell's method of direct search was used. The bulkhead is designed to satisfy the requirements of Det norske Veritas (a ship-classification society). The problem involves twelve variables related to thickness of the plates, the shape of the corrugations and the positions of the horizontal girders supporting the corrugated plates. The number of constraints is $M = 33$.

Table 9.1   Variations in response-surface parameters (Example 9.5)

| Initial penalty factor $r_1$ | Initial penalty-term percentage $p$ | Factors of reduction $t_k$, $k = 2, \ldots, K$ [a] | | |
|---|---|---|---|---|
| | | $K = 3$ | $K = 4$ | $K = 5$ |
| 0·1 | 5·4 | 126·5 | 25·2 | 11·3 |
| 0·3 | 16·1 | 219·0 | 36·3 | 14·8 |
| 1·0 | 54·0 | 400·0 | 54·2 | 20·0 |
| 3·0 | 161·0 | 691·0 | 78·0 | 25·6 |
| 10·0 | 540·0 | 1265·0 | 116·5 | 35·6 |

[a] $K$ = number of response surfaces.

Table 9.2   Number of function evaluations $N_f$

| Initial penalty factor $r_1$ | Number of response surfaces $K$ | | |
|---|---|---|---|
| | 3 | 4 | 5 |
| 0·1 | 1201 | 1280 | 1655 |
| 0·3 | 1128 | 1018 | 1497 |
| 1·0 | 772 | 1232 | 1636 |
| 3·0 | 1848 | 1678 | 1831 |
| 10·0 | 2007 | 2067 | 2190 |

Table 9.3   Obtained minima of objective function $W_{min}$

| Initial penalty factor $r_1$ | Number of response surfaces $K$ | | |
|---|---|---|---|
| | 3 | 4 | 5 |
| 0·1 | 174·082 | 173·683 | 173·673 |
| 0·3 | 174·479 | 174·049 | 173·557 |
| 1·0 | 174·943 | 174·353 | 173·788 |
| 3·0 | 188·924 | 174·125 | 173·551 |
| 10·0 | 184·222 | 173·949 | 173·668 |

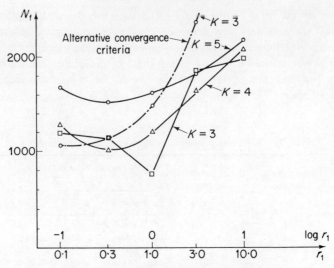

**Figure 9.12** Number of function evaluations $N_f$

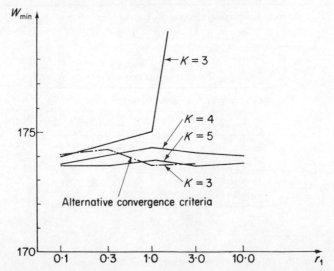

**Figure 9.13** Value of objective function $W_{min}$ at the obtained minimum point

*With one and the same starting point, fifteen different runs have been performed, systematically varying $r_1$, $K$ and $t_k$, $k = 2, \ldots, K$, as shown in Table 9.1. The variations are performed so that the value of the final response factor is identical in all cases. The results obtained are shown in Tables 9.2 and 9.3 and in Figures 9.12 and 9.13. It is seen from the figures that little is gained by reducing the number of response surfaces below four, or by increasing $t_k$ beyond, say, 50.*

### 9.5.4 Extrapolations

Fiacco and McCormick[4] have shown that the minima obtained for decreasing values of the response factor $r$ may be considered as continuous functions of $r$, i.e.

$$\xi_n = x_n(r), \qquad n = 1, \ldots, N \tag{9.35}$$

with continuous derivatives at $r = 0$. If one could determine the functions (9.35), one would then obtain the optimum of the objective function at $x_n(0)$, $n = 1, \ldots, N$.

An examination of results obtained from numerical examples also shows that, if the values of the variables obtained after the unconstrained minimizations on consecutive response surfaces are plotted as functions of the $r$ values, smooth parabolic curves are obtained in most cases. This is shown schematically in Figure 9.14.

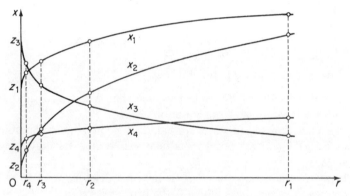

**Figure 9.14** Unconstrained minima as a function of $r$

Having carried out the minimizations on a few of the response surfaces, it now seems reasonable to fit parabolic curves through the points of solution and extrapolate toward $r = 0$ to obtain an improved starting point for the following sequences of the search. Fiacco and McCormick suggest, among others, an extrapolation formula of the following type:

$$x_n(r) = A_n + B_n r^{\frac{1}{2}} \tag{9.36}$$

to be fitted through two points, or alternatively

$$x_n(r) = A_n + \sum_{i=1}^{k-1} B_{ni} r^{i/2} \tag{9.37}$$

if $k$ points are available. The extrapolated approximation to the solution obviously is

$$x_n(0) = A_n. \tag{9.38}$$

This solution must be checked for feasibility before it can be accepted.

Lund[13] suggests a modified procedure which has been used with considerable success and is incorporated in a well established optimization program[16]. Having found the minima $x_{1n}$ and $x_{2n}$ of two response surfaces corresponding to $r_1$ and $r_2$, extrapolation curves of the following type are tried:

$$x_n = z_{1n} + z_{2n}r^\beta, \qquad n = 1, \ldots, N \qquad (9.39)$$

where $\beta$ is a preselected constant.

The correspondingly extrapolated values of constraints and objective functions are, respectively:

$$g_m(\mathbf{z}_1) \quad \text{and} \quad f(\mathbf{z}_1) \qquad (9.40)$$

where

$$\mathbf{z}_1 = [z_{11}, z_{12}, \ldots, z_{1N}]^{\mathrm{T}}. \qquad (9.41)$$

From equation (9.39), it follows that

$$z_{1n} = \frac{r_1^\beta x_{2n} - r_2^\beta x_{1n}}{r_1^\beta - r_2^\beta}, \qquad n = 1, \ldots, N. \qquad (9.42)$$

The following procedure is now followed:

(1) A trial value $\beta_1$ of $\beta$ is selected such that $0 < \beta_1 < 1$, and the $z_{1n}$ values are computed from equation (9.42).

(2) The constraints $g_m(\mathbf{z}_1)$ are evaluated. If point $\mathbf{z}_1$ is acceptable, it is used as the starting point for a final search on a last (fifth) response surface.

(3) If point $\mathbf{z}_1$ is unacceptable, $\beta_1$ is next increased by a value of $\Delta\beta$. If the new $\beta$ value is less than 1·0, the process is repeated from Step (1). If no acceptable point is found for $\beta$ values less than 1·0, the search enters the third response surface in the ordinary way, and a similar attempt to extrapolate can be made when the minimum point for the third response surface is available.

Lund suggests selecting 0·4 as a starting value of $\beta_1$, with $\Delta\beta = 0·1$. Usually one will obtain an acceptable point for $\beta = 0·6$–$0·7$. The extrapolation technique presented here has been found to yield a reduction in the number of function evaluations of approximately 30 per cent.

### 9.5.5 Convergence criteria and step sizes

In the Sumt method one needs criteria for the termination of the iterative search processes on the following three levels:

(a) For each unidirectional search.
(b) For unconstrained minimization of each response surface.
(c) For the final completion of the search.

A number of criteria can be developed, based on changes in the values of the variables or the functions $W(\mathbf{x})$ or $\phi(\mathbf{x}, r_k)$. Such criteria may be, for instance:

$$|x_n'' - x_n'| \leqslant \varepsilon_1, \qquad n = 1, \ldots, N \tag{9.43a}$$

$$\left[ \sum_{n=1}^{N} (x_n'' - x_n')^2 \right]^{\frac{1}{2}} \leqslant \varepsilon_2 \tag{9.43b}$$

$$\left| \frac{W''(\mathbf{x}) - W'(\mathbf{x})}{W'(\mathbf{x})} \right| \leqslant \varepsilon_3 \tag{9.43c}$$

$$\left| \frac{\phi''(\mathbf{x}, r) - \phi'(\mathbf{x}, r)}{\phi'(\mathbf{x})} \right| \leqslant \varepsilon_4 \tag{9.43d}$$

$$|\mathbf{g}| = |\nabla \phi(\mathbf{x}, r)| = \left\{ \sum_{n=1}^{N} \left[ \frac{\partial \phi(\mathbf{x}, r)}{\partial x_n} \right]^2 \right\}^{\frac{1}{2}} \leqslant \varepsilon_5 \tag{9.43e}$$

where the superscripts $''$ and $'$ refer to two suitable points of comparison and

$$\mathbf{g} = \left[ \frac{\partial \phi}{\partial x_1}, \frac{\partial \phi}{\partial x_2}, \ldots, \frac{\partial \phi}{\partial x_N} \right]^{\mathrm{T}}.$$

The inclusion of upper and lower limits on the number of steps in the iterations may also be advisable.

In the unidirectional search employing the method described in Section 9.4, a value of $\varepsilon_5 = 0 \cdot 1$ has been used for all response surfaces. An upper limit is also put on the number of points investigated.

In the minimization of each response surface with direct search methods, a value of $\varepsilon_1 = 0 \cdot 005$ is generally used for the first response surface (with scaled variables). This value is typically reduced by a factor $C_k = 0 \cdot 4$ each time a new response surface is entered. The differences $|x_n'' - x_n'|$ are taken between two consecutive cycles consisting of $N + 1$ unidirectional searches. Figure 9.15 shows that, in the optimization of the corrugated bulkhead described in Example 9.5, a factor of reduction $C_k = 0 \cdot 5$ yielded fewer function evaluations for five response surfaces. One should also prescribe a minimum number of cycles to avoid a premature termination of the search if the process temporarily gets into a deadlock.

It is not really necessary to require that the minimum of each response surface is determined with high accuracy, as long as it is only used as starting point for a search on the following surface. This fact is of great importance in structural optimization, since it justifies the approximate methods of structural analysis to be employed in the search, as described in Section 9.6 of this chapter.

A higher accuracy of the response-surface minima may be required if they are to be used as a basis for extrapolations as described in Section 9.5.4.

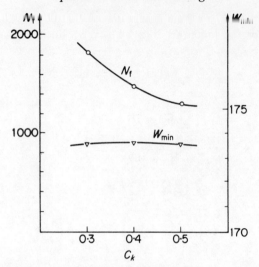

**Figure 9.15** Influence of variations in the factor $C_k$ on the number of function evaluations $N_f$ and the obtained solution $W_{min}$ (corrugated-bulkhead example with $K = 5, r_1 = 0.3$ and $C_s = 0.4$)

The final termination of the search process usually takes place after a predetermined number of response surfaces or alternatively when some suitable criteria of the types listed in equations (9.43a–d) are satisfied for two consecutive response surfaces.

Fiacco and McCormick[4] have deduced the following absolute bounds on the true minimum value of the objective function $W(\xi)$, based on a dual formulation of the minimization problem [penalty function of equation (9.2)]:

$$W(\mathbf{x}) - r_k \sum_{m=1}^{M} \frac{1}{g_m(\mathbf{x})} \leqslant W(\xi) \leqslant W(\mathbf{x}). \qquad (9.44)$$

This may alternatively be written in the following way:

$$2W(\mathbf{x}) - \phi(\mathbf{x}, r_k) \leqslant W(\xi) \leqslant W(\mathbf{x}). \qquad (9.45)$$

Here $W(\mathbf{x})$ and $\phi(\mathbf{x}, r_k)$ are the function values in the optimum point of the current response surface.

For small values of $r_k$, equation (9.45) can apparently be relied on only if the minimum of the response surface $\phi(\mathbf{x}, r_k)$ is determined with high accuracy. Figure 9.16 shows the upper and lower bounds as found for the five response surfaces of one of the corrugated-bulkhead examples.

The initial step length in the unidirectional searches is typically selected as 0.05 for the scaled variables. This value should also be reduced each time a new response surface is entered, for instance by a constant factor $C_s$ which may typically be selected in the range 0.3–0.5. Figure 9.17 shows the influence

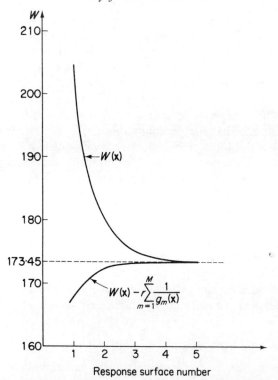

**Figure 9.16** Upper and lower bounds of the optimum value determined by means of results from consecutive response surfaces ($C_k = 0\cdot4$, $C_s = 0\cdot5$)

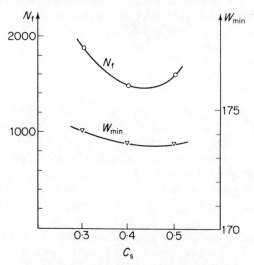

**Figure 9.17** Influence of variations in the factor $C_s$ on the number of function evaluations $N_f$ and the obtained solution $W_{min}$ (corrugated-bulkhead example with $K = 5$, $r_1 = 0\cdot3$ and $C_k = 0\cdot4$)

of the factor $C_s$ on the search efficiency for the corrugated-bulkhead problem treated in Example 9.5. In Figures 9.12 and 9.13 the fully drawn curves have been obtained with $C_s = C_k = 0.4$. Figure 9.13 clearly shows that this has not yielded satisfactory results for the case with $K = 3$.

The 'dash–dot' lines drawn for $K = 3$ are obtained using $C_k = C_s = 0.253$. Thereby the convergence criterion for the last response surface has been brought down to the same level as with $K = 4$. This example is just a reminder of the fact that the optimum selection of the number of response surfaces $K$, the response-factor-reduction coefficients $t_k$, the step sizes and the convergence criteria is an involved problem that requires the simultaneous consideration of all these items. If few response surfaces are selected, correspondingly small values of $C_k$ and $C_s$ should be selected to reach a desired accuracy in the final solution. The accuracies attained in the numerical results presented here may be beyond those required in practical design work. An efficient way of reducing central-processing-unit times will be to relax the convergence criteria accordingly.

The problem of selecting suitable step sizes in the determination of gradient directions for the variable metric method by means of finite differences is discussed at some length by Kavlie and Moe[12].

## 9.6 Approximate analyses

In the optimization of complex statically indeterminate structures the numercial work involved in the structural analyses usually predominates. It therefore becomes mandatory to reduce this work as much as possible if one wants to solve anything but the simplest academic problems within acceptable computing times. In a straightforward application of the Sumt method combined with Powell's method of direct search and a Golden Section method of unidirectional search, one may typically expect that the number of designs to investigate is

$$N_a = 40N(N + 1). \tag{9.46}$$

This number may be reduced somewhat by means of the unidirectional search method described in Section 9.4. The required number of analyses also depends, of course, on the accuracy prescribed, as well as other details of the search routines used. In a structure with six variables, which, for a practical problem, is a very low number, this suggests, at best, a need to analyse the structure almost one thousand times. It is fortunately possible to reduce this number very considerably by the introduction of a so-called 'behaviour-model' technique[19].

The idea behind the behaviour-model concept is simply to create a strictly mathematical and approximate model of the structural behaviour, and to perform the optimization on this mathematical model. One may, for instance, make a Taylor-series expansion of the behaviour in terms of stress resultants or stresses as functions of the variables. For practical reasons the model should be as simple as possible. Second-degree behaviour models, with no crosscoupling terms, have been applied with success[20], i.e.

$$B_l(\mathbf{x}) = a_{0l} + \mathbf{a}_{1l}^{\mathrm{T}}(\mathbf{x} - \mathbf{x}^0) + \tfrac{1}{2}(\mathbf{x} - \mathbf{x}^0)^{\mathrm{T}}\mathbf{H}'_l(\mathbf{x} - \mathbf{x}^0) \qquad (9.47)$$

where

$$B_l(\mathbf{x}) = \text{behaviour for quantity number } l$$
$$\mathbf{H}'_l = \text{a diagonal matrix.}$$

The model may alternatively be written as follows:

$$B_l(\mathbf{x}) = a_{0l} + \sum_{n=1}^{N} \{a_{1ln}(x_n - x_{0n}) + a_{2ln}(x_n - x_{0n})^2\}. \qquad (9.48)$$

The $2n + 1$ coefficients $a_{0l}$, $a_{1ln}$ and $a_{2ln}$ may be calculated from known behaviour values $B_l(\mathbf{x})$ in $2n + 1$ points. This is demonstrated for a two-dimensional case in Figure 9.18. The behaviour values are calculated by

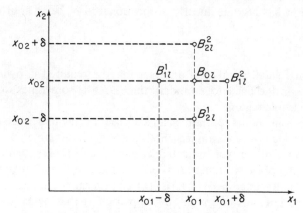

**Figure 9.18**  Points for evaluation of behaviour model

means of complete analyses for the reference point $(x_{01}, x_{02})$ as well as for small perturbations $\delta$ in each of the variables to either side of the reference point. The coefficients for the behaviour model [equation (9.48)] are then:

$$\left.\begin{array}{l} a_{0l} = B_{0l} \\[2mm] a_{1ln} = \dfrac{B_{nl}^2 - B_{nl}^1}{2\delta}, \qquad n = 1, 2 \\[4mm] a_{2ln} = \dfrac{B_{nl}^1 - 2B_{0l} + B_{nl}^2}{2\delta^2}, \qquad n = 1, 2. \end{array}\right\} \qquad (9.49)$$

Such behaviour models are, of course, inaccurate except in the vicinity of the reference points used for model construction. The rate of convergence of the Sumt method is, fortunately, not very sensitive to precision in the evaluation of the minima of consecutive response surfaces, since these are only used as starting points for the next surfaces.

The question now arises as to which quantities to select for such modelling. One may, for instance, select:

   (i) Stress resultants (bending moments, shearing forces, etc.).
   (ii) Stresses, or even:
   (iii) Constraint functions $g_m(\mathbf{x})$ and object function $W(\mathbf{x})$.

The relative merits of these alternatives are discussed by Lund[15,20]. Numerical tests[21] show that the number of complete analyses for a frame design example with six nodal points, six members and six variables could, by means of alternatives (i) and (ii), be brought down from 1199 to 56–84, depending on the program parameters used.

It is gratifying that the required number of complete analyses, which, according to equation (9.46), is proportional to the square of the number of variables, now has become linearly proportional to $N$, being approximately equal to

$$N_a = K(2N + 1). \tag{9.50}$$

An additional small number of analyses may be required if, owing to the inaccuracies in the behaviour models, the search is sometimes carried out into the infeasible region.

The behaviour-model technique can also be implemented as a completely general method of constrained non-linear programming. Polynomial approximations of the type indicated by equation (9.47) are then generated directly for the objective function $W(\mathbf{x})$ and the constraint functions $g_m(\mathbf{x})$, and a new penalty function $\phi_1(\mathbf{x}, r_k)$ is created by means of these polynomials. A very important advantage in this approach is that general analytical expressions may then be derived for the gradient vector $\mathbf{g}$ and the Hessian matrix $\mathbf{H}$ of the $\phi_1(\mathbf{x}, r_k)$ function. The unconstrained minimum may then be found by means of the modified Newton–Raphson method. A simplified flow chart of this procedure is shown in Figure 9.19.

In this approach the number of search steps is very moderate. It has therefore been found possible to make complete analyses in each step, while the behaviour models are only used in the generation of search directions:

$$\mathbf{s} = -\mathbf{H}^{-1}\mathbf{g}. \tag{9.51}$$

This procedure has not been thoroughly tested, but some promising results have been obtained in exploratory investigations. The corrugated-bulkhead problem described in Example 9.5 was selected for this study. The program

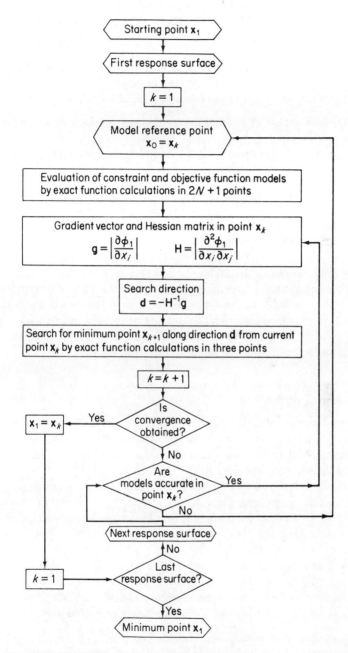

**Figure 9.19** Flow chart for general optimization program using behaviour models for constraints and objective function

using constraint and objective-function models yielded a minimum-weight solution of 173·46 ton with 421 complete analyses. Five response surfaces were used. These numbers should be compared to those of Tables 9.2 and 9.3 to judge the efficiency of the method.

The efficiency of this new approach in terms of number of analyses is, of course, partly obtained at the expense of larger time consumption in the search routines. The main time-consuming items are:

(i) Determination of coefficients in the polynomials of the behaviour models.
(ii) Evaluation of gradient vector $\mathbf{g}$ and Hessian matrix $\mathbf{H}$.
(iii) Inversion of the Hessian matrix.

For the corrugated-bulkhead example just mentioned, the analyses were quite simple. In this case the central-processing-unit time was halved by the new approach. For problems involving more complex analyses, the time reduction is expected to be much more significant.

While the method of behaviour models effectively reduces the required number of analyses, it still remains important to improve the efficiency of the methods of analysis and to take advantage of the fact that approximate analyses of modified structures may be performed with much less computational effort than complete new analyses. For further information on this subject reference is made to a recent publication by Kavlie and Powell[22], and to a method recently published by Fox and Miura[23].

### 9.7 Practical applications

Penalty-function methods have been used to a considerable extent in the optimization of ship structures. The problems treated include optimization of corrugated bulkheads[18], tanker midship sections[17], decks of car carriers[24], two- and three-dimensional frames[7,12,15], and grillages[14].

Code constraints as well as general stress and deflection constraints have been used. The optimization has been based partly on weight and partly on cost criteria. In some cases minimum-cost structures have been found to be radically different from those corresponding to minimum weight[17,24].

**Example 9.6.** *A typical midship section through a tanker is shown in Figure 9.20. The stiffened panels shown in the figure are supported on transverse frames at regular intervals a. Assuming that the principal dimensions of the ship are given, the following free variables remain to be determined:*

$s_d$ = *spacing of longitudinals in deck and bottom*
$s_s$ = *spacing of longitudinals in ship sides and longitudinal bulkheads*
$t_d$ = *thickness of plates in deck*

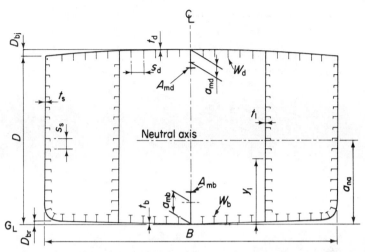

Reproduced from J. Moe and S. Lund, 'Cost and weight minimization of structures with special emphasis on longitudinal strength members of tankers', *RINA Trans.*, **110**, 43–70 (1968) by permission of The Royal Institution of Naval Architects

**Figure 9.20** Midship section of tanker

$t_b$ = *thickness of plates in bottom*
$t_s$ = *thickness of plates in ship sides*
$W_d$ = *section modulus of deck longitudinals*
$W_b$ = *section modulus of bottom longitudinals.*

*Plate thicknesses in longitudinal bulkheads and section moduli of longitudinals on sides and bulkheads are calculated as functions of the free variables.*

*A minimum-cost design is sought, while complying with the constraints imposed by the rules of the relevant ship-classification society (in this case Det norske Veritas).*

*Applying the Sumt method in conjunction with Powell's method of direct search, this problem is solved with 15–20 s central-processing-unit time on a Univac 1108. A few of the results obtained during some parametric studies are shown in Figures 9.21 and 9.22. The cost factor k expresses the relationship between cost of labour (per hour) and cost of steel (per tonne).*

Optimization programs of this type have been found to be valuable tools in design as well as in teaching and research. It may be quite important to the designers that the optimization programs are highly efficient automated design routines that allow them to investigate numerous alternative designs. In teaching and research, the programs may be used to perform parametric studies that yield valuable insight into the cost performances of various competing structural configurations.

An attempt to incorporate several optimization programs into an integrated design package intended to be administered interactively from a remote

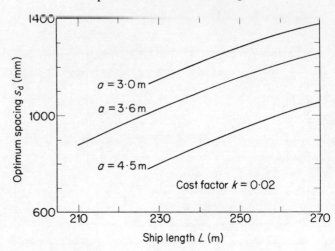

Reproduced from J. Moe and S. Lund, 'Cost and weight minimization of structures with special emphasis on longitudinal strength members of tankers', *RINA Trans.*, **110**, 43–70 (1968) by permission of The Royal Institution of Naval Architects

**Figure 9.21**    Optimum spacing ($s_d$) of longitudinals in deck and bottom for various web spacings
*a* and ship lengths *L*

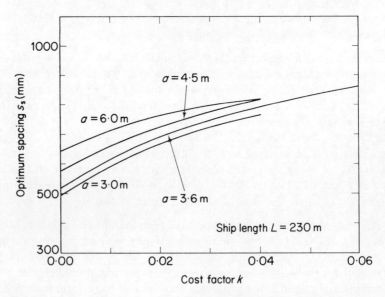

Reproduced from J. Moe and S. Lund, 'Cost and weight minimization of structures with special emphasis on longitudinal strength members of tankers', *RINA Trans.*, **110**, 43–70 (1968) by permission of The Royal Institution of Naval Architects

**Figure 9.22**    Optimum spacing ($s_s$) of longitudinals in ship sides and longitudinal bulkheads
for various web spacings *a*

terminal is described in Reference 25. Applications of the penalty method to other types of structural problems have been described by Moses *et al.*[26], while Bracken and McCormick[18] present a more general review of applications in various fields.

## 9.8 Closure

The penalty-function technique seems to be gaining increasing popularity in the field of structural optimization, and the volume of literature on practical applications in this field is increasing rapidly.

The main advantage of the method seem to be:

(i) General applicability to widely different types of problems.
(ii) Ease of programming. The optimization program can be built on a modular basis so that the practical user only needs to formulate and program his own problem, while a general-purpose optimization program is used more or less as a 'black box'[16].
(iii) The method is robust in the sense that it works well even with approximate information. This permits one to use approximate and time-saving methods of analysis during major portions of the search.
(iv) By means of the extended penalty-function technique or other available techniques, the method is also able to accept infeasible starting points. This may be a quite important feature when dealing with complex design problems for which feasible starting points are not readily available.
(v) Some success has also been reported on the use of the penalty-function technique in discrete non-linear programming[27,28].
(vi) Although the method is already used with success in practical design work, there are great potentials for further improvements. There is hardly any program working today that includes even the majority of the known features that would have to be incorporated in an 'optimum' version. Much also remains to be done to increase our knowledge concerning optimum choices of search parameters for different types of problems.

## References

1. R. Courant, 'Variational methods for the solution of problems of equilibrium and vibrations', *Bull. Am. Math. Soc.*, **49**, 1–23 (1943).
2. R. Frisch, 'The logarithmic potential method of convex programming', Memorandum 13, University Institute of Economics, Oslo, 1955.
3. C. W. Carroll, 'The created response surface technique for optimizing nonlinear restrained systems', *Operations Res.*, **9**, No. 2, 169–184 (1961).
4. A. V. Fiacco and G. P. McCormick, *Nonlinear Programming, Sequential Unconstrained Minimization Techniques*, Wiley, New York, 1968.

5. C. M. Ablow and G. Brigham, 'An analog solution of programming problems', *Operations Res.*, **3**, 388–394 (1955).
6. M. J. D. Powell, 'An efficient method for finding the minimum of a function of several variables without calculating derivatives', *Computer J.*, **7**, No. 3, 155–162 (1964).
7. D. Kavlie, 'Optimum design of statically indeterminate structures', Ph.D. Thesis, University of California, Berkeley, 1971.
8. R. L. Fox, 'Unconstrained minimization approaches to constrained problems', in *Structural Design Applications of Mathematical Programming Techniques* (Eds. G. Pope and L. Schmit), AGARDograph 149, 1971, Chap. 6.
9. G. Zoutendijk, 'Nonlinear programming: A numerical survey', *J. SIAM Control*, **4**, 194–210 (1966).
10. R. Fletcher and M. J. D. Powell, 'A rapidly convergent decent method for minimization', *Computer J.*, **6**, 163–168 (1963).
11. J. Kowalik, 'Nonlinear programming procedures and design optimization', Acta Polytechnica Scandinavica, Math. and Computing Machinery Series, No. 13, Trondheim, 1966.
12. D. Kavlie and J. Moe, 'Automated design of frame structures', *J. Struct. Div., ASCE*, **97**, No. ST1, 33–62 (1971).
13. S. Lund, 'Experiences with direct search methods for nonlinear structural optimization', Nord-DATA-69, Stockholm, June 1969.
14. D. Kavlie and J. Moe, 'Application of nonlinear programming to optimum grillage design with nonconvex sets of variables', *Int. J. Num. Methods Eng.*, **1**, No. 4, 351–378 (1969).
15. S. Lund, 'Frame design example', in *Optimization and Automated Design of Structures* (Eds. J. Moe and K. M. Gisvold), Division of Ship Structures, Meddelelse SK/M21, Trondheim, 1971, Chap. 7.4.
16. S. Lund, 'SKOPT—direct search program for optimization of nonlinear functions with nonlinear constraints', Users Manual SK/P13, Division of Ship Structures, Norges Tekniske Høgskole, Trondheim, 1971.
17. J. Moe and S. Lund, 'Cost and weight minimization of structures with special emphasis on longitudinal strength members of tankers', *Trans. Royal Inst. Naval Arch.*, **110**, No. 1, 43–70 (1968).
18. J. Bracken and G. P. McCormick, *Selected Applications of Nonlinear Programming*, Wiley, New York, 1968.
19. J. Moe, 'Design of ship structures by means of nonlinear programming techniques', *Int. Shipb. Progress*, **17**, 69–86 (1970).
20. S. Lund, 'Behaviour model technique', in *Optimization and Automated Design of Structures* (Eds. J. Moe and K. M. Gisvold), Division of Ship Structures, Meddelelse SK/M21, Trondheim, 1971, Chap. 7.2.
21. J. Moe and K. M. Gisvold (Eds.), *Optimization and Automated Design of Structures*, Division of Ship Structures, Meddelelse SK/M21, Trondheim, 1971.
22. D. Kavlie and G. H. Powell, 'Efficient reanalysis of modified structures', *J. Struct. Div., ASCE*, **97**, No. ST1, 377–392 (1971).
23. R. L. Fox and H. Miura, 'An approximate analysis technique for design calculations', *AIAA J.*, **9**, 177–179 (1971).
24. D. Kavlie, J. Kowalik and J. Moe, 'Design optimization using a general nonlinear programming method', *European Shipbuilding*, **15**, No. 4 (1966).
25. J. Moe, 'Integrated design of tanker structures', in *Proceedings of the International Symposium on Numerical and Computer Methods in Structural Mechanics, The University of Illinois, Urbana, Sept. 1971*, Academic Press, to be published.

26. F. Moses, R. L. Fox and G. Goble, 'Mathematical programming applications in structural design', in *Computer-Aided Engineering* (Ed. G. L. M. Gladwell), S.M. Study No. 5, University of Waterloo Press, 1971.
27. P. V. Marcal and R. A. Gellatly, 'Application of the created response surface technique to structural optimization', *Proceedings of the 2nd Conference on Matrix Methods in Structural Mechanics, Oct. 1968*, AFFDL-TR-68-150.
28. K. M. Gisvold, 'Discrete programming', in *Optimization and Automated Design of Structures* (Eds. J. Moe and K. M. Gisvold), Division of Ship Structures, Meddelelse SK/M21, Trondheim, 1971, Chap. 8.1.

*Chapter 10*

# Dynamic Programming and Structural Optimization

*Andrew C. Palmer*

## 10.1 Introduction

Dynamic programming has only relatively recently been applied to structural optimization. It is a mathematical approach to the theory of multi-stage decision processes. In a typical problem, a sequence of decisions has to be made, and these decisions affect the state of a system, and perhaps its response to future decisions. By the choice of decisions some result is achieved. Some results are better than others (in some definable way), and one wishes to achieve the best result, and therefore the best sequence of decisions. Typical decision sequences arise in planning an investment program, deciding whether or not to continue a so-far unsuccessful research project, routing a pipeline across a mountain range, or deciding how to play a hand of bridge.

A decision sequence will be called a *policy*; the decision sequence leading to the best result is the *optimal policy*. The classical approach to the problem is to think of the result as some function of all the decisions, and then somehow to optimize with respect to these decisions. If the decision sequence is at all long there are very many alternative policies, and the dimension of the optimization problem is very large. Dynamic programming escapes from this difficulty by thinking instead of what has to be decided at each step in the sequence. If we have at each step some way of looking at the current state of the system and deciding what to do in consequence, we do not need at that step to consider the whole chain of past and future decisions, except in so far as they are reflected by the current state. Dynamic programming rests on the systematic exploitation of this idea.

Dynamic programming was developed by Bellman in the early 1950s and energetically applied and developed in an extensive series of books and papers[1-4]. It has been widely applied in operations research and economics (see, for example, Beckmann[5] and White[6]). It is known in mathematical control theory, but seems to have gone almost unnoticed in other areas of engineering. The first application to structural optimization seems to be due to Kalaba[7].

179

180    *Optimum Structural Design*

Because dynamic programming is less well known than linear and non-linear programming, this chapter begins with a discussion of a simple problem that illustrates the few central ideas of the theory. It then goes on to a simple structural problem, which can be developed to exemplify most of the techniques which have so far been used in structural applications. Generalizations of this problem illustrate the limitations of dynamic programming, which is not by any means a method that is applicable to any structure. A review of existing work on structural applications comes next, and is followed by a discussion of possible lines along which the theory might be developed.

## 10.2 A network problem

A problem posed by Bellman and Dreyfus[2] illustrates the ideas of dynamic programming, although it is not a structural problem at all. Figure 10.1 shows a rectangular grid of lines, on which nodes can be referred to by coordinates $(i, j)$. Written next to each line segment is a positive number, which we interpret as the time required to move along the segment from one end to the other. We consider the problem of finding a path through the network, starting at the origin $(0, 0)$ and finishing at point A $(3, 3)$, such that the sum of the numbers encountered on the way is a minimum. This is the minimum-time path from $(0, 0)$ to A. So, for example, one path is $(0, 0)$, $(1, 0)$, $(1, 1)$, $(2, 1)$, $(2, 2)$, $(3, 2)$, $(3, 3)$; the total time by that path is $3 + 5 + 1 + 4$

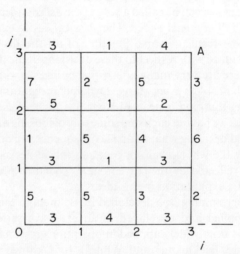

From *Applied Dynamic Programming*, by Richard E. Bellman and Stuart A. Dreyfus (published by Princeton University Press) copyright © 1962 by the Rand Corporation: Fig. 65 on p. 208, in adapted form. Reprinted by permission of Princeton University Press

Figure 10.1

# Dynamic Programming and Structural Optimization 181

$+ 2 + 3 = 18$, but is not necessarily optimal. There is one restriction on the paths: each move must be either 'up' in the positive $j$ direction or 'across' in the positive $i$ direction, but backtracking from $(i, j)$ to $(i - 1, j)$ or $(i, j - 1)$ is not allowed.

One central idea to be used is the concept of *embedding*. In this instance the particular problem of finding a minimum-time path to A from $(0, 0)$ is embedded in the family of problems of finding minimum-time paths to A from a general point $(i, j)$. In keeping with this, define $W(i, j)$ as the minimum time to A from $(i, j)$; in addition, let $t(i, j; k, l)$ be the time from $(i, j)$ to the neighbouring node $(k, l)$. A second central concept is that of sequential decision: the problem of deciding on a complete path is imagined to be broken up into a sequence of simpler decisions, to be made one by one, and the structure of this decision sequence is then studied.

Imagine that the path has reached node $(i, j)$. The next move will take it either to $(i + 1, j)$ or to $(i, j + 1)$. If the next move is to $(i + 1, j)$, and the path from $(i, j)$ is to be a minimum-time path, then the continuation from $(i + 1, j)$ to A must be along the minimum-time path from $(i + 1, j)$; in that case the total time from $(i, j)$ is the time to get to $(i + 1, j)$ plus the time from there to A by a minimum-time path, i.e.

$$t(i, j; i + 1, j) + W(i + 1, j) \quad [\text{move across to } (i + 1, j)].$$

If, on the other hand, the next move from $(i, j)$ is to $(i, j + 1)$, then the total time from $(i, j)$ to A, moving first to $(i, j + 1)$ and thence by a minimum- time path to $A$, is

$$t(i, j; i, j + 1) + W(i, j + 1) \quad [\text{move up to } (i, j + 1)].$$

Note that it is not yet known which is the optimal move from $(i, j)$, and neither is it known whether or not $(i, j)$ is on the optimal path from $(0, 0)$ to A. The only idea used so far is that, if a path should arrive at some point, the optimal continuation from there is along the minimum-time path.

The alternative moves from $(i, j)$ each give a total time to A. Since there is no other allowed choice, one of the moves must give a minimum time, and that minimum time is by definition $W(i, j)$; therefore

$$W(i, j) = \min \begin{cases} t(i, j; i + 1, j) + W(i + 1, j) \\ t(i, j; i, j + 1) + W(i, j + 1) \end{cases} \quad \begin{array}{l} (\text{move across}) \\ (\text{move up}). \end{array} \quad (10.1)$$

A solution can be found by the systematic use of this equation. One by one we construct $W$ functions that will later be needed on the right-hand side of the equation, and at each step record the optimal decision. The calculation can be carried out on the diagram itself (Figure 10.2). Since $W(i + 1, j)$ and $W(i, j + 1)$ are needed to calculate $W(i, j)$, the calculation starts at A and works downward and to the left.

Optimum Structural Design

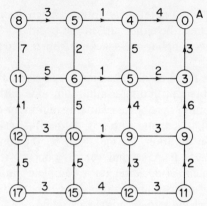

From *Applied Dynamic Programming*, by Richard E. Bellman and Stuart A. Dreyfus (published by Princeton University Press) copyright © 1962 by the Rand Corporation: Fig. 66 on p. 210, in adapted form. Reprinted by permission of Princeton University Press

Figure 10.2

By definition, $W(3, 3) = 0$. From $(2, 3)$ the only allowed move is across, and so $W(2, 3) = 4$. Values of $W$ are written in the circle centred on the corresponding node, and an arrow indicates the direction of the minimum-time path to A from that node. From $(3, 2)$ the only allowed move is up, and so $W(3,2) = 3$. From $(2, 2)$ there is a choice of move; from equation (10.1)

$$W(2, 2) = \min \begin{Bmatrix} t(2, 2\,;3, 2) + W(3, 2) \\ t(2, 2\,;2, 3) + W(2, 3) \end{Bmatrix} = \min \begin{Bmatrix} 2 + 3 \\ 5 + 4 \end{Bmatrix} = 5 \quad (10.2)$$

and the optimal move from $(2, 2)$ leads to $(3, 2)$. At $(1, 3)$ there is again no choice of move, so $W(1, 2) = 1 + 4 = 5$, and next this value and $W(2, 2)$ can be used to find $W(1, 2)$, and so on. Equation (10.1) is applied at each node in turn until the whole network has been covered. Each node now has a value of $W$ and an arrow showing the optimal move from that node.

The original special problem was to find the minimum-time path from $(0, 0)$ to A. The minimum time is $W(0, 0) = 17$. The optimal path is now constructed by starting at $(0, 0)$ and following the arrows from one node to the next, from $(0, 0)$ to $(0, 1)$, then through $(0, 2)$, $(1, 2)$, $(2, 2)$ and $(3, 2)$ to $(3, 3)$. Minimum-time paths from other points can be constructed in the same way.

Although dynamic programming can be applied to a much wider class of problems than those of paths through networks, many problems can be reformulated in this way, although they may at first sight look very different. Bellman and Dreyfus[2] describe, for instance, an application to the search for an optimal procedure for bringing a fighter aircraft initially cruising at a certain height and speed to another height and speed, taking account of the complex dependence of thrust and drag on speed and height. A simple structural application is considered next.

## 10.3   A simple structural problem

Consider the problem of optimizing the layout of a plane pin-jointed cantilever truss of the type shown in Figure 10.3. It consists of several panels, numbered from the left 1 through $n$, each of unit horizontal extent, each except the first with the same $\times$ bracing. Coordinates $x$ and $y$ define points on the truss, which is symmetrical about the $x$ axis. At the left-hand end it carries a unit vertical load, and at the right it is supported by a rigid foundation.

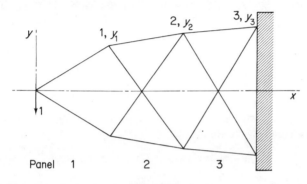

**Figure 10.3**

Since the truss is statically determinate, the axial force in each bar is determined by the layout. The layout is defined by the ordinates $y_1, \ldots, y_n$ of the nodes in the upper chord, since the $x$ coordinates for the nodes are known, and an optimal layout therefore requires an optimal choice of these ordinates. It can easily be shown that the forces in the bars belonging to panel $i$ depend only on $y_{i-1}$ and $y_i$, and not on the configurations of other panels. It will be assumed that some method exists by which an individual bar can be designed, given its length and the axial force it must transmit, and that a cost can be assigned to each such design. The most natural way of assigning a cost to the whole structure is to sum the costs of individual bars. Since $y_{i-1}$ and $y_i$ determine both the forces in the bars of panel $i$ and the lengths of those bars, the cost of that panel is a function of those two variables; denote this cost $w_i(y_{i-1}, y_i)$. The total cost of the first $i$ panels is therefore

$$w_1(0, y_1) + w_2(y_1, y_2) + \cdots + w_i(y_{i-1}, y_i). \tag{10.3}$$

Imagine a continuous path drawn on Figure 10.3 from a point on the line $x = i$ to the origin, the path consisting of a sequence of line segments, the first from $(i, y_i)$ to $(i - 1, y_{i-1})$, the next from $(i - 1, y_{i-1})$ to $(i - 2, y_{i-2})$, and so on, until finally it goes from $(1, y_1)$ to $(0, 0)$. A cost $w_j(y_{j-1}, y_j)$, which is a function of the segment number $j$, the position of the left-hand end (defined by $y_{j-1}$) and the position of the right-hand end (defined by $y_j$), can be assigned to each segment. The path itself marks out a possible layout for the upper chord of the truss, which we have seen to define the design. The minimum-cost path minimizes expression (10.3), and defines an optimal design for the first $i$ panels if the upper chord passes through $(i, y_i)$. An analogy with the network problem considered earlier becomes obvious.

Let $W_i(y_i)$ be the minimum of expression (10.3) with respect to $y_1, \ldots, y_{i-1}$, the minimum cost of the first $i$ panels if their right-hand end is fixed at $(i, y_i)$, and the cost of the minimum-cost path from $(i, y_i)$ to $(0, 0)$. Suppose that for a fixed value of $y_i$ we make an arbitrary choice of $y_{i-1}$. The cost assigned to the line segment from $(i, y_i)$ to $(i - 1, y_{i-1})$ is $w_i(y_{i-1}, y_i)$, and the cost of the remainder of the path by an optimal continuation from $(i - 1, y_{i-1})$ to $(0, 0)$ is $W_{i-1}(y_{i-1})$. An arbitrary choice of $y_{i-1}$ makes the total cost from $(i, y_i)$ to $(0, 0)$

$$w_i(y_{i-1}, y_i) + W_{i-1}(y_{i-1}). \tag{10.4}$$

Minimizing this expression with respect to $y_{i-1}$ locates the minimum cost of all paths from $(i, y_i)$ to $(0, 0)$, and so

$$w_i(y_i) = \min_{y_{i-1}} \{w_i(y_{i-1}, y_i) + W_{i-1}(y_{i-1})\}. \tag{10.5}$$

This equation is analogous to equation (10.1) in the network problem. It can also be derived by a more formal mathematical argument, and this is done in Appendix 10A.

One method of finding an optimal design by using equation (10.5) is like the one used in the network problem. There the allowable paths were limited to a finite set of alternatives, along fixed grid lines. Now suppose each of the $y_i$ limited to a finite number of alternatives. In this instance, let each of the $y_i$ take one of the values 0·2, 0·4, 0·6, 0·8 and 1. Figure 10.4 shows all possible

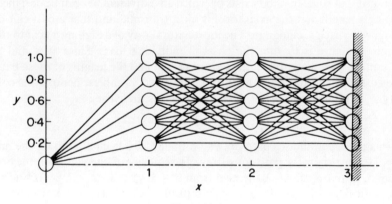

**Figure 10.4**

paths; there are altogether 125 of them, each one corresponding to a different design. A cost $w_i(y_{i-1}, y_i)$ has to be assigned to each segment. Adopting the simplest of all models of bar costs, we let the cost of each bar be the product of its length and the absolute value of the axial force it transmits (as it would be if the allowable stress were unity in tension or compression, and if the material cost one unit per volume). It can easily be shown that the cost $w_i$ of panel $i$ is then

$$w_i(y_{i-1}, y_i) = \begin{cases} i(1 + y_{i-1}^2)/y_i - (i - 2)y_i & i > 1, \; y_i/i < y_{i-1}/(i-1) \\ (i-1)(1 + y_i^2)/y_{i-1} - (i+1)y_{i-1} & i > 1, \; y_i/i > y_{i-1}/(i-1) \end{cases}$$

$$(10.6)$$

$$w_1(0, y_1) = y_1 + 1/y_1 \tag{10.7}$$

The calculation procedure is very like that used before. If only the first panel of the truss is considered, then

$$W_1(y_1) = w_1(0, y_1) \tag{10.8}$$

and can be immediately calculated from equation (10.7). In Figure 10.5 values of $W_1(y_1)$ are written next to the corresponding nodes for the different values of $y_1$. Exactly as in the network problem, the calculation started from the 'target' A and worked step by step towards the origin; here it starts at the load at $(0, 0)$ and works panel by panel towards the foundation. Moving on to panel 2, let $y_2$ be $0.8$. In Figure 10.5 alternative paths from $(2, 0.8)$ to $(1, y_1)$ are

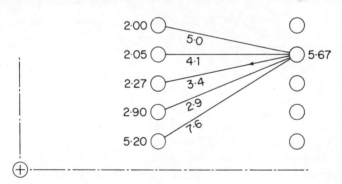

**Figure 10.5**

marked with corresponding values of $w_2(y_1, 0.8)$ calculated from equation (10.6). Specializing equation (10.5):

$$W_2(0.8) = \min_{y_1} \{w_2(y_1, 0.8) + W_1(y_1)\} \tag{10.9}$$

and is, by inspection, $3.4 + 2.27 = 5.67$ when $y_1 = 0.6$. The corresponding path is labelled by an arrow.

This comparison is now repeated for the other four allowed values of $y_2$. The results are shown in Figure 10.6. Moving to panel 3, and making these comparisons again, results for three panels are given in Figure 10.7. The choice of an optimal path is now immediate. If $y_3$ can be as large as 1·0, the cost is 9·59, and the optimal design has $y_2 = 0\cdot8$, $y_1 = 0\cdot6$. If $y_3$ is restricted to 0·4, for instance, the cost is 16·57 and $y_2 = y_1 = 0\cdot6$.

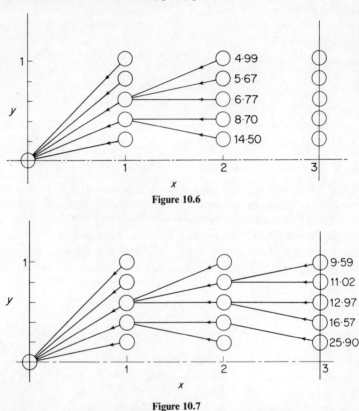

Figure 10.6

Figure 10.7

An optimal design has been achieved without any linearization of the panel cost functions $w_i$. The form of the function is indeed immaterial. In equation (10.6) it is already non-linear and not everywhere continuously differentiable. It would be a trivial matter to introduce into this calculation a design procedure for individual bars which took account of buckling, or which limited the maximum length of bars, or which assigned costs in a more sophisticated way. A computer program can easily be written to carry out a calculation of this kind. Even if there were more panels, and the $y_i$ could take many more values, the amount of calculation would remain small. A solution by dynamic programming requires not nearly so many calculations as a simple enumeration and comparison of all possible designs. Consider a cantilever truss with $n$ panels and $m$ alternative values for each of the $n$ ordinates $y_i$. There are $m^n$ distinct designs to be compared in an enumerative solution, and enumeration requires that the cost for each design be evaluated, by the addition of $n$ distinct $g_i$ values, and that the smallest cost then be found. Dynamic programming, on the other hand, requires $(n - 1)m^2$ additions of two values, and $(n - 1)m^2$ comparisons to determine the smallest among $n$ alternatives. If the computing effort needed to add together $p$ numbers is $(p - 1)$ units, and the effort to find the smallest among $q$ numbers is $(q - 1)$ units, then the 'advantage' of dynamic programming, the ratio of the effort used by enumeration to that used by dynamic programming, is

$$\frac{2m^n(n - 1)}{n(n - 1)m^2} = 2m^{n-2}n^{-1} \tag{10.10}$$

which is much greater than 1 unless $m$ and $n$ are small.

The explanation for this advantage is that dynamic programming only examines designs that contain optimal designs as substructures. When the designs for panel $i$ are being compared, by the choice of $y_{i-1}$ as a function of $y_i$, this is naturally not done without thinking of the effect of the design of other panels, but the presence of $W_{i-1}(y_{i-1})$ in equation (10.5) ensures that all the designs being considered include optimal designs for the first $(i - 1)$ panels. Similarly, in the optimal-path problem of Figures 10.1 and 10.2, when the optimal decision at $(i, j)$ is being considered, the only paths examined are those which have optimal continuations from the node reached as a result of the decision. Bellman called this the *principle of optimality*[1]:

> 'An optimal policy has the property that whatever the initial state and initial decision are, the remaining decisions must constitute an optimal policy with regard to the state resulting from the first decision.'

In the context of allocation problems, Aris put it less formally[8]:

> 'If you don't do the best you can with what you happen to have got, you will never do the best you might have done with what you should have had'.

In many problems this principle makes it possible to write down immediately the key equation corresponding to equations (10.1) or (10.5).

## 10.4 Computational questions

Examine first of all the problem of constructing a solution to equation (10.5), the expression of the principle of optimality. A similar equation appears in almost all applications of dynamic programming, and the same problems recur. The first part of the solution requires the step-by-step construction of the functions $W_i$, at each step constructing a function that will be used in the next step in the right-hand side of the equation. What is therefore needed is a means of storing a function $W_i(y_i)$ so that a value of the function can be calculated for any value of the argument. One solution has been devised already: it limited each $y_i$ to a small set of alternative values, and tabulated $W_i(y_i)$ for each possible argument. It is natural to look next for a way of allowing $y_i$ to take any value in some continuous range.

In particularly simple problems $W_i$ can be represented analytically, but even in the truss problem the moderate complexity of the $w_i$ functions soon leads to extremely complex expressions for $W_i$. One of the great advantages of dynamic programming is precisely that it allows objective functions to take complex forms, not necessarily smooth or continuous or able to be represented analytically, and so analytic representation of the $W_i$ functions is rarely appropriate.

The next possibility is to calculate $W_i(y_i)$ for values at regular intervals of its argument, store the results as a table, and obtain intermediate values by interpolation. As long as the function to be represented is a function of only a single variable, a satisfactory representation usually needs only a modest amount of storage. This method is extremely simple to program. Another possibility is to use polynomial representation, originally suggested by Bellman and Dreyfus[9]. In the simplest form $W_i$ is expressed as a polynomial:

$$W_i(y_i) = \sum_{j=0}^{k} a_{ij} y_i^j \qquad (10.11)$$

and the $k$ coefficients $a_{ij}$ are stored instead of values of the function. It is almost always better first to normalize the allowable range of $y_i$ to $(-1, 1)$, so that, if $y_i$ lies in the range $(a, b)$, we define

$$z = -1 + 2(y_i - a)/(b - a) \qquad (10.12)$$

and then let

$$W_i = \sum_{j=0}^{k} a_{ij} \phi_j(z) \qquad (10.13)$$

where the $\phi_i$ are polynomials orthonormal over $(-1, 1)$, such as Chebyshev and Legendre polynomials. General applications of this idea to dynamic programming are described by Bellman, Kalaba and Kotkin[10], and applications in mechanics are described by Palmer[11], Barras[12] and Cammaert[13].

The other computational problem in solving equation (10.5) is that of carrying out the minimization. If each of the $y_i$ can take only a limited number of values, the function to be minimized itself has a limited number of values, and the minimum can be found by inspection or by straightforward search.

If the $y_i$ vary continuously, the minimum can be found by one of the well known minimization techniques[14]. It can often be shown that the function to be minimized is unimodal or convex, and this can be used to make the search more efficient.

## 10.5 Multidimensional problems

In the truss problem described above each of the nodes had a prescribed horizontal position, and only their vertical position coordinates $y_i$ had to be optimized. If there are $m$ alternative values of $y_i$, $m^2$ values of $w_i(y_{i-1}, y_i) + W_{i-1}(y_{i-1})$ have to be computed for each value of $i$; in the simple example worked out in Section 10.3 $m = 5$.

Now consider a generalization of the problem. Suppose the truss to remain symmetrical about the $x$ axis, but now let both the $y_i$ and $x_i$ coordinates of the nodes on the upper chord be free to be optimized. The four corners of panel $i$ are now at $(x_{i-1}, y_{i-1})$, $(x_{i-1}, -y_{i-1})$, $(x_i, -y_i)$ and $(x_i, y_i)$, and the cost of the panel is a function $w_i(x_{i-1}, y_{i-1}; x_i, y_i)$. It can then be shown that the principle of optimality is expressed by

$$W_i(x_i, y_i) = \min_{x_{i-1}, y_{i-1}} \{w_i(x_{i-1}, y_{i-1}; x_i, y_i) + W_{i-1}(x_{i-1}, y_{i-1})\} \quad (10.14)$$

where $W_i(x_i, y_i)$ is the minimum cost of an $i$-panel truss whose right-hand end is supported at $(x_i, y_i)$ and $(x_i, -y_i)$. The problem can still be solved in the same way, but the computational problem is more serious. The $W_i$ are now functions of two variables, and the minimization has to be over two variables rather than one. It is still possible to solve by letting $x_i$ and $y_i$ each take one of a limited range of allowed values, and considering all combinations of $x_{i-1}$ and $y_{i-1}$, but there are now many more combinations. If $y_i$ and $x_i$ each have $m$ possible values, $W_i(x_i, y_i)$ has to be determined for $m^2$ combinations of its argument variables, and this requires $m^4$ calculations of

$$w_i(x_{i-1}, y_{i-1}; x_i, y_i) + W_{i-1}(x_{i-1}, y_{i-1})$$

for each value of $i$. The generalization has increased the amount of calculation by a factor of $m^2$.

This exponential growth in the amount of calculation is a consequence of what Bellman called the 'curse of dimensionality'. It was shown earlier that dynamic programming escapes from the multidimensional problem of considering all decisions in a policy simultaneously, and it does this by examining how the decisions at each stage depend on the current state. What determines the amount of computation is not primarily the length of the decision sequence, but the number of variables required to describe the current state, and the dimension of the problem is the number of these 'state variables'. In the simplest version of the truss problem there is one state variable $y$. At the end of $i - 1$ panels $y_{i-1}$ locates the ends of that part of the truss, at the end of $i$ panels $y_i$ locates the ends of that part, and so on. When we are considering how to design panel $i$ for a given value of $y_i$, only one variable $y_{i-1}$ is at the time relevant to the description of the first $i - 1$ panels. However, in the generalization there are two state variables $x_i$ and $y_i$, and the amount of calculation is much increased. It will be seen later that a further generalization to the general asymmetric plane $\times$ braced truss requires four state variables. The curse of dimensionality compels us to be as stingy as we can in the choice of state variables. It is sometimes possible to reduce the number of state variables by a reformulation of the problem[11]. In multidimensional problems so much time is required to calculate each value of the $W_i$ function that we are driven to methods of function representation that need as few input values of the function as possible. Polynomial representation is one of these methods.

The computational difficulty of multidimensional problems is the chief obstacle to the wider use of dynamic programming. A great deal of attention is being given to overcoming it, and a discussion of possible developments follows later in this chapter.

### 10.6    A review of previous work

Kalaba[7] considered a structure consisting of a number of elements, each of which has a certain probability of failure which is a known function of its cross-sectional area. The failure of one element leads to the failure of the whole structure. The problem is to assign material to different elements in such a way as to minimize the structure weight and yet attain a given overall failure probability. If $L_m$ is the length of the $m$th element, $\rho_m$ the specific weight of its material, $P_{f_m}(A_m)$ its failure probability if its cross-sectional area is $A_m$, and if $W_i(p)$ is the smallest weight of the first $i$ elements that gives an overall probability of failure within those elements of $p$, it can be shown from the principle of optimality that

$$W_i(P_f) = \min_{A_i} \left( \rho_i L_i A_i + W_{i-1}\{[P_f - P_{f_i}(A_i)]/[1 - P_{f_i}(A_i)]\} \right). \quad (10.15)$$

A solution can be found in the usual way.

Razani and Goble[15] and Goble and DeSantis[16] investigated the optimal proportioning of plate girder flanges composed of a number of sections, each made from a plate of uniform thickness. If there are many sections the thickness of the flange can follow the variation in bending moment, so that steel is saved, but the cost of splices is increased. The optimal policy is the solution of a smoothing problem of the general type:

$$\text{minimize} \quad \sum_{k=1}^{n} g_k(x_k - r_k) + \sum_{k=1}^{n} h_k(x_k - x_{k-1}) \quad (10.16)$$

where $\{r_k\}$ is a given sequence, $g(x)$ and $h(x)$ are given functions, $x_0$ is given and $\{x_k\}$ is a minimizing sequence to be determined. Bellman[1] discusses the solution of these problems, which often arise in operations research.

Porter-Goff[17–19] investigated the optimal layout of a cantilever and a two-hinged arch designed by simple plastic theory. He went on to study the optimal design of a pin-jointed framework supporting a point load and connected to a circular foundation (Figure 10.8). The weight of each bar was assumed to be proportional to the product of its length and the absolute value of the force it transmits. An absolute-minimum-weight Michell solution to this problem is known[20]. Porter-Goff investigated frameworks of the general

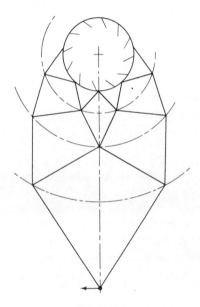

**Figure 10.8**

layout shown in Figure 10.8, and used dynamic programming to find minimum-weight designs within this class. They were found to have weights roughly 5 per cent above the weights of the Michell solutions. A similar analysis was used to optimize a class of symmetrical cantilever trusses supported at two points, and to find optimal layouts for a symmetrical bridge truss. The general ideas of this analysis are those used in the simple truss problem described earlier.

Palmer[21] applied dynamic programming to the optimal plastic design of continuous beams and rigid-jointed frames. He considered a beam continuous over a number of spans of known length, each span uniform in section and carrying a known load. It follows from the lower-bound theorem of plastic design[22] that the beam will not collapse if it is so proportioned that its limit moment is nowhere less than the limit moment in a statically admissible bending-moment distribution in equilibrium with the loads. The maximum bending moment in a single span is a function of the moments over the supports at either end of the span, and they therefore determine the design and cost of the span. The cost of the whole beam is a function of all the moments over the supports, and the optimization problem therefore becomes a problem of the optimal choice of those moments. It turns out to have a mathematical formulation identical with the truss problem and leading to equation (10.5). It can be extended to optimal elastic design[23], where the moments and beam stiffnesses have to obey additional conditions expressing compatibility.

This idea can be applied to the plastic design of frames. In a rectangular frame with $p$ storeys and $q$ bays, the number of state variables is $q + 1$ (one for each column) if the dynamic-programming procedure moves down the frame, and is $p$ (one for each beam) if the procedure moves across the frame. The minimum number of state variables is thus min $(q + 1, p)$, and the method is only feasible if this is small. Cammaert[13] has recently made a detailed study of dynamic-programming design of frames with particular attention to the computational problem. He is also able to extend his analysis to include effects not taken into account in simple plastic design, such as the effect of axial force on limit moment and the effect of deflection at the top of the frame. A text by Heyman[24] includes an introduction to dynamic-programming analysis of beams.

Palmer and Sheppard[25,26] studied the optimal layout of cantilever trusses. Their analysis of the symmetrical × braced truss is that outlined earlier in the paper. They then studied the general cantilever truss, in which the load direction is no longer parallel to the foundation and the truss is no longer

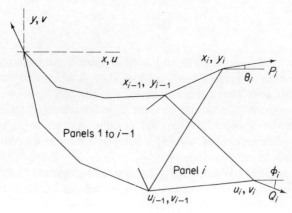

**Figure 10.9**

symmetric. An extra complexity appears when the assumption of symmetry is dropped. In the analysis of the symmetrical truss it was stated that the forces in the bars of the $i$th panel depend on the positions of the corners of that panel. It might be expected that this would still be true if the truss were asymmetric. This is not so, as can be seen by considering the forces acting on the first $i$ panels of such a truss (Figure 10.9). Four variables $P_i, Q_i, \theta_i$ and $\phi_i$ define the forces supporting that part of the truss, and if they are known the axial forces in the four bars of panel $i$ can be found. Three equilibrium equations express the equilibrium of the first $i$ panels as a whole. Only if the whole truss is symmetric does an extra symmetry condition enable all four forces to be found. If the truss is asymmetric, an additional state variable, such as $\theta_i$, is necessary. If panel points on the upper chord are located by $(x_i, y_i)$, and those on the lower chord are located by $(u_i, v_i)$, so that the four corners of panel $i$ are at $(x_{i-1}, y_{i-1}), (u_{i-1}, v_{i-1}), (u_i, v_i)$ and $(x_i, y_i)$, it turns out that the principle of optimality is expressed by:

$$W_i(x_i, y_i, u_i, v_i, \theta_i)$$

$$= \min_{x_{i-1}, y_{i-1}, u_{i-1}, v_{i-1}} \{w_i + W_{i-1}(x_{i-1}, y_{i-1}, u_{i-1}, v_{i-1}, \theta_{i-1})\} \quad (10.17)$$

where $w_i(x_{i-1}, y_{i-1}, u_{i-1}, v_{i-1}; x_i, y_i, u_i, v_i, \theta_i)$ is the cost of panel $i$ and $W_i(x_i, y_i, u_i, v_i, \theta_i)$ is the cost of the first $i$ panels if their right-hand end is supported at $(x_i, y_i)$ and $(u_i, v_i)$, the reaction at $(x_i, y_i)$ being directed at $\theta_i$ to the horizontal. The cost is not minimized over $\theta_{i-1}$, because $\theta_{i-1}$ is a function of $x_{i-1}, y_{i-1}, u_{i-1}, v_{i-1}, x_i, y_i, u_i, v_i$ and $\theta_i$.

These methods have recently been applied to the more clearly practical problem of synthetizing optimal designs for the transmission towers ('pylons') that support overhead electric lines. An effort was made to include as many as possible of the constraints that govern the design of real towers, and to escape from the concentration on simple and over-idealized structures that has characterized much previous work on optimal design. It is important to distinguish between two quite different questions that such a study might ask. One might take a broad view, and look for the best possible design concept. In a particular situation, is the best tower one constructed in the now conventional way, a lattice-braced steel space frame from angle sections bolted together, or might it be better to use a steel-tube mast or a timber tower? One can instead ask the less ambitious question of whether or not the conventional design can be systematically optimized in detail. In this study the second point of view was taken; the wider question remains open.

The general configuration of the tower is shown in Figure 10.10. Attention was concentrated on the part of the tower below the lowest cross arm, although the method can be used to optimize the design of the remainder of the tower as well. The height of the lowest cross arm is fixed by the required clearance between the ground and the suspended cables. Neither the width

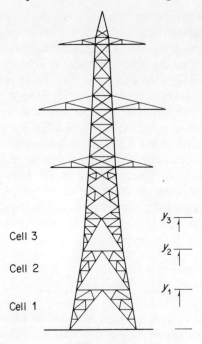

**Figure 10.10**

at the cross arm nor the width at the base are fixed, but it turned out to be simplest to assign each of these widths a fixed value, to optimize the remainder of the design, and then to examine the effect of altering the widths. The tower is divided into a number of 'cells' (the three-dimensional counterpart of panels) and dynamic programming is used to determine the optimal number of cells between the cross arm and the ground, the levels which fix the top and bottom of each cell, and the optimal type of bracing to be used in each cell, this type being chosen from six topologically distinct bracing configurations.

Suppose that the heights of the top and bottom of each cell are $y_i$ and $y_{i-1}$ (Figure 10.10). In general, a lattice-braced tower is not statically determinate, and so the forces within the bars of cell $i$ depend on the layout and dimensions of all the bars in the tower. Though it would be possible to take indeterminacy into account by an appropriate choice of additional state variables, to do so in this case would lead to unacceptable computation times, for the reasons discussed earlier. It can, however, be argued that the stresses within cell $i$ depend primarily on the stress resultants transmitted across a horizontal section of the tower at level $y_i$ and only to a much smaller extent on the detailed arrangement of other cells. If the bars of cell $i$ are designed so that they can sustain forces which equilibrate the stress resultants, the resulting design is safe under the conditions of the lower-bound theorem of plastic design. This argument, which is developed in more detail in the original papers[26,27], is reinforced by the evidence of field tests and of previous satisfactory experience of its use by tower designers. It then follows that for any bracing configuration of cell $i$ the forces within the bars of that cell depend only on the prescribed external loads on the tower and on the heights $y_{i-1}$ and $y_i$. The force analysis can be repeated for each loading condition. The extreme values taken by the axial force in each bar, in whichever loading condition is critical for that bar, determine the design for that bar, through an appropriate design process for individual bars. Each bar can then be costed, and the costs added to determine a cell cost. The analysis can be repeated for each allowed bracing configuration. The minimum cost of cell $i$ can then be found as a function of $y_{i-1}$ and $y_i$. The problem of determining the optimal design of the tower then becomes formally identical to the truss problem discussed earlier, although, of course, it is more complex in detail.

The basic configuration of transmission towers has remained unchanged for many years, and their detail design has been substantially refined by traditional methods. No dramatic further improvement is to be hoped for, but it was nonetheless found that a dynamic-programming approach led to further reductions in structural weight, of the order of 4 per cent, without any significant increase in structural complexity. Because large numbers of towers are built to a single design, even such small savings are economically worthwhile, and more than repay the extra design costs.

### 10.7  Possible future development

It was natural for dynamic programming to be applied first to problems in which the number of state variables was very small. This led to a concentration on long, narrow structures, such as beams and cantilevers, in which the interaction between different parts is rather simple, and therefore can be expressed by a few state variables. The advantages of dynamic programming are then dramatic: the analysis for the optimal design of an $n$-panel symmetric cantilever truss, for instance, is reduced from a $2n$-dimensional non-linear-programming problem to a two-dimensional dynamic-programming problem. If the technique is to be more widely applicable, however, attention must be given to ways of overcoming the curse of dimensionality. Larson[28] reviewed computational aspects of dynamic programming, against the rather different background of control theory, and among the techniques he listed the following look particularly promising in structural problems.

*Direct iteration*  Each state variable is assumed to take one of a small number of values spaced on a regular grid. An approximate optimal policy is determined in the usual way. This approximate optimal policy is then used to select an improved quantization of the state variables, which now take values on a new regular grid centred on the approximate values found in the first iteration. In the truss problem, for instance, we might want to locate the ordinates more accurately than on the $0.2$, $0.4$, $0.6$, $0.8$, $1.0$ grid. The first iteration would use this grid; as we have seen, it finds the optimal value of $y_2$ to be $0.8$ and that of $y_i$ to be $0.6$. A second iteration could then be carried out by exactly the same procedure, but with $y_2$ being limited to values $0.70, 0.75, 0.80$, $0.85$ and $0.90$, and $y_1$ being limited to $0.50$, $0.55$, $0.60$, $0.65$ and $0.70$. A new optimal policy would then be found, and the process could be repeated.

*Successive approximation*    A trial optimal policy is assumed. All the state variables except one are then held constant at their values in the trial policy, and a one-variable dynamic-programming procedure is then used to optimize the policy with respect to that one variable. That variable is then held fixed, a second variable is released, and the procedure is repeated to optimize with respect to the second variable. The process is repeated cyclically. In this way an $n$-dimensional problem becomes a sequence of one-dimensional problems. Convergence is not assured, but can be shown to occur under quite wide conditions[29].

*State increment dynamic programming*    Restrictions are imposed on the change in a state variable allowed to occur in the passage from one step in the procedure to the next. In the truss problem, for instance, the change from $y_i$ to $y_{i-1}$ might be limited to $0.2$, so that if $y_2$ were $0.6$, $y_1$ could only be $0.4$, $0.6$ or $0.8$. This would reduce the amount of calculation in the ratio $25:9$, and, in fact, Figure 10.7 shows that it would make no difference to the solutions for $i = 3$. The technique is useful in control applications, where the restriction in the change of a state variable is achieved by restricting the step length, and has been extensively discussed by Larson[30].

*Polynomial approximation*

*Lagrangian multipliers*    The number of state variables can sometimes be reduced by using Lagrangian multipliers, just as it can in non-linear programming.

These techniques have been little used in structural problems as yet. Sheppard[26] used direct iteration to refine transmission-tower designs, and Cammaert[13] found that successive approximation substantially reduced the computation time needed for optimal plastic design of multibay multistorey frames.

Dynamic programming does not lend itself to the construction of general-purpose computer programs suitable for a wide range of distinct problems. Here it can be contrasted with linear programming. It is far from easy to write an efficient linear-programming subroutine, but once it has been written it can be used to solve a wide range of problems. In dynamic programming, on the other hand, each problem usually requires a new formulation, but once the appropriate formulation has been found the preparation of a computer program is usually quite straightforward.

## 10.8 Closure

Dynamic programming is well suited to the optimal design of certain kinds of structure, in general those in which the interaction between different parts is rather simple. It is much less suitable for highly redundant structures with complex interactions, and is not a general-purpose method. It does not require the objective function to be a continuous or smooth or linear function of the parameters describing the design. Its main drawback is the amount of computation necessary if there are more than a few state variables expressing the complexity of the interaction, but several promising techniques for avoiding this difficulty have still to be explored.

### APPENDIX 10A. DERIVATION OF EQUATION (10.5)

The formal derivation of equation (10.5) is as follows. By definition

$$W_i(y_i) = \min_{y_1, y_2, \ldots, y_{i-1}} \{w_1(0, y_1) + \cdots + w_{i-1}(y_{i-2}, y_{i-1}) + w_i(y_{i-1}, y_i)\}. \tag{10A.1}$$

The minimization can be split into two stages, an inner minimization with respect to $y_1, y_2, \ldots, y_{i-2}$, and an outer minimization with respect to $y_{i-1}$, so that

$$W_i(y_i) = \min_{y_{i-1}} \{ \min_{y_1, \ldots, y_{i-2}} [w_1(0, y_1) + \cdots + w_{i-2}(y_{i-2}, y_{i-1}) + w_i(y_{i-1}, y_i)]\}$$

$$\tag{10A.2}$$

The term $w_i(y_{i-1}, y_i)$ is not affected by the inner minimization, and so equation (10A.2) can be rearranged as

$$W_i(y_i) = \min_{y_{i-1}} \{w_i(y_{i-1}, y_i) + \min_{y_1, \ldots, y_{i-2}} [w_1(0, y_1) + \cdots + w_{i-2}(y_{i-2}, y_{i-1})]\}.$$

$$\tag{10A.3}$$

The second term in the square brackets is exactly analogous to the right-hand side of equation (10A.1), the only difference being that $i$ has been replaced by $i - 1$. Accordingly

$$W_i(y_i) = \min_{y_{i-1}} [w_i(y_{i-1}, y_i) + W_{i-1}(y_{i-1})]. \tag{10A.4}$$

## References

1. R. E. Bellman, *Dynamic Programming*, Princeton University Press, Princeton, N.J., 1957.

# 198      *Optimum Structural Design*

2. R. E. Bellman and S. E. Dreyfus, *Applied Dynamic Programming*, Princeton University Press, Princeton, N.J., 1962.
3. R. E. Bellman, *Adaptive Control Processes: A Guided Tour*, Princeton University Press, Princeton, N.J., 1961.
4. R. E. Bellman, 'The theory of dynamic programming', *Bulletin of the American Mathematical Society*, **60**, 503–516 (1954).
5. M. J. Beckmann, *Dynamic Programming of Economic Decisions*, Springer-Verlag, Berlin, 1968.
6. D. J. White, *Dynamic Programming*, Oliver and Boyd, Edinburgh, 1969.
7. R. Kalaba, 'Design of minimum-weight structures for given reliability and cost', *J. Aero. Sci.*, **29**, 355–356 (1962).
8. R. Aris, *Dynamic Programming*, Colloquium lectures in pure and applied science, No. 8, Socony–Mobil Oil Company Inc., Dallas, Texas, 1965.
9. R. E. Bellman and S. E. Dreyfus, 'Functional approximation and dynamic programming', *Mathematical Tables and Other Aids to Computation*, **13**, 247–251 (1959).
10. R. E. Bellman, R. Kalaba and B. Kotkin, 'Polynomial approximation—a new computational approach in dynamic programming', *Mathematics of Computation*, **17**, 155–161 (1963).
11. A. C. Palmer, 'Limit analysis of cylindrical shells by dynamic programming', *Int. J. Solids and Struct.*, **5**, 289–302 (1969).
12. R. Barras, 'Limit analysis of spherical shells by dynamic programming', University of Cambridge, Department of Engineering, Technical Report CUED/C—STRUCT/TR 1, 1969.
13. A. B. Cammaert, 'The optical design of multi-storey frames using mathematical programming', *Ph.D. Dissertation*, Cambridge University, 1971.
14. D. J. Wilde, *Optimum Seeking Methods*, Prentice-Hall, Englewood Cliffs, N.J., 1964.
15. R. Razani and G. G. Goble, 'Optimum design of constant-depth plate girders', *J. Struct. Div. ASCE*, **92**, No. ST2, 253–281 (1966).
16. G. G. Goble and P. V. DeSantis, 'Optimum design of mixed steel composite girders', *J. Struct. Div. ASCE*, **92**, No. ST6, 25–43 (1966).
17. R. F. D. Porter-Goff, 'Decision theory and the shape of structures', *J. Royal Aero. Soc.*, **70**, 448–452 (1966).
18. R. F. D. Porter-Goff, 'Shape as a structural design parameter', *Ph.D. Dissertation*, University of Leicester, 1968.
19. R. F. D. Porter-Goff, 'Dynamic programming for optimal structural design', University of Leicester Engineering Department, Report 69-4, 1969.
20. A. G. M. Michell, 'The limits of economy of material in framed structures', *Phil. Mag. (Series 6)*, **8**, 589–597 (1904).
21. A. C. Palmer, 'Optimum structure design by dynamic programming', *J. Struct. Div. ASCE*, **94**, No. ST8, 1887–1906 (1968).
22. J. F. Baker and J. Heyman, *Plastic Design of Frames*, Vol. I, Cambridge University Press, Cambridge, 1969.
23. A. C. Palmer, unpublished.
24. J. Heyman, *Plastic Design of Frames*, Vol. II, Cambridge University Press, Cambridge, 1971.
25. A. C. Palmer and D. J. Sheppard, 'Optimizing the shape of pin-jointed structures', *Proc. Inst. Civil Engrs.*, **47**, 363–376 (1970).
26. D. J. Sheppard, 'Structural design optimization by dynamic programming', *Ph.D. Dissertation*, Cambridge University, 1970.

27. D. J. Sheppard and A. C. Palmer, 'Optimal design of transmission towers by dynamic programming', University of Cambridge, Department of Engineering, Technical Report CUED/C—STRUCT/TR 21, 1971 [to appear in *Computers and Structures* (1972)].
28. R. E. Larson, 'A survey of dynamic programming computational procedures', *IEEE Transactions on Automatic Control*, **AC-12**, 767–774 (1967).
29. R. E. Larson and A. J. Korsak, 'A dynamic programming successive approximations technique with convergence proofs', *Automatica*, **6**, 245–260 (1970).
30. R. E. Larson, *State Increment Dynamic Programming*, American Elsevier, New York, 1968.

*Chapter 11*

# Discrete Variables in Structural Optimization

*A. Cella and K. Soosaar*

## 11.1 Introduction

The first papers on structural optimization by the use of mathematical programming raised a great interest in the development of the related numerical techniques. As noted in Chapter 1, after ten years of intensive research a wide range of algorithms has been developed and tested, but systematic applications of structural optimization to actual design problems have been scarce.

One reason for the gap existing between academic research and professional practice is the inadequacy of the mathematical models in representing costs and variables of the design problem. In the mathematical model, for instance, the commonly used weight objective function neglects the often predominant costs of fabrication and erection of the structure. The geometric configuration and the material distribution in the structure have rarely appeared among the variables of the model. The stiffness distribution is usually the variable, and it is assumed that the field of available member sizes and wall thicknesses is continuous; in reality, the choice is always limited to a discrete set that often is quite small because of secondary design constraints. Thus the need to revise some of the original axioms of structural optimization, to formulate different models and algorithms, and, particularly, to adopt more flexible design procedures, has been increasingly recognized.

Discrete-variable models were first introduced into structural design by Toakley[1], who assigned a Boolean variable to each of the available profiles. Toakley solved discrete linear problems such as the optimal plastic design of frames and the optimal design of statically determinate trusses under deflection constraints. Gomory's first algorithm (see Reference 2) was used; unfortunately the performance of the algorithm is quite unpredictable, except in very small problems. A simple combinatorial scheme to provide upper bounds on the optimal value of the objective function was tested. Reinschmidt[3], on the same path as Toakley, but with a more sophisticated approach, compared the implicit enumeration algorithm with Balintfy's approximate method[4] in the design of a simple plastic frame. He also optimized elastic trusses under stress, deflection and material-selection constraints. The problem was linearized by a first-order Taylor-series expansion of the constraints, and then made discrete by implicit enumeration. The convergence of the procedure and the strategy to reach the global optimum are discussed extensively.

Marcal and Gellatly[5] have ingeniously applied the penalty-function method to compute discrete optima from a continuous local optimum. Substantial reductions in computations were obtained by embedding the algorithm into the structural problem. The pilot program was confined to small trusses.

These discrete-variable algorithms could be called 'pseudodiscrete', because they ultimately 'round' a discrete optimum from a previously computed continuous optimal solution. An all-integer algorithm designed for structural optimization problems has been developed by Cella[6,7] and is described in this chapter. Both material distribution[3] and geometric configuration have been included into discrete-variable models; the latter are described by Professor Maier in the next chapter. Because of the flexibility of the discrete-variable approach, extensions of the present models into more complete ones seem promising, particularly from the point of view of the numerical solution.

This chapter is organized as follows. First, in Section 11.2, a typical structural optimization problem in discrete variables is formulated and discussed. The all-integer algorithm already referred to is described in Section 11.3. Section 11.4 is dedicated to the analysis of different design procedures associated with discrete algorithms. A number of design applications are described in Section 11.4. Appendix 11A gives a general branch-and-bound procedure.

## 11.2   A discrete optimal-design problem

A typical, fairly general, structural optimization problem in discrete variables can be formulated as follows: compute the least-cost configuration for a structure of a general kind, idealized as an assembly of bars and finite elements, and subjected to static loads. The geometry of the structure and the partition into finite elements are given; the variable is the stiffness distribution across the structure. The properties of each member are selected from a table containing the properties of available profiles, and the thickness of each element from a table of available sizes. The length and the content of the tables may vary from problem to problem.

The variables are either the areas of the members (that are entries to the other properties in the table) or the thickness of the elements, both indicated as $x_j$. In the remainder of this chapter, the term 'elements' will refer to both members and finite elements. The objective function $W$ is the total cost of the structure, computed as the sum of the element costs, which are each, in turn, proportional to the weight of the element. Using the nomenclature of prior chapters, the minimization problem is therefore

$$\min \mathbf{c}^T \mathbf{x} \tag{11.1}$$

subject to constraints as discussed in the following.

A fundamental constraint stems from the discreteness of the problem. Thus

$$\mathbf{x} \in T \tag{11.2}$$

where $T$ is the table of available member areas or finite-element thicknesses. Three types of constraints on the elastic behaviour of the structure, in the form of linear inequalities, are taken into account:

(a) Stress constraints: the maximum absolute value of the element principal stresses, calculated at the centre of each element, must be equal to or lower than the allowable stress. In a member, the maximum stresses computed at a number of critical cross-sections are compared and the largest are selected; the allowable stresses are computed from existing codes (in the present case the AISC code[8]). Thus

$$\bar{\boldsymbol{\sigma}} - \boldsymbol{\sigma} \geqslant 0 \tag{11.3}$$

where $\bar{\boldsymbol{\sigma}}$ and $\boldsymbol{\sigma}$ are the vectors of allowable and calculated stresses, respectively.

(b) Displacement constraints:

$$\bar{\boldsymbol{\Delta}} - \boldsymbol{\Delta} \geqslant 0 \tag{11.4}$$

where $\bar{\boldsymbol{\Delta}}$ and $\boldsymbol{\Delta}$ are the vectors of allowable and calculated node point displacements, respectively.

(c) Stiffness constraints: for given nodes, specified terms $(k_j)$ of the global stiffness matrix must be equal to or lower than the allowable values $(\bar{k}_j)$. Thus

$$\bar{\mathbf{k}} - \mathbf{k} \geqslant 0 \tag{11.5}$$

where $\bar{\mathbf{k}}$ and $\mathbf{k}$ represent the appropriate vector listings of the terms in question.

An additional type of constraint, in the form of a linear equality, allows the designer to collect elements into 'equivalence classes'. In an equivalence class the constraints are checked for each element, but the variations in the element sizes are the same for all the elements of the class. Such constraint is appropriate for continua, and for structures that show symmetries in the configuration but not in the loading conditions. For this type of constraint, we write

$$x_j^k - x_l^k = 0, \qquad k = 1, \dots, n_e \tag{11.6}$$

*Optimum Structural Design*

for all $j$, $l \in E^k$, where $n_e$ is the total number of equivalence classes and $E^k$ is a set containing the indices $j$ of the elements belonging to an equivalence class.

In a discrete-variable approach the behaviour constraints are in implicit form, i.e. the dependence of stresses or deflections on the variables $x_i$ is not expressed in closed form. Each constraint is checked on a 'pass-or-fail' basis at every step of the algorithm. With the constraints in implicit form the mathematical model becomes quite simple and achieves the flexibility of handling fairly complex constraints.

## 11.3  A branch-and-bound algorithm

The foregoing optimization problem has been solved by a special branch-and-bound algorithm[7]. The term branch and bound is loosely associated with a family of combinatorial algorithms characterized by the existence of procedures for the systematic elimination of non-optimal solutions. While the number of combinations increases exponentially with the number of variables, the elimination scheme keeps within reasonable bounds the number of combinations to be explored.

For an overall picture of branch-and-bound methods one may consult Lawler and Wood[9] or Beckmann *et al.*[10], and, for more sophisticated algorithms, Balas[11], Geoffrion[12] and Abadie[13]. A general procedure for branch-and-bound methods, developed by Herve[14] and Roy[15], is briefly reviewed in Appendix 11A. The core of the algorithm that we will now consider is a procedure called diagonal enumeration that is described using the framework and notation of the general procedure in Appendix 11A.

### 11.3.1  Diagonal enumeration

In the general procedure the set $Y$ of feasible solutions is assumed to be known. When the constraints are in implicit form, however, the algorithm must simultaneously explore and define $Y$. Repetitive checks of the constraints are required, implying a number of complete analyses (or reanalyses) of the structure. The efficiency of the algorithm thus depends on minimizing the number of checks.

The following assumptions underlie the algorithm:

(a) The structure has linearly elastic behaviour.
(b) The partition into finite elements is adequate in defining the behaviour constraints with the required accuracy.
(c) The element properties in the tables are ordered with increasing stiffness.
(d) The objective function is linear, or, at most, convex in the design variables $x_i$.

Diagonal enumeration begins by setting $Y$ equal to $I^P$, the set of all possible combinations of element properties. The optimal solution, under the previous assumptions, is known to be on the 'boundary' of the feasible region. Diagonal enumeration explores piecewise a limited band around the boundary, and converges to a 'local optimum', starting from a feasible solution, in a finite number of steps. The quotes emphasize that, in discrete variables, the distance between solutions is not defined. The use of the Cartesian metric of the space $R^n$ would be misleading, as in the case of uneven scattering of the discrete solutions in $R^n$. Boundary points and local optima can be introduced in discrete variables, either on the basis of the topological definition of neighbourhood, or with an appropriate non-Cartesian metric (see Reference 16).

Diagonal enumeration cycles on the following operations:

(1) Given a feasible solution $x^f$, the subset $Y^f$ of discrete solutions with element properties smaller than in $x^f$ is defined as:

$$Y^f = \{x^p\}$$

$$x_i^p \leqslant x_i^f, \qquad i = 1, \dots, n.$$

In Figure 11.1(a), $x^f$ is the point A, $Y^f = Y^o$ is the set of points included between A, M, N and P and indicated as [AMNP].

(2) The subset $Y^f$ is searched, by successive partitions, for the suboptimal solution $(x^f)^o$. The search must be exhaustive; an efficient strategy is one that eliminates a large number of solutions in $Y^f$. When the objective function is linear in $x$ and has a positive coefficient, and when the constraints have the convex shape as in Figure 11.1, it seems most convenient to search by decreasing the element properties one at a time, in order to maximize the penetration into $Y^f$. The geometric analogue is a movement from $x^f$ along the diagonal of $Y^f$; and thus the procedure is termed diagonal enumeration.

(3) The suboptimal solution $(x^f)^o$ is perturbed by increasing the element properties; $x^p$ is obtained. From $x^p$, a subset $Y^p$ is defined as in step (1) and searched as in step (2). If $(x^p)^o = (x^f)^o$, $(x^p)^o$ is called a local optimum; the radius of $(x^p)^o$ (see Reference 16) is the radius of the perturbation.

In the example of Figure 11.1(a), the search procedure of step (2) is performed in the following stages:

$$\text{separation of } Y^o = [AMNP] \text{ from } Y$$

$$p = 0; Z^o = \{Y^o, Y^1\} \qquad \text{where } Y^1 = Y - Y^o.$$

$Y^o$ is partitioned by decreasing the element property of $x_2$:

$$p = 1; Y^o = \{Y^2, Y^3\} \qquad \text{where } Y^2 = [BQNP], Y^3 = [AMQB]$$

$$Z^1 = \{Y^2, Y^3, Y^1\}.$$

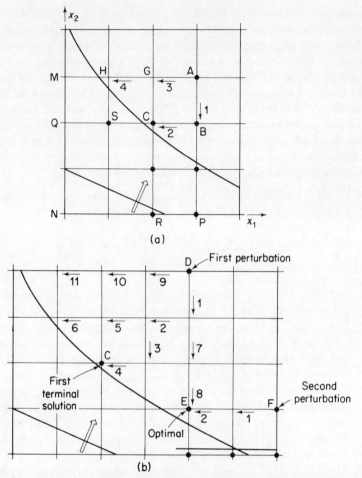

(a)

(b)

**Figure 11.1** Method of diagonal enumeration: (a) search of a terminal solution and (b) perturbation and local optimum

Partition of $Y^2$ by decreasing the element properties of $x_1$:

$$p = 2; \quad Y^2 = \{Y^4, Y^5\} \quad \text{where } Y^4 = [\text{CQNR}], \; Y^5 = [\text{BCRP}]$$

$$Z^2 = \{Y^4, Y^5, Y^3, Y^1\}.$$

The subset $Y^4$ cannot be further partitioned. Thus it is a terminal. Point C represents a suboptimal solution $(x^4)^\circ$ and $Y^4$ is eliminated. Also the next subset $Y^5$ is a terminal, with $f(x^5)^\circ \geqslant (x^4)^\circ$; $Y^5$ is eliminated. The subset $Y^3$ is partitioned:

$$p = 3, \, Y^3 = \{Y^6, Y^7\} \text{ where } Y^6 = [\text{GMQC}], \; Y^7 = [\text{AGCB}].$$

The subset $Y^7$ is eliminated because all of its solutions have higher cost than $(x^4)^o$. In diagonal enumeration, such bounding is embedded in the scheme of enumeration of the solutions.

$$Z^3 = \{Y^6, Y^1\}.$$

$$p = 4, Y^6 = \{Y^8, Y^9\} \text{ where } Y^8 = [HMQS], Y^9 = [GHSC].$$

$Y^8$ and $Y^9$ are terminal and do not improve on the suboptimal solution $(x^4)^o$. Finally $Z^4 = Y^1$; the search of $Y^o$ is completed. In Figure 11.1(a) the dotted points show where the constraints were checked, and thus structural analysis (or reanalysis) performed.

In Figure 11.1(b) for step 3, starting from point C with a perturbation of two steps, a local optimum of radius two is reached in point F.

### 11.3.2 The path of least cost increase

Often a feasible solution in the neighbourhood of a local optimum is not known. A heuristic procedure, called a least-cost-increase procedure, is then used. Starting from a random infeasible solution, the procedure generates a feasible solution near the boundary. One additional assumption, that the structure has normal behaviour with respect to stresses, is made, i.e. the increase in the stiffness of an element produces a decrease in the stress level throughout the structure, and vice versa for a decrease in stiffness.

The infeasible solution sets a lower bound on the minimum value of the objective function; the procedure consists in satisfying one constraint at a time, selecting at every step the constraint that requires the least increase in the cost function. The corresponding path of movement 'jumps' from one constraint to another, as against the fine-grain search of diagonal enumeration.

The direction of motion at each step varies with the type of constraint. In the case of stress constraints, postulating normal behaviour, the least cost increase consists in stiffening the overstressed element, because of the predominance of the local effects. Here an element whose properties lower the stress level below the allowable values is picked from the table. The effect of the redistribution of stresses on the choice of the element is neglected. Notice that discrete distributions allow for some fluctuation.

A discrete-solution space is presented in Figure 11.2 for the classical three-bar truss problem shown in Figure 2.1. Starting from A', the path lies in A'B'C'. First $x_2$ is increased to satisfy the stress constraint on member 2 due to load 2. Then $x_1$ is increased to satisfy the constraint on member 1 due to load 2. The constraint near points B' and B corresponds to hybrid behaviour; even in that case the procedure is not 'trapped'.

For deflection constraints, say on node $j$, the influence of element-property variation on the deflection $u_j$ on node $j$ is computed and assembled into a

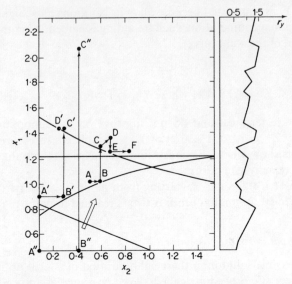

Reproduced by permission from A. Cella and R. D. Logcher, 'Automated optimum design from discrete components', *J. Struct. Div.*, *ASCE*, **97**, No. ST1, 175–190 (January 1971)

**Figure 11.2**   Discrete-solution space for three-bar truss (Figure 2.1)

vector

$$g(u_j) = \frac{\Delta u_j}{\Delta x_i}, \qquad i = 1, \ldots, n.$$

g is the discrete analogue of the gradient.

The path for the simple frame of Figure 11.3 is shown in Figure 11.4. The starting point 1 violates both constraints. First the constraint on the *x* component of the deflection of node 5 is satisfied; because of the non-

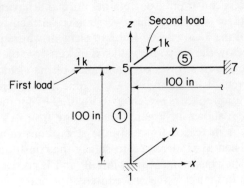

**Figure 11.3**   Plane frame

linearity of the constraints more than one step may be required. Then the constraint on the $y$ component is satisfied. Point 4 is a feasible solution.

The procedure works for stiffness constraints as for deflection constraints. The difference is that only the elements surrounding node $j$ contribute to the vector **g** of sensitivity coefficients. The path corresponding to the problem in Figure 11.3 is shown in Figure 11.4.

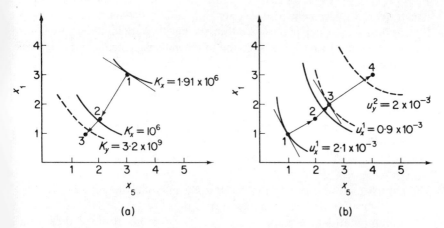

**Figure 11.4** Constraint surfaces for plane frame: (a) stiffness and (b) deflection

### 11.3.3 Reanalysis and sensitivity coefficients

It is a well established fact that a perturbation in the stiffness of an element produces perturbations in the displacements equal to the displacements induced into the structure by an appropriate system of self-equilibrated loads. The perturbed stiffness equations have, in standard notation[17], the form:

$$(\mathbf{K} + \delta\mathbf{K})(\mathbf{\Delta} + \delta\mathbf{\Delta}) = \mathbf{P} + \delta\mathbf{P} \qquad (11.7)$$

where $\delta\mathbf{K}$ and $\delta\mathbf{P}$ are the known perturbations in the global stiffness matrix and $\delta\mathbf{P}$ is the node-equivalent vector of the dead weight. $\delta\mathbf{\Delta}$ is the unknown variation in the displacements owing to the perturbation in the stiffness. Eliminating the equation $\mathbf{K}\mathbf{\Delta} = \mathbf{P}$ from equation (11.7), one obtains

$$(\mathbf{K} + \delta\mathbf{K})\,\delta\mathbf{\Delta} \approx \mathbf{K}\,\delta\mathbf{\Delta} = \delta\mathbf{P} - \delta\mathbf{K}\mathbf{\Delta} = \mathbf{P}_1. \qquad (11.8)$$

Equation (11.8) provides a simple reanalysis procedure that has been adopted in the implementation of the algorithm. The choice of equation (11.8) as against more sophisticated procedures[18] lay in the particular handling of data performed by STRUDL II, on which the algorithm is implemented. The running time for reanalysis is about 30 per cent of the time for complete analysis.

When in the presence of deflection constraints, the sensitivity coefficients for deflections, i.e. the variations in a particular node deflection owing to perturbations in the element stiffness, need to be computed. Using Betti's law the sensitivity coefficients can be computed without performing a cycle of reanalyses using equation (11.8).

At the beginning of the process, the original load vector $P$ is expanded into $P_q = [P, Q]$ where $Q$ is a matrix of unit load vectors corresponding to the $q$ constrained displacement components. Solving the system of equations for $P_q$, one has, at each step of the algorithm, the matrix $\Delta_q$ of joint displacements owing to the unit loads in the direction of the $q$ displacement components of the $j$ constrained nodes.

By Betti's law, $\Delta_q^T$ is the matrix of the displacements in the direction of the $q$ displacement components owing to unit loads applied to the nodes of the structures. From equation (11.8), one obtains the variation in the constrained displacement components owing to a perturbation $\Delta K$ in the stiffness of an element as

$$\delta \Delta_q \, \Delta_q^T P_1. \tag{11.9}$$

The approximation is due to the fact that $\Delta_q^T$ refers to the unmodified stiffness configuration (see also Chapter 7, p. 114).

For computational efficiency, the number $q$ of constrained deflection components should not be too large; that is often the case in actual design problems. The implementation of the algorithm, however, does not limit the number of deflection constraints.

## 11.4 Suboptimization

Discrete-variable algorithms of combinatorial type are fairly efficient in the exhaustive (and partly implicit) search of a limited subset. They lack, however, the fast convergence from a random start down to the neighbourhood of a local optimum, a feature that is typical of gradient or penalty-function methods. Hence there is a need to integrate discrete algorithms with procedures that provide a 'good' starting solution. References 1 and 3 have suggested, as a starting point, the continuous optimal solution computed via linear programming. Here, the least cost increase was adopted. The efficiency of the entire optimization process depends substantially on the 'quality' of the starting solution. The continuous local optimum, being unique (at least in the neighbourhood of a discrete optimum), is not always good, depending on the problem in hand. The output of the least cost increase, on the other hand, can be controlled by replacing a random start by an educated guess. In the three-bar-truss problem in Figure 11.2 one may see the improvement of the first feasible solutions $C''$, $C'$, and $C$, starting from $A''$, $A'$ and $A$, respectively.

A more significant example is the optimization for total cost, subject to stress constraints, of the tapered plane frame in Figure 11.5. Starting the least-cost-increase procedure with a poor guess, a poor feasible solution that requires a substantial search before reaching a discrete optimum [see

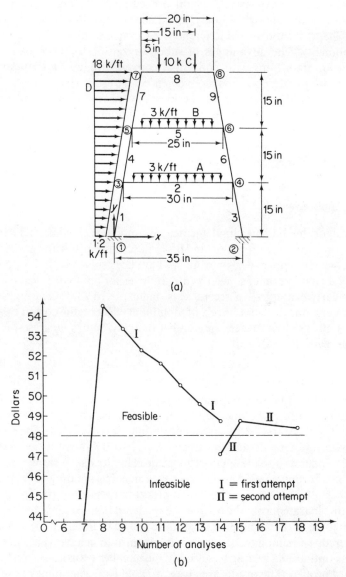

(a)

(b)

Reproduced by permission from A. Cella and R. D. Logcher, 'Automated optimum design from discrete components', *J. Struct. Div.*, *ASCE*, **97**, No. ST1, 175–190 (January 1971)

**Figure 11.5**   Tapered frame: (a) geometry and loads and (b) solution history

Figure 11.5(b)] is obtained. Restarting with a good solution (easy to derive by inspection of the member design output) the convergence is immediate.

In complex structures, or in continua where a fine mesh partition into elements is required, the use of fully automated optimization procedures proves to be too expensive. Several procedures have been developed for piecewise optimal design, loosely called suboptimization. Such procedures are in Chapter 9, and also in Reference 19. Complete automation is sacrificed for optimality. The advantages of suboptimization, however, can be substantial for algorithms with good final convergence, and are questionable for algorithms with poor final convergence.

Discrete-variable algorithms, besides having good final convergence, provide the flexibility of performing several suboptimizations with different meshes of discrete solutions. At each step the computed suboptimal solution becomes the initial solution for the following step, while the new mesh is refined around the previous suboptimal solution.

## 11.5 Applications

The algorithm has been implemented on a modified version of STRUDL II and is presently operational on an IBM 360/65 equipped with two 2314 disk drives. Several types of design problems have been solved with the algorithm. The total running time is given as a parameter for computational efficiency for the early examples. In recent tests under the MVT Operating System, the running time does not have a predominant effect on the cost of computation; as intrinsic parameter, the number of analyses (or reanalyses) is given, together with the total cost.

### 11.5.1  The hammerhead of a radiotelescope

A large radar and radiotelescope is supported on a cantilevered hammerhead subjected to stress and deflection constraints. The end portion of the hammerhead, controlled by stress constraints, was analysed. The structure is a space frame of 72 steel members ($E = 29,000 \text{ kip/in}^2$, $\sigma_y = 36 \text{ kip/in}^2$, $v = 0.3$) under two loading conditions: in-plane loads, as shown in Figure 11.6, and a torque applied at the cross-section corresponding to joints 5 and 6 in Figure 11.7(a). A piecewise design procedure was adopted. The in-plane loads are separated from the torque and analysed first; because of symmetry, only one side of the hammerhead (the one shown in Figure 11.6) needs to be considered. The simplified structure was assumed to behave as a plane truss, and was optimized for weight, selecting the member properties from a set of standard profiles. The results are given in Table 11.1. The shape of the structure suggests that the torque is relevant to the statically indeterminate portion of the hammerhead shown in Figure 11.6. Such a 40-member space frame

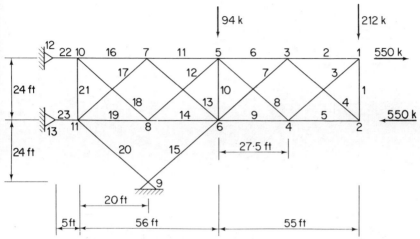

Reproduced by permission from A. Cella and R. D. Logcher, 'Automated optimum design from discrete components', *J. Struct. Div.*, *ASCE*, **97**, No. ST1, 175–190 (January 1971)

**Figure 11.6** Hammerhead, plane truss

Table 11.1   Solution of the hammerhead plane truss
(see Figure 11.6)

| Members | Lower bound | Optimal |
|---|---|---|
| 1 | 8 WF 24 | 8 WF 24 |
| 2 | 36 WF 150 | 36 WF 160 |
| 3 | 10 WF 45 | 12 WF 53 |
| 4 | 14 D 17·2 | 16 WF 40 |
| 5 | 14 WF 119 | 14 WF 150 |
| 6 | 36 WF 182 | 33 WF 200 |
| 7 | 8 WF 24 | 10 WF 45 |
| 8 | 16 WF 40 | 18 WF 45 |
| 9 | 14 WF 193 | 14 WF 211 |
| 10 | 8 WF 35 | 14 WF 61 |
| 11 | 33 WF 220 | 36 WF 230 |
| 12 | 12 JR 11·8 | 12 JR 11·8 |
| 13 | 10 WF 45 | 12 WF 53 |
| 14 | 14 WF 74 | 14 WF 87 |
| 15 | 14 WF 150 | 14 WF 158 |
| 16 | 14 WF 150 | 14 WF 176 |
| 17 | 16 B 26 | 18 WF 45 |
| 18 | 8 WF 24 | 8 WF 24 |
| 19 | 14 WF 61 | 14 WF 87 |
| 20 | 18 WF 45 | 18 WF 50 |
| 21 | 8 JR 6·5 | 8 JR 6·5 |
| Volume (in³) | 181,340 | 213,720 |
| Time (min) | | 3·4 |

Reproduced by permission from A. Cella and R. D. Logcher, 'Automated optimum design from discrete components', *J. Struct. Div.*, *ASCE*, **97**, No. ST1, 175–190 (January 1971)

214      *Optimum Structural Design*

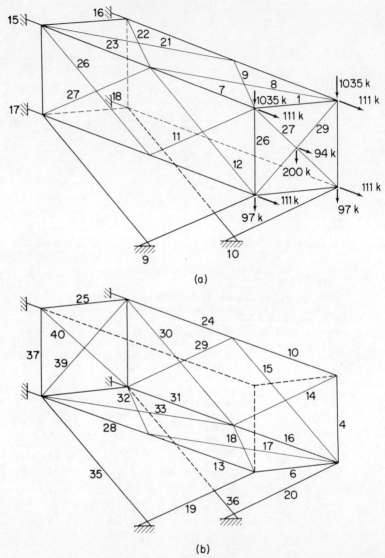

(a)

(b)

Reproduced by permission from A. Cella and R. D. Logcher, 'Automated optimum design from discrete components', *J. Struct. Div., ASCE*, **97**, No. ST1, 175–190 (January 1971)

**Figure 11.7**   Hammerhead, space truss: (a) exterior members and (b) interior and additional exterior members

was optimized with respect to weight, using the optimal truss configuration as a starting point. The results are shown in Table 11.2.

A near-optimal configuration for the hammerhead can be assembled from the two previous suboptimizations, and could be reused as a starting point for a further step of optimization.

Table 11.2   Solution of the hammerhead space frame
(see Figure 11.7)

| Members | Lower bound | Optimal |
|---|---|---|
| 1, 6 | 8 WF 35 | 10 WF 45 |
| 2, 3 | 14 WF 61 | 14 WF 74 |
| 4, 5 | 8 WF 24 | 12 WF 53 |
| 7, 10 | 33 WF 220 | 36 WF 230 |
| 8, 9 | 12 JR 11·8 | 12 JR 11·8 |
| 11, 14 | 14 B 17·2 | 16 B 26 |
| 12, 15 | 21 WF 55 | 26 WF 68 |
| 13, 16 | 14 WF 119 | 14 WF 136 |
| 17, 18 | 10 JR 9 | 10 JR 9 |
| 19, 20 | 14 WF 237 | 14 WF 264 |
| 21, 22 | 10 JR 9 | 10 JR 9 |
| 23, 24 | 36 WF 182 | 36 WF 182 |
| 25, 34 | 10 JR 9 | 10 JR 9 |
| 26, 30 | 8 WF 35 | 8 WF 35 |
| 27, 29 | 16 WF 40 | 18 WF 45 |
| 28, 31 | 14 WF 74 | 14 WF 74 |
| 32, 33 | 10 JR 9 | 10 JR 9 |
| 35, 36 | 16 WF 40 | 18 WF 45 |
| 37, 38 | 10 JR 9 | 10 WF 9 |
| 39, 40 | 10 JR 9 | 10 WF 9 |
| Volume (in³) | 224,980 | 320,620 |
| Time (min) | | 10·9 |

Reproduced by permission from A. Cella and R. D. Logcher, 'Automated optimum design from discrete components', *J. Struct. Div. ASCE*, **97**, No. ST1, 175–190 (January 1971)

## 11.5.2   Supporting structure for an optical mirror

In the design of the 120 in optical mirror for the NASA Large Space Telescope[20], one of the aims was to define the optimal structural substrate for the reflecting surface of a diffraction-limited telescope subjected to both gravity-field test procedures and in-orbit thermal perturbations.

Given a finite-element partition of the mirror, as in Figure 11.8(a), the optimal displacements of the support nodes C and D are computed with a separate procedure. The optimality criterion is the minimization of the root-mean-square error, that is, the least-square error of the surface distortion. The optimal $z$ displacements are $u_z^C = 0·188 \times 10^{-3}$ in and $u_z^D = 0·198 \times 10^{-3}$ in.

The supporting structure is designed for minimum weight under the previous deflection constraints. The members are 'I' beams, as in Figure 11.8(b), with their overall dimensions fixed; the thickness of the plates is variable. Because the constraints are in the form of equalities as opposed to inequalities, a very fine mesh of discrete solutions would be necessary to obtain the displacement within the required accuracy. Instead, the design

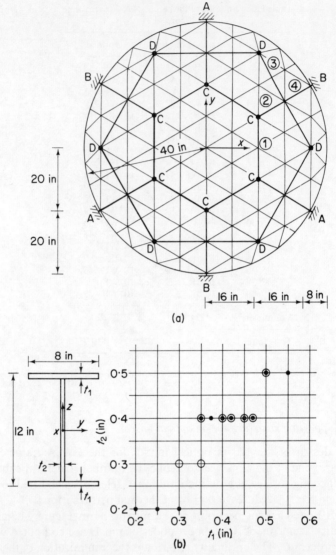

**Figure 11.8**  Mirror support structure: (a) layout and (b) member properties

procedure adopted is to perform several runs with different mesh sizes; from the suboptimal solution obtained in a run, the mesh is refined for the next run in the area where a local optimum is expected to fall. In Figure 11.8(b) the large circles indicate the discrete solutions available at the first run; the dots show the discrete solutions for the final run, which were $t_1 = 0.375$ in, $t_2 = 0.4$ in for member 1, $t_1 = 0.3$ in, $t_2 = 0.2$ in for member 2,

$t_1 = 0.6$ in, $t_2 = 0.5$ in for member 3, and $t_1 = t_2 = 0.2$ in for member 4. The corresponding support displacements are $u_z^C = 0.185 \times 10^{-3}$ in, $u_z^D = 0.203 \times 10^{-3}$ in, with a few per cent error from the required values.

### 11.5.3 Box-shaped beam

The structure shown in Figure 11.9 was optimized for weight in two steps, using two different finite-element partitions. The procedure is the same as in the previous example. The difference, however, is in the fact that the principal stresses computed in the slab of the building module vary smoothly with the refinement of the partition, while on the loaded sides of the box the variation

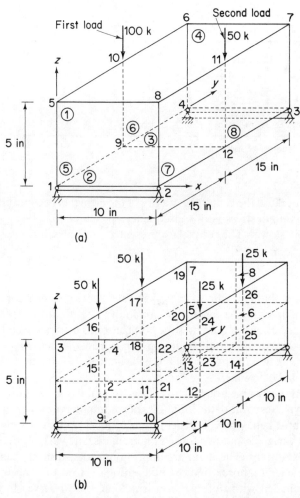

**Figure 11.9** Box-shaped beam: (a) elementary partition and (b) final partition

*Optimum Structural Design*

Table 11.3   Solution of box-shaped beam—elementary mesh

| Elements | Initial thickness (in) | Optimal thickness (in) |
|----------|-----------------------|------------------------|
| 1, 4 | 0·5 | 0·1 |
| 2, 3 | 0·5 | 0·5 |
| 5, 6 | 1·0 | 1·0 |
| 7, 8 | 1·0 | 0·5 |

Number of analyses = 5
Cost of computation = $25·14

Table 11.4   Solution of box-shaped beam—final mesh

| Elements | Initial thickness (in) | Optimal thickness (in) |
|----------|-----------------------|------------------------|
| 1, 2, 3, 4, 5, 6, 7, 8 | 0·1 | 0·1 |
| 9, 10, 11, 12, 13, 14 | 0·5 | 0·4 |
| 15, 16, 17, 18, 19, 20 | 1·0 | 2·0 |
| 21, 22, 23, 24, 25, 26 | 0·5 | 0·9 |

Number of analyses = 7

is sharper, because the centres of gravity of the elements become closer to the areas of maximum stress. As a consequence, the search for a local optimum is considerably slower. The results are presented in Tables 11.3 and 11.4.

## 11.6   Closure

The results obtained with discrete-variable algorithms of a combinatorial type are encouraging as far as the possibility of optimizing medium-scale structures, either directly, or by suboptimization. The most useful feature of such algorithms is the good final convergence in the neighbourhood of a local optimum, making possible a piecewise-optimal design process with fair computational efficiency.

The discrete-variable approach is closer to real design conditions and, in addition, bypasses some of the loopholes of continuous variable optimization, such as hybrid behaviour. The definition of local optima is looser than in continuous variables, but this feature offers, in discrete variables, a number of equivalent solutions. The main assumption introduced in problem (11.1) is the finite-element approximation of continua. How that approximation can be reconciled with optimization is a topic that deserves further basic research and extensive numerical tests.

As for the general outlook, one may conclude that optimal design can be achieved in several non-trivial design problems; its price might be the replacement of full automation with a piecewise procedure. On the other hand, the third computer revolution prophesied in the early sixties has not arrived yet. While waiting, we can still make good use of the skill and of the ingenuity of the designer in an integrated, partly automated approach.

*Acknowledgment* The authors are very grateful to A. Cornell, R. D. Logcher and K. Reinschmidt of the MIT Civil Engineering Department for the important contributions they made to several stages of the present work.

The research was partly funded by DSR Project 55-34900 sponsored by Goddard Space Flight Center of NASA through contract NAS 5-21542.

APPENDIX 11A:
A GENERAL PROCEDURE FOR
BRANCH-AND-BOUND ALGORITHMS

Herve[14] and, more recently, Roy[15] proposed a general procedure that encompasses several types of branch-and-bound algorithms. The procedure is broadly applicable to decision-making problems, where the solutions are related only by ordering, or subjected to multiple objective functions. A brief review of the procedure is given here, restricted, for simplicity, to the treatment of integer programming problems.

**11A.1 Definitions**

The following are preliminary definitions:

$U$ = the set of solutions; in all-integer problems $U = I^m$; in mixed-integer problems $U = R^n \times I^P$

$Y \in U$ = the subset of admissible solutions

$W$ = the objective function; $Y \to R$

$\delta$ = a 'threshold' function, $R \times R \to [0, 1]$

$$\begin{aligned} \delta(r, r') = 0 \quad &\text{if } r \leqslant r' \\ \delta(r, r') = 1 \quad &\text{if } r > r' \end{aligned} \qquad \text{for } r, r' \in R$$

$Y^n$ = a sequence of subsets of $Y$ obtained by successive partitions; $n$ is the sequential number of the partition

$\phi(n)$ = a function that estimates the minimum of $W$ on $Y^n$; $Y^n \to R$

$S(n)$ = the set of all subsets obtained partitioning $Y^n$

$Z^P$ = the set of all subsets $Y^n$ available for partition at the $p$th step of the procedure; in $Z^P$ the subsets $Y^n$ are ordered with respect to the values of $(n)$

$Z_0^P$ = the first subset in $Z^P$

The aim is to isolate, by successive partitions of the subsets in $Z^P$, a solution $\mathbf{x}^o$ such that either

$$W^o = W(\mathbf{x}^o) < W(\mathbf{x}) \qquad \text{for all } \mathbf{x} \text{ in } Y$$

or

$$\delta[W^o, W(\mathbf{x}^o)] = 0.$$

In the first case, $\mathbf{x}^o$ is an optimal solution for the objective function $W$; in the second case, it is a 'satisfactory' solution with respect to the threshold function $\delta$. The procedure is divided into the following operations: partition of $Y^n$ into subsets (branch); evaluation through $\phi$ of the minimum value of $W$ on the subsets (bound); selection of the subset to be partitioned next; elimination of the subsets that cannot contain the optimal solution $\mathbf{x}^o$. The procedure is often represented in the form of an arborescence (see Figure 11A.1) where a node represents a subset $Y^n$ and an arc connects $Y^n$ with one of the 'descendent' subsets of $S(n)$. At the side of each node is written the estimate value $\phi(n)$ of the minimum.

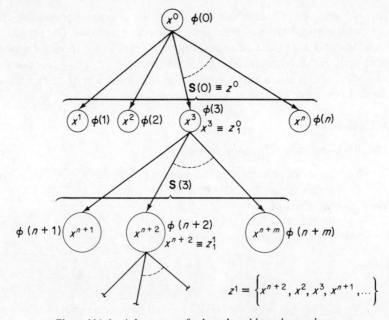

**Figure 11A.1**   Arborescence for branch-and-bound procedure

## 11A.2   The procedure

The solution of an integer programming problem is obtained in $p$ steps, preceded by an *initialization*:

$$p = 0, \qquad z^o = \{Y\}, \qquad z^o_0 = Y.$$

At each step, the following operations are performed:

*Partition and evaluation* The first subset $z_0^p$ is partitioned according to a given rule that is specific to the algorithm. In the subsets 'descending' from $z_0^p$, the minimum is evaluated by the function $\phi$, which is also specific to the algorithm. Two situations can occur:

(a) $z_0^p$ is a 'terminal', i.e. a subset in which either the optimal solution $(x^p)^\circ$ can be computed exactly, or there are no solutions. In the first case the conditions for completion are checked; in the second case one more step is performed with $Z^{p+1} = Z^p - z_0^p$.

(b) $z_0^p$ is not a terminal; $z_0^p$ is partitioned and an estimate of the minimum is computed by the function $\phi$, for each of the descendants of $z_0^p$ in $S(p)$.

*Elimination* From $S(p)$, the subsets $Y^n$ that are either empty terminals or are such that $\phi(n) \geqslant W(x^\circ)$ are eliminated. At every step of the algorithm, $x^\circ$ is a known feasible solution of the problem. Calling $\hat{S}(p)$ the set $S(p)$ after the eliminations, the available subsets for the next step are given by

$$Z^{p+1} = (Z^p - \{z_0^p\}) \cup \hat{S}(p).$$

*Completion* When $W(x^\circ)^p \leqslant \phi(\bar{z}^1)$, where $\bar{z}^1$ is the second subset in $\bar{Z}^n$, $\bar{x}^\circ$ is the optimal solution $x^\circ$. Otherwise, the next step begins by partitioning $\bar{z}^1$.

Notice that, substituting the threshold function $\delta$ for the objective function $W$, the procedure terminates with a 'satisfactory' solution $\bar{x}^\circ$.


## 11A.3 Convergence

The procedure converges to the optimal solution in a finite number of steps if the rules for partition and for evaluation that are specific to the algorithm satisfy the following axioms:

For partition:

*Conservation*

$$Y^n = \bigcup_{r \in S(n)} Y^r, \quad n = 1, \ldots, p.$$

In other words, no subset of $Y$ can be lost.

*Termination* After a finite number of partitions, a terminal is reached.

For evaluation:

*Lower bound* The function $\phi$ is such that $\phi(n) \leqslant \min_{x \in Y} W(x)$

*Coincidence* For non-empty terminals $\hat{Y}_n$

$$\hat{\phi} = W(\hat{x}^\circ)$$

where $\hat{x}^\circ$ is the optimal solution in $\hat{Y}$. Using the threshold function, the condition of coincidence becomes

$$\delta[\hat{\phi}, W(\hat{x}^\circ)] = 0.$$

222　　　　　　　　*Optimum Structural Design*

## References

1. A. R. Toakley, 'Optimum design using available sections', *J. Struct. Div. ASCE,* **94**, No. ST5, 1219–1241 (1968).
2. G. Greenberg, *Integer Programming*, Academic Press, 1970.
3. K. F. Reinschmidt, 'Discrete structural optimization', *J. Struct. Div. ASCE,* **97**, No. ST1, 133–156 (1971).
4. J. L. Balintfy, 'Menu planning by computer', *Comm. ACM,* **7**, No. 4, 255–259 (1964).
5. P. V. Marcal and R. A. Gellatly, 'Application of the created response surface technique to structural optimization', *Proc. 2nd Conf. on Matrix Meth. in Struct. Mech., Wright Patterson AFB, 1968.*
6. A. Cella and R. D. Logcher, 'Automated optimum design from discrete components', *J. Struct. Div. ASCE,* **97**, No. ST1, 175–190 (1971).
7. A. Cella and K. Soosaar, 'Optimization of structures by finite element methods', MIT C. S. Draper Laboratory Report E 2570, Cambridge, Mass., March 1971.
8. *Manual of Steel Construction*, 7th ed., AISC, 1969.
9. E. L. Lawler and D. F. Wood, 'Branch and bound methods: a survey', *Operations Research,* **14**, 699–719 (1968).
10. M. Beckmann, *Einfuhring in die Methode Branche and Bound*, Springer, Berlin, 1968.
11. E. Balas, 'Discrete programming by the filter method', *Operations Research,* **13**, 517–546 (1965).
12. A. M. Geoffrion, 'Integer programming by implicit enumeration and Balas' method', *SIAM Review,* **9**, 178–190 (1967).
13. J. Abadie, 'Une methode arborescente pour les programmes nonlineaires partiellment discrets', *Revue d'Inf. et de Rech. Oper.,* **3**, V-3, 25–50 (1969).
14. P. Herve, 'Les procedures arborescentes d'optimisation', *Revue d'Ing. et de Rech. Oper.,* **14**, 69–80 (1968).
15. B. Roy, 'Procedure d'exploration par separation et evaluation', *Revue d'Ing. et de Rech. Oper.,* **3**, V-1, 61–90 (1966).
16. A. Cella, 'Properties of discrete optima in structural optimization', *J. Struct. Div. ASCE,* **98**, No. ST3, 787–792 (1972).
17. O. C. Zienkiewicz, *The Finite Element Method in Engineering Sciences*, McGraw-Hill, New York, 1971.
18. R. J. Melosh, 'Optimal analysis of structures', *Computers and Structures,* **I**, 181–190 (1971).
19. U. Kirsch, M. Reiss and U. Shamir, 'Optimum design by partitioning into substructures', *J. Struct. Div. ASCE,* **98**, No. ST1, 249–267 (1972).
20. K. Soosaar, 'Design of optical mirror structures', MIT C. S. Draper Laboratory Report R-673, 1971.

*Chapter 12*

# Limit Design in the Absence of a Given Layout: A Finite-element, Zero–One Programming Problem*

G. Maier

## 12.1 Introduction

The problem discussed in this chapter can be formulated as follows. A three-dimensional structure, composed of a perfectly plastic material obeying Drucker's postulate (i.e. 'normality' and 'convexity'[1]), must be designed so that:

(a) It is fixed to a given surface or set of surfaces S″ and contained in a prescribed 'available region' $R_a$ (geometric constraints).
(b) It can carry given live loads in the form of surface tractions on an assigned surface or surfaces S′, and its own weight, i.e. body forces $F_B$ per unit volume (behaviour constraints).
(c) The material consumption is the least possible (the merit function is weight or volume).

The peculiar feature of this limit-design problem is the absence of any layout of axes or middle surfaces to which cross-sectional properties or thicknesses are referenced. Shape optimization of structures in this sense was pioneered by Mitchell[2] and has been intensively studied recently (see, for example, Reference 3). However, the present investigation is particularly related to and motivated by References 4–7, in which optimality conditions, either sufficient or necessary, were established for the design problem stated above. These criteria, however interesting they may be from the theoretical standpoint, are, admittedly, incapable of leading to the solution except in special simple cases. Therefore further research, especially on numerical techniques, has been advocated by Prager[7]. This chapter is intended as a contribution in this direction.

Using the tensor-field description of continuum mechanics, a 'minimax' characterization of the optimal structure resting on the kinematic theorem of limit analysis is discussed first. The problem is subsequently tackled by an

---

* This chapter has appeared in the *Journal of Structural Mechanics*.

alternative, engineering-oriented approach, based on a finite-element discretization of the available space; in this context the static and kinematic formulations turn out to be dual structured problems in mixed zero–one programming. Balas's duality theory of integer programming is used in the subsequent discussion. Implicit enumeration algorithms are suggested as efficient tools for numerical solutions, and interpreted in mechanical terms. Various ways of reducing the size of the problem in practical engineering situations are pointed out.

## 12.2 The continuum approach

### *12.2.1 Formulation by the lower-bound theorem (static)*

Let $R$ be a symbol for the generic structure, i.e. for a space region filled with material [Figure 12.1(a)]. Its volume is measured by the scalar $V$, its boundary is formed by the given loaded surface $S'$, the given fixity surface $S''$ or part of it, and the free surface $S_f$. Each $S'$, $S''$, $S_f$ may represent a set of disjointed surfaces, rather than a single surface. Some volumes, like E in Figure 12.1(a), might have to be left free for surrounding structures or installations, thus reducing the available region $R_a$. The symbol R refers to all design variables, i.e. all the geometrical entities adopted in order to define a structure or the free part $S_f$ of its boundary. With this convention, the optimal (minimum-weight) structure $R_0$ can be formally characterized as follows:

$$\min_{R,\sigma_{hk}} V \equiv \int_R dV \qquad (12.1)$$

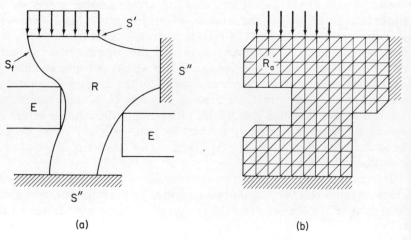

(a)                                              (b)

**Figure 12.1**

subject to

$$\frac{\partial \sigma_{hk}}{\partial x_k} + F_{B_h} = 0 \text{ in R} \tag{12.2}$$

$$\sigma_{hk} n_k = \begin{cases} F_{L_h} \text{ on S}' \\ 0 \text{ on S}_f \end{cases} \tag{12.3}$$

$$f(\sigma_{hk}) \leqslant 0 \text{ in R} \tag{12.4}$$

$$\text{R} \in \text{R}_a. \tag{12.5}$$

In the preceding formulae $\sigma_{hk}$ represents the stress tensor, whose components are assumed to be piecewise differentiable functions of the coordinates $x_1$, $x_2$, $x_3$, and whose possible discontinuities are assumed to obey equilibrium (tractions normal to discontinuity surfaces are continuous across them); $n_k$ are direction cosines of the outward normal to the boundary of R; $F_{B_h}$ and $F_{L_h}$ are body forces per unit volume and surface tractions, respectively; $f$ is the (convex) plastic potential or yield function of the material; the index summation convention holds. The behaviour constraint (capacity of carrying the external loads) is expressed, on the basis of the static (lower-bound) theorem of limit analysis[8], by the equilibrium equations (12.2) and (12.3) and by the plastic conformity requirement [expression (12.4)]; condition (12.5) reflects the geometrical constraint limiting the space that may be occupied by the structure.

### 12.2.2   Formulation by the upper-bound theorem (kinematic)

It is easy to show that the optimal structure $R_o$ can also be defined by the minimax of a functional, in which kinematic variables as well as the design variables intervene. To be precise:

$$\min_{R} \max_{\dot{u}_h} W \equiv \int_R \{1 + \alpha[F_{B_h}\dot{u}_h - D(\dot{\varepsilon}_{hk})]\} \, dV + \alpha \int_{S'} F_{L_h}\dot{u}_h \, dS \tag{12.6}$$

subject to:

$$\dot{u}_h = 0 \text{ on S}'' \tag{12.7}$$

$$\dot{\varepsilon}_{hk} = \frac{1}{2}\left(\frac{\partial \dot{u}_h}{\partial x_k} + \frac{\partial \dot{u}_k}{\partial x_h}\right) \text{ in R} \tag{12.8}$$

$$\text{R} \in \text{R}_a \tag{12.9}$$

where $\alpha$ is an arbitrary positive constant; $\dot{u}_h$ represents a continuous, piecewise-differentiable field of velocities; $\dot{\varepsilon}_{hk}$ represents the relevant plastic strain rates; and $D$ is the dissipation rate density.* The assumed convexity

---

* No explicit mention needs to be made of the plastic incompressibility $\dot{\varepsilon}_{hh} = 0$. Because of the normality property, this implied unboundedness of the yield locus and its violation leads, through equation (12.15), to $D \to \infty$, which is ruled out by the maximization in definition (12.6).

of the yield surface and normality to it of the strain rate vector (briefly, the admitted validity of Drucker's postulate[1] for the material behaviour), imply that $D$ is uniquely defined by $\dot\varepsilon_{hk}$[9]. Since the stressless state is an admissible stress state (i.e. the origin of the $\sigma_{hk}$ axes is not external to the elastic range), $D$ cannot be negative.

A proof of the above minimax characterization of the optimal design can be derived from the kinematic (upper-bound) theorem of limit analysis[1]. This can be expressed by the following statement, where $s$ is the safety factor against plastic collapse; for given external forces $F_{L_h}$, $F_{B_h}$, collapse impends ($s = 1$) or has taken place previously ($s < 1$) if, and only if, a compatible velocity field $\dot u_i$ can be found such that the rate at which work is done by the external forces equals or exceeds that at which energy is dissipated by plastic strains.

The difference between external-force power and dissipated power is:

$$\Pi = \int_{S'} F_{L_h}\dot u_h \, dS + \int_R F_{B_h}\dot u_h \, dV - \int_R D(\dot\varepsilon_{hk}) \, dV \qquad (12.10)$$

and intervenes in expression (12.6):

$$W(R, \dot u_h) = V(R) + \alpha\Pi(R, \dot u_h). \qquad (12.11)$$

For any given R, let $\Pi_m(R)$ be the maximum of $\Pi$ with respect to all possible $\dot u_h$ under the geometric compatibility constraint [which implies: equation (12.7); continuity and differentiability of $\dot u_i$ so that equation (12.8) may have a meaning; use of equation (12.8) in defining $D$]. By virtue of the above-stated limit theorem, one observes that:

(a) $\Pi_m(R) \to +\infty$ characterizes any structure R incapable of carrying the loads $F_{L_h}, F_{B_h}$, i.e. any R with a safety factor $s < 1$. Let $\Sigma_1$ be the class of all these 'unsafe' designs contained in $R_a$.

(b) $\Pi_m(R) = 0$ characterizes any R which satisfies the behaviour constraint to carry the loads, i.e. which exhibits $s \geqslant 1$. Let $\Sigma_2$ be the class of all these 'safe' designs contained in $R_a$. The class $\Sigma_2$ can be thought of as the union of the two following subclasses:

(b') $\Sigma_2'$ of all 'strictly safe' R, i.e. with $s > 1$. For these designs $\Pi_m(R) = 0$ is attained only for $\dot u_h \equiv 0$, since any velocity field implies an excess of the dissipation over the external power ($\Pi < 0$).

(b'') $\Sigma_2''$ of all R at collapse, i.e. with $s = 1$. For these there exists some field $\dot u_i \not\equiv 0$ corresponding to $\Pi_m(R) = 0$ and defined to within a positive multiplier. In fact any 'collapse mechanism' $\dot u_h^c$ fulfils both compatibility and the energy balance $\Pi = 0$ (i.e. it defines a 'kinematically admissible situation').

The values $\Pi_m(R)$ generated by the maximization of $W$ with respect to the kinematic variables $\dot u_h$ play the role of a 'penalty functional' in the subsequent minimization with respect to the design variables. It restricts this minimization to the class $\Sigma_2$ of the safe designs, since:

$$\max_{\dot{u}_h} W = V(\mathbf{R}) + \Pi_m(\mathbf{R}) \begin{cases} = V(\mathbf{R}) \text{ over } \Sigma_2 \\ \rightarrow +\infty \text{ over } \Sigma_1. \end{cases} \tag{12.12}$$

Thus, provided the class $\Sigma_2$ is not empty, its member of minimum volume, i.e. the/any safe and geometrically possible structure of minimum volume ($\mathbf{R}_o$ with $V_o \equiv W_o$) is shown to solve the minimax problem stated at the beginning of this section. Moreover, $\mathbf{R}_o$ belongs to the subclass $\Sigma_2''$ of structures at impending collapse ($s = 1$) and hence will be associated with some collapse mechanisms $\dot{u}_h^c$. In fact, if $s > 1$:

$$s \int_{S'} F_{L_h} \dot{u}_h \, dS + s \int_{R_o} F_{B_h} \dot{u}_h \, dV - \int_{R_o} D(\dot{\varepsilon}_{hk}) \, dV \leqslant 0 \tag{12.13}$$

for any $\dot{u}_h$ complying with equations (12.7) and (12.8). It would then be possible to subtract from $\mathbf{R}_o$ some infinitesimal part $\delta\mathbf{R}$ such that, for any $\dot{u}_h$ complying with equations (12.7) and (12.8):

$$\int_{S'} F_{L_h} \dot{u}_h \, dS + \int_{R_o} F_{B_h} \dot{u}_h \, dV - \int_{\delta R} F_{B_h} \dot{u}_h \, dV$$
$$- \int_{R_o} D(\dot{\varepsilon}_{hk}) \, dV + \int_{\delta R} D(\dot{\varepsilon}_{hk}) \, dV \leqslant 0. \tag{12.14}$$

The structure $\mathbf{R} = \mathbf{R}_o - \delta\mathbf{R} \in \mathbf{R}_a$ would be safe with $V < V_o$, in contradiction to the meaning of $\mathbf{R}_o$. Note that this argument fails when a continuous variation of $\mathbf{R}$ is not admitted, as in the subsequent sections.

### 12.2.3 Limitations

In both the static and kinematic formulations considered so far, the design problem in question appears to be intractable in most practical situations. The basic common difficulty arises in the quantitative definition and manipulation of the design variables symbolized by $\mathbf{R}$. The latter minimax formulation involves the peculiar difficulty of the suboptimization which supplies the term $D(\dot{\varepsilon}_{hk})$ of functional $W$. In fact, by Hill's maximum principle, or as a consequence of Drucker's postulate[9], the dissipation rate for given strain rates and convex yield surfaces is defined as

$$D(\dot{\varepsilon}_{hk}) = \max \sigma_{hk} \dot{\varepsilon}_{hk} \quad \text{subject to } f(\sigma_{hk}) \leqslant 0 \tag{12.15}$$

i.e. as the optimum of a constrained convex programming problem in the stress components. These, however, do not otherwise intervene in the kinematic characterization of $\mathbf{R}_o$. New formulations in algebraic terms, based on a systematic preliminary discretization, seem at this point necessary for setting up concepts and solution methods of practical interest.

## 12.3 A discrete approach

### 12.3.1 Finite elements

A frequently adopted discrete model of three-dimensional inelastic continua consists of an assembly of tetrahedral finite elements, within each of which the strain and stress fields are required to be constant through the choice of linear 'shape functions' for displacements[10,11]. From now on in this chapter, bold-face letters will indicate matrices or column vectors. Superscript $i$ refers to the generic $i$th finite element. Let $V^i$ be its volume; let $\sigma^i$ and $\varepsilon^i$ denote the vectors of the six independent stress and strain components [the strain components being taken according to the 'engineering definition' so that $(\sigma^i)^T \varepsilon^i V^i$ represents the internal work rate performed in the element]. We indicate the vector of the six 'natural' generalized strains (i.e. elongations of the tetrahedron edges) by $\mathbf{p}^i$ and indicate the vector of the corresponding natural generalized stresses (i.e. six pairs of forces acting at the vertices along the edges[10] by $\mathbf{Q}^i$. Thus defined, generalized strains and stresses are intrinsic quantities (unaffected by rigid-body motions and equilibrated, respectively) related to the above physical strains and stresses through the transformations:

$$\mathbf{p}^i = \mathbf{T}^i \varepsilon^i \qquad \sigma^i = \frac{1}{V^i}(\mathbf{T}^i)^T \mathbf{Q}^i \tag{12.16}$$

where $\mathbf{T}^i$ is a non-singular matrix, easily evaluated from the vertex coordinates (or from the direction cosines and lengths of the edges). Through equations (12.16), the description of the material in the superposed $\sigma^i$, $\varepsilon^i$ spaces is transformed into the description of the element behaviour in the superposed $\mathbf{Q}^i$, $\mathbf{p}^i$ spaces, the essential features (normality and convexity) being preserved. It is convenient to piecewise linearize the element yield locus. For this purpose, $y$ linear yield functions are introduced:

$$\phi_j^i = (\mathbf{N}_j^i)^T \mathbf{Q}^i - K_j^i, \qquad j = 1, \ldots, y. \tag{12.17}$$

By properly choosing vectors $\mathbf{N}_j^i$ and positive constants $K_j^i$ and condensing them in a matrix $\mathbf{N}^i \equiv [\mathbf{N}_1^i \cdots \mathbf{N}_y^i]$ and in a vector $(\mathbf{K}^i)^T \equiv [K_1^i \cdots K_y^i]$, respectively, the approximate polyhedrical admissible range in the $\mathbf{Q}^i$ space is defined by the inequality:

$$\phi^i = (\mathbf{N}^i)^T \mathbf{Q}^i - \mathbf{K}^i \leqslant \mathbf{O} \tag{12.18}$$

where $\mathbf{O}$ stands for a vector with zero entries and the inequality sign applies to each pair of corresponding components. The normality property of the plastic strain rates can be expressed accordingly, in the form:

$$\dot{\mathbf{p}}^i = \mathbf{N}^i \dot{\lambda}^i \qquad \dot{\lambda}^i \geqslant \mathbf{O} \tag{12.19}$$

$(\dot{\lambda}^i)^T \equiv [\dot{\lambda}_1^i \ldots \dot{\lambda}_y^i]$ being the vector of the plastic multiplier rates. A yielding mode $j$ can be activated ($\lambda_j^i > 0$) only if the relevant plane contains the stress point ($\phi_j^i = 0$); this fact is simply expressed by the non-linear equation:

$$(\boldsymbol{\phi}^i)^\mathrm{T}\boldsymbol{\lambda}^i = 0. \qquad (12.20)$$

### 12.3.2 Zero–one description of structure

Consider a pattern of $m$ tetrahedra covering the available region $R_a$ and associate to each a Boolean variable $\xi^i$ which assumes the values 0 or 1, depending on whether the $i$th tetrahedron is either empty or filled with material [see Figure 12.1(b)]. The class $\Sigma$ of all possible structures in $R_a$ is restricted by approximation to the $2^m$ possible ways of filling completely some of the $m$ elements of the assigned pattern. Any structure R in $\Sigma$ is defined by a vector

$$[\xi]^\mathrm{T} \equiv [\xi^1 \cdots \xi^m], \qquad \xi^i = (0, 1) \qquad (12.21)$$

and its volume by the scalar product

$$V = [\mathbf{V}]^\mathrm{T} . \boldsymbol{\xi} \qquad (12.22)$$

$\mathbf{V}$ being the $m$-vector whose components are the volumes of the tetrahedra, taken in a prescribed order. Let $m$ be the number of the nodes of the finite-element assembly which are free, i.e. do not belong to the fixity surface $S''$. Let the relevant $3n$ components of nodal velocities and nodal external forces, referred to a global reference frame, be collected in the vectors $\dot{\mathbf{u}}$ and $\mathbf{F}$, respectively. $\mathbf{F}$ includes two parts: the live load vector $\mathbf{F}_L$, whose non-zero entries concern the nodes on the given loaded surface $S'$ and are obtained by suitable discretization of the given surface tractions $F_{L_h}$; the self-weight vector $\mathbf{F}_B$, which clearly depends on $\boldsymbol{\xi}$. In order to specify this dependence, consider the vector $\mathbf{G}^i$ of the $3 \times 4$ components (in the global reference axes) of the nodal forces which are 'equivalent' in the work sense to the body forces $F_{B_h}$ acting throughout element $i$[11]. Through the virtual-work principle:

$$\mathbf{G}^i = \xi^i \int [\mathbf{M}^i(x_h)]^\mathrm{T}[F_{B_h}^i] \, \mathrm{d}V \qquad (12.23)$$

where: $\mathbf{M}^i$ represents the $3 \times 12$ matrix of the (linear) shape functions for the displacements in element $i$, the integration extends over the volume of element $i$ and $[F_{B_h}^i]$ is the vector of the three-body force components per unit volume. The supervector $[\mathbf{G}]^\mathrm{T} \equiv \{[\mathbf{G}^1]^\mathrm{T} \cdots [\mathbf{G}^m]^\mathrm{T}\}$ supplies the elemental representation of the body-forces distribution for the whole element assembly over $R_a$; from equation (12.23):

$$\mathbf{G} = \mathrm{diag}\left\{ \int [\mathbf{M}^i(x_h)]^\mathrm{T}[F_{B_h}^i] \, \mathrm{d}V \right\}\boldsymbol{\xi} \qquad (12.24)$$

whence, if $\mathbf{B}$ stands for the Boolean, topological matrix that performs the force composition at each node:

$$\mathbf{F_B} = \mathbf{BG} = \mathbf{B}\,\text{diag}\left\{\int [\mathbf{M}^i(x_h)]^{\mathrm{T}}[F^i_{B_h}]\,\mathrm{d}V\right\}\boldsymbol{\xi} \equiv \mathbf{P}\boldsymbol{\xi}. \qquad (12.25)$$

The last equality in expression (12.25) defines the 'self-weight matrix' $\mathbf{P}$, which depends on the geometry of the finite-element pattern over $\mathbf{R}_a$ and on the specific weight of the given material.

We now form the supervectors

$$[\dot{\mathbf{p}}]^{\mathrm{T}} \equiv \{[\dot{\mathbf{p}}^1]^{\mathrm{T}}\cdots[\dot{\mathbf{p}}^m]^{\mathrm{T}}\} \qquad [\mathbf{Q}]^{\mathrm{T}} \equiv \{[\mathbf{Q}^1]^{\mathrm{T}}\cdots[\mathbf{Q}^m]^{\mathrm{T}}\} \qquad (12.26)$$

which define strains and stresses throughout $\mathbf{R}_a$. Thus compatibility and equilibrium can be expressed for the whole assembly in the form:

$$\dot{\mathbf{p}} = \mathbf{C}\dot{\mathbf{u}} \qquad (12.27)$$

$$[\mathbf{C}]^{\mathrm{T}}\mathbf{Q} = \mathbf{F_L} + \mathbf{P}\boldsymbol{\xi} \qquad (12.28)$$

where the 'compatibility matrix' $\mathbf{C}$ depends on the geometry of the discrete model of $\mathbf{R}_a$.

Let us reconsider the vector inequality (12.18) which defines the admissible stress domain in the $\mathbf{Q}^i$ space of element $i$. It is always possible to make this domain bounded, by adding, if necessary, some fictitious yielding plane far enough from the origin of the $Q^i_r$ axes in order not to alter significantly the description of the plastic behaviour. Thus, by introducing in expression (12.18) the Boolean variable $\xi^i$ as a multiplier of the constant vector $\mathbf{K}^i$, the inequality

$$\boldsymbol{\phi}^i = [\mathbf{N}^i]^{\mathrm{T}}\mathbf{Q}^i - \mathbf{K}^i\xi^i \leqslant \mathbf{O} \qquad (12.29)$$

for $\xi^i = 0$ makes the admissible domain shrink to the origin, i.e. element $i$ is incapable of carrying any stresses. In other terms, this artifice deprives empty elements of any 'strength'. By means of new symbols:

$$[\boldsymbol{\phi}]^{\mathrm{T}} \equiv [\boldsymbol{\phi}^1]^{\mathrm{T}}\cdots[\boldsymbol{\phi}^m]^{\mathrm{T}}, \qquad [\boldsymbol{\lambda}]^{\mathrm{T}} \equiv [\dot{\boldsymbol{\lambda}}^1]^{\mathrm{T}}\cdots[\dot{\boldsymbol{\lambda}}^m]^{\mathrm{T}} \qquad (12.30)$$

$$\mathbf{N} \equiv \text{diag}\,[\mathbf{N}^1]^{\mathrm{T}}\cdots[\mathbf{N}^m]^{\mathrm{T}}, \qquad \mathbf{K} \equiv \text{diag}\,[\mathbf{K}^1\cdots\mathbf{K}^m] \qquad (12.31)$$

the element constitutive laws [expressions (12.29), (12.19), (12.20)] can be condensed into a compact formulation concerning all tetrahedra in $\mathbf{R}_a$:

$$\boldsymbol{\phi} = [\mathbf{N}]^{\mathrm{T}}\mathbf{Q} - \mathbf{K}\boldsymbol{\xi}, \qquad \boldsymbol{\phi} \leqslant \mathbf{O} \qquad (12.32)$$

$$\dot{\mathbf{p}} = \mathbf{N}\boldsymbol{\lambda}, \qquad \boldsymbol{\lambda} \geqslant \mathbf{O}, [\boldsymbol{\phi}]^{\mathrm{T}}\boldsymbol{\lambda} = 0. \qquad (12.33)$$

*12.3.3   Static and kinematic discretized problems*

On the basis of the above discretization, the design problem in question can be reformulated as follows, simply by transferring to the discrete model the concepts examined in Section 12.2.

Static formulation:

$$\min_{\xi Q} V = [\mathbf{V}]^T \xi \qquad (12.34)$$

subject to:

$$
\begin{aligned}
[\mathbf{C}]^T \mathbf{Q} - \mathbf{P}\xi &= \mathbf{F}_L \\
[\mathbf{N}]^T \mathbf{Q} - \mathbf{K}\xi &\leqslant \mathbf{O}, \qquad \xi^i = (0, 1).
\end{aligned}
\qquad (12.35)
$$

Kinematic formulation:

$$\min_{\xi} \max_{\dot{u}} W = [\mathbf{V}]^T \xi + \alpha\{[\mathbf{F}_i]^T \dot{\mathbf{u}} + [\xi]^T [\mathbf{P}]^T \dot{\mathbf{u}} - [\xi]^T \mathbf{D}(\dot{p})\} \qquad (12.36)$$

subject to:

$$\dot{\mathbf{p}} = \mathbf{C}\dot{\mathbf{u}}, \qquad \xi^i = (0, 1). \qquad (12.37)$$

The last term in curly brackets in equation (12.36) represents the total rate dissipation and makes use of the new $m$-vector

$$[\mathbf{D}]^T \equiv [D^1 \cdots D^m] \qquad (12.38)$$

whose component $D^i(\dot{\mathbf{p}}^i)$ is the dissipation rate produced by the generalized strain rates $\dot{\mathbf{p}}^i$ in the $i$th finite element when it is filled with material ($\xi^i = 1$). For any $\dot{\mathbf{p}}^i, D^i$ is defined as the optimal value of the linear-programming problem:

$$D^i(\dot{\mathbf{p}}^i) = \max_{\mathbf{Q}^i} [\dot{\mathbf{p}}^i]^T \mathbf{Q}^i \qquad \text{subject to:} [\mathbf{N}^i]^T \mathbf{Q}^i \leqslant \mathbf{K}^i. \qquad (12.39)$$

In the present context of piecewise linear yield loci, expression (12.39) expresses the principle of maximum work and is the counterpart of expression (12.15). The formal dualization of the linear program [expression (12.39)][12], and the mechanical interpretation of the dual variable, lead to the following alternative definition of $D^i$:

$$D^i(\dot{\mathbf{p}}^i) = \min_{\lambda^i} [\mathbf{K}^i]^T \lambda^i \qquad \text{subject to:} \mathbf{N}^i \lambda^i = \dot{\mathbf{p}}^i, \lambda^i \geqslant \mathbf{O}. \qquad (12.40)$$

This minimum characterization of the dissipated power has a theoretical interest of its own, as the dual of Hill's principle, and was first established in Reference 13. It is very useful for present purposes, since it removes the difficulty pointed out in Section 12.2.3 with reference to the continuous kinematic approach.

In fact, by condensing all linear programs [expression (12.40)] into a single one, the total dissipated power can be expressed in the form:

$$[\xi]^T \mathbf{D}(\dot{\mathbf{p}}) = \min_{\lambda} [\xi]^T [\mathbf{K}]^T \lambda \qquad \text{subject to:} \mathbf{N}\lambda = \dot{\mathbf{p}}, \lambda \geqslant \mathbf{O}. \qquad (12.41)$$

Taking account of expression (12.41) in equation (12.17), and substituting

232 Optimum Structural Design

equation (12.33), which appears among the constraints of expression (12.41), into equation (12.37), the kinematic characterization of the optimal structure becomes:

$$\min_{\xi} \max_{\dot{u}\dot{\lambda}} W = [\mathbf{V}]^T \xi + \alpha\{[\mathbf{F_L}]^T \dot{u} + [\xi]^T [\mathbf{P}]^T \dot{u} - [\xi]^T [\mathbf{K}]^T \dot{\lambda}\} \quad (12.42)$$

subject to:

$$\mathbf{N}\dot{\lambda} = \mathbf{C}\dot{u}, \quad \dot{\lambda} \geqslant \mathbf{O}, \quad \xi^i = (0, 1). \quad (12.43)$$

Thus a minimax problem is obtained, which (in contrast to that of Section 12.2.2), very advantageously, no longer implies a subordinate optimization with respect to static variables. Together with the design variables $\xi$, only kinematic variables intervene here, namely both the free nodal velocities $\dot{u}$ and the sign-constrained plastic multiplier rates $\dot{\lambda}$.

### 12.3.4 Relation of continuum and its discretization

It seems appropriate now to compare the solution of the discretized problem (superscript d) to that of the continuous problem (superscipt c). Suppose first that the set $\Sigma(m)$ of all $2^m$ geometrically possible $(R \in R_a)$ designs be the same in both cases, i.e. that the boundary of any $R$ in the continuous approach consists exclusively of tetrahedron faces of the $m$-finite-element pattern adopted for the discrete approach. Any design of the 'unsafe' class $\Sigma_1^d(m)$ in the latter approach belongs also to the unsafe class $\Sigma_1^c(m)$ in the former, since compatibility, including continuity of the velocity field, is everywhere obeyed throughout the finite-element pattern (whereas equilibrium is ensured only at the nodes). From $\Sigma_1^d(m) \in \Sigma_1^c(m)$, the relation $\Sigma_2^d(m) \in \Sigma_2^d(m)$ follows for the 'safe' classes which include the optimal structures. Therefore the solution $R_o^d(m)$ to the discrete problem involves an approximation on the unsafe side (i.e. its ability to carrying the loads is not rigorously guaranteed) and supplies a lower bound to the minimum volume obtainable by the continuous approach within the same set $\Sigma(m)$:

$$V_o^d(m) \leqslant V_o^c(m). \quad (12.44)$$

Neither $R_o^c(m)$ nor $R_o^d(m)$ are necessarily at collapse. Both tend to the solution $R_o^c$ of the truly continuous formulation of Section 12.2 as $m \to \infty$, i.e. as more and more nodes are added in $R_a$ in order to make the tetrahedral pattern finer. This process broadens the minimization domain of design variables in both cases, but the maximization domain only in the latter ($R_o^d$) case. Therefore, $V_o^c(m)$ tends to $V_o^c$ from above,

$$V_o^c(m) \geqslant V_o^c \quad (12.45)$$

but an analogous conclusion cannot be drawn for $V_o^d(m)$.

## 12.4   Shape optimization as an integer programming problem

### 12.4.1   Integer programming

The static formulation [expressions (12.34) and (12.35)] of the discretized design problem turns out to be, from the mathematical standpoint, a special problem in integer programming, namely a zero–one mixed linear program (the term mixed refers to the presence of continuous variables, besides the Boolean ones). A general theory of integer, linear and non-linear programming has been recently developed by Balas[14–16]. One of its main results can be specialized to the present context as follows. Consider the problems:

$$\min_{\mathbf{x}^1 \mathbf{x}^2} \Omega^P \equiv [\mathbf{c}^1]^T \mathbf{x}^1 + [\mathbf{c}^2]^T \mathbf{x}^2 \tag{12.46}$$

subject to:

$$\mathbf{A}^1 \mathbf{x}^1 + \mathbf{A}^2 \mathbf{x}^2 + \mathbf{y} = \mathbf{b}, \qquad x_i^1 = (0, 1), \qquad \mathbf{x}^2 \geqslant \mathbf{O}, \qquad \mathbf{y} \geqslant \mathbf{O} \tag{12.47}$$

$$\min_{\mathbf{x}^1} \max_{\mathbf{v}} \Omega^D \equiv [\mathbf{x}^1]^T \{\mathbf{c}^1 + [\mathbf{A}^1]^T \mathbf{v}\} - [\mathbf{b}]^T \mathbf{v} \tag{12.48}$$

subject to:

$$\mathbf{A}^2 \mathbf{v} - \mathbf{w} = -\mathbf{c}^2, \qquad x_i^1 = (0, 1), \mathbf{v} \geqslant \mathbf{O}, \mathbf{w} \geqslant \mathbf{O} \tag{12.49}$$

where $\mathbf{c}^1$, $\mathbf{c}^2$ and $\mathbf{b}$ are given vectors, $\mathbf{A}^1$ and $\mathbf{A}^2$ are given matrices, and $\mathbf{x}^1$, $\mathbf{x}^2$, $\mathbf{y}$, $\mathbf{v}$ and $\mathbf{w}$ are variable vectors ($\mathbf{y}$ and $\mathbf{w}$ are 'slack' variable vectors). If expressions (12.46) and (12.47) have an optimal solution $\mathbf{x}_o^1, \mathbf{x}_o^2, \mathbf{y}_o$, there exists a solution $\mathbf{x}_o^1, \mathbf{v}_o, \mathbf{w}_o$ to expressions (12.48) and (12.49) with:

$$\Omega^P(\mathbf{x}_o^1, \mathbf{x}_o^2) = \Omega^D(\mathbf{x}_o^1, \mathbf{v}_o)$$

$$[\mathbf{v}_o]^T \mathbf{y}_o = 0 \qquad [\mathbf{w}]_o^T \mathbf{x}_o^2 = 0$$

$$[\mathbf{c}^2]^T \mathbf{x}_o^2 + [\mathbf{v}_o]^T (\mathbf{b} - \mathbf{A}^1 \mathbf{x}_o^1) = 0.$$

Also the converse of this statement is true (the solution to the latter problem gives a solution to the former). Because of the above (and other) mutual ties, the problem posed by expressions (12.46) and (12.47) is called 'primal' and that posed by expressions (12.48) and (12.49) is called 'dual'.* When the number of components in $\mathbf{x}^1$ and $\mathbf{c}^1$ and the number of columns in $\mathbf{A}^1$ vanishes, the problems posed by equations (12.46) and (12.47) and (12.48) and (12.49) and the relevant duality theorem reduce to a pair of dual problems in 'continuous' linear programming and to their main duality properties, respectively.

---

* In the general theory of Balas, the pair of dual problems exhibits a formal symmetry, which makes evident the involutory property of their duality (the dual of the dual is the primal) but which is lost through the present specialization.

234          *Optimum Structural Design*

### 12.4.2   Duality of static and kinematic forms

The static formulation [expressions (12.34) and (12.35)] of the design problem can be easily given the form of expressions (12.46) and (12.47). Since **U** is a vector with $3n$ components all equal to one, let the equilibrium equations in expression (12.35) be replaced by the equivalent inequalities:

$$[C]^T Q - P\xi \leqslant F_L, \qquad -[U]^T[C]^T Q + [U]^T P\xi \leqslant -[U]^T F_L. \quad (12.53)$$

Let the $6m$ free variables **Q** be replaced by $6m + 1$ sign-constrained variables $\{[Q^+]^T : Q'\}$ through the relations:

$$Q = Q^+ - U'Q', \qquad Q^+ \geqslant O, Q' \geqslant 0 \quad (12.54)$$

where **U'** denotes the $6n$-vector of unit components. Thus, using among the slack variables the plastic potentials $\phi$ according to equation (12.32), we can write, instead of expressions (12.34) and (12.35):

$$\min_{\xi Q^+} V = [V]^T \xi \quad (12.55)$$

subject to:

$$
\begin{bmatrix} -P \\ \cdots \\ [U]^T P \\ \cdots \\ -K \end{bmatrix} \xi +
\begin{bmatrix} [C]^T & : & -[C]^T U' \\ \cdots & & \cdots \\ -[U]^T[C]^T & : & [U]^T C U' \\ \cdots & & \cdots \\ [N]^T & : & -N U' \end{bmatrix}
\begin{bmatrix} Q^+ \\ \cdots \\ Q' \end{bmatrix} -
\begin{bmatrix} F_L \\ \cdots \\ -[U]^T F_L \\ \cdots \\ O \end{bmatrix}
$$

$$
= -\begin{bmatrix} \eta \\ \cdots \\ \zeta \\ \cdots \\ -\phi \end{bmatrix} \leqslant O \quad (12.56)
$$

$$\xi^i = (0, 1); \qquad Q^+ \geqslant O, Q' \geqslant 0. \quad (12.56a)$$

A comparison between expressions (12.46) and (12.47) and (12.55) and (12.56) establishes the identifications (e.g. $x^1 \equiv \xi$, $c^1 \equiv V$, $c^2 \equiv O$, etc.) which have to be introduced in expressions (12.48) and (12.49) in order to obtain formally the dual to the problem expressed in expressions (12.55) and (12.56). Thus, in particular, the constraints (12.49) become:

$$[C : -CU : N]v = C[I : -U]v^1 + Nv^2 = O, \qquad v \geqslant O \quad (12.57)$$

where **I** means identity matrix and **v** has been suitably partitioned. Equation (12.57), by the mechanical meaning of **C** and **N**, suggests the following mechanical interpretation of the dual variables **v**:

$$[I : -\ddot{u}]v^1 \equiv -\dot{u}\alpha, \qquad v^2 \equiv \lambda\alpha, \quad (12.58)$$

where $\alpha$ is a positive constant scalar, with arbitrary value and suitable dimensions in order to make homogeneous the addends in equation (12.48) ($\dot{u}$ and $\dot{\lambda}$ are velocities). By means of the aforementioned identification and expression (12.58), the dual problem [expressions (12.48) and (12.49)] turns out to be identical to the kinematic formulation [expressions (12.42) and (12.43)] of the design problem. It is a remarkable fact that straightforward formal application of duality theory in integer programming avoids the use of both the upper-bound theorem of limit analysis and the extremum properties of the specific dissipation power. Indeed, it leads directly to the alternative minimax characterization of the optimal structure in the space of the kinematic and design variables only.

The duality property expressed by equation (12.50) shows that, at the/a solution by the dual, kinematic approach:

$$[\mathbf{F}_L]^T\dot{\mathbf{u}}_o + [\mathbf{\xi}_o]^T\mathbf{P}\dot{\mathbf{u}}_o - [\mathbf{\xi}_o]^T[\mathbf{K}]^T\dot{\lambda}_o = 0. \tag{12.59}$$

Through equations (12.58) and the identification $[\mathbf{y}]^T \equiv \{[\mathbf{\eta}]^T \vdots \zeta \vdots [\mathbf{\phi}]^T\}$ obtained by comparing equation (12.56) to equation (12.47), equation (12.51a) implies that:

$$[\mathbf{\phi}_o]^T\dot{\lambda}_o = 0. \tag{12.60}$$

Equations (12.59) and (12.60) show that, if the solution to the dual includes non-zero vectors $\dot{\mathbf{u}}_o$, $\dot{\lambda}_o$, these define an actual collapse mechanism for the optimal structure $\mathbf{\xi}_o$, i.e. a situation which is both kinematically and statically admissible. In fact, compatibility is expressed by the constraints, and the energy balance is expressed by equation (12.59); $\dot{\lambda}_o$ is associated through the constitutive laws, in accordance with equation (12.60), with stresses $\mathbf{Q}_o$ which are part of the solution to the primal problem and, hence, fulfil equilibrium and conformity.

### 12.4.3   Implicit enumeration techniques

Among the many numerical techniques devised in linear integer programming, the most promising for the present problems seem to be the 'implicit enumeration' algorithms[14-16], which preserve the structure of the continuous part of the problem. These algorithms can be described, in the present terminology, as procedures for systematically enumerating a small number of structures $\mathbf{\xi}$ in the geometrically possible class $\Sigma$, and examining them in such a way as to ensure that all the $2^m$ members of $\Sigma$ have been implicitly examined. In other terms, in the graph of all possible structures, each of which corresponds to a node, the procedure constructs a tree by means of certain tests such that, if not passed, a node and all its descendents are abandoned and never considered again[14]. The dual problem [expressions (12.48) and (12.49)] lends itself better than the primal to the application of the method, because Boolean variables intervene only in its objective. Details can be found in the literature quoted. We specialize below to the present case and interpret in mechanical terms the generation of the constraints to be satisfied by any $\xi$ candidate for optimality in Balas's implicit enumeration algorithm[15,16]. For any fixed structure $\xi_r$, expressions (12.42) and (12.43) become an ordinary linear program $L(\xi_r)$, for whose optimal value $W_m(\xi_r)$ there are two alternatives (corresponding to those of Section 12.2.2):

(a) $W_m(\xi_r) \to \infty$. Then $\xi_r$ belongs to the unsafe class $\Sigma_1$. A 'feasible ray' can be found, i.e. a mechanism [complying with conditions (12.43)] $\dot{u}_r, \dot{\lambda}_r$ such that the external power exceeds or equals the dissipation:

$$[\mathbf{F_L}]^T + [\xi_r]^T [\mathbf{P}]^T \dot{u}_r \geqslant [\xi_r]^T [\mathbf{K}]^T \dot{\lambda}_r. \qquad (12.61)$$

(b) $W_m(\xi_r) = [\mathbf{V}]^T \xi_r$. Then $\xi_r$ belongs to the safe class $\Sigma_2$, and $\dot{u}_r = \mathbf{O}$, $\dot{\lambda}_r = \mathbf{O}$ is an optimal vector for $L(\xi_r)$. If this is not the only optimal vector, a mechanism can be obtained as in (a), which, however, satisfies expression (12.55) as an equality and represents a collapse mechanism for $\xi_r$.

A typical iteration of the procedure[15] can be outlined in mechanical terms as follows:

(i) When $s$ structures $\xi_r$ have been examined, the $s_b$ of kind (b) supply the test constraint:

$$[\mathbf{V}]^T \xi < \min_r [\mathbf{V}]^T \xi_r, \qquad r = 1, \ldots, s_b. \qquad (12.62)$$

The $s_a$ of kind (a) and possibly some $\bar{s}_b$ of (b) supply the constraints:

$$[\xi]^T [\mathbf{K}]^T \dot{\lambda}_r \geqslant [\mathbf{F}]^T \dot{u}_r + [\xi]^T \mathbf{P}]^T \dot{u}_r, \qquad r = 1, \ldots, s_a + \bar{s}_b. \qquad (12.63)$$

(ii) Using implicit enumeration, find a $\xi$ satisfying expression (12.62), i.e. with volume less than those of all the previous $s_b$ safe structures, and satisfying (12.63), i.e. such that the dissipated power exceeds the external power for all the previously considered mechanisms. If there is no such $\xi$, the last considered $\xi_r$ of kind (b) is optimal (if $s_b = 0$ no solution exists). Otherwise go to (i), update the test constraints (12.62) and (12.63) and continue. Clearly, a $\xi$ infeasible for the current constraints cannot become feasible later, and hence is to be never reconsidered.

## 12.5  On the size of the mathematical-programming problem

The mathematical models to which the discretized design problem has been reduced exhibit computationally favourable structures. In fact the matrices involved, $\mathbf{A}^1$ and $\mathbf{A}^2$, are formed, as shown in equation (12.55), by the following submatrices: the $6m \times m$ self-weight matrix $\mathbf{P}$ and the $6m \times 3n$ compatibility matrix $\mathbf{C}$, which are both sparse by their very nature; the $ym \times m$ plastic constraints matrix $\mathbf{K}$ and the $6m \times ym$ outward normals matrix $\mathbf{N}$, which are both block diagonal (each block is $y \times 1$ in the former, $6 \times y$ in the latter). However, in most practical situations, the numbers $m$ of finite elements, $n$ of free nodes and $y$ of yielding planes for each element are such that large-scale programming problems arise. Some ways of reducing the computational effort are suggested below. The number of the Boolean variables, which strongly affects the computational difficulty, can be sometimes kept reasonably small as follows.

(a) The finite-element assembly covering the available region $\mathbf{R}_a$ is divided into $\bar{m}$ subregions and in each of these all elements are obliged to be simultaneously either filled with material or empty, through the relation:

$$\xi = \mathbf{M}\bar{\xi}, \qquad \bar{\xi}^i = (0, 1), \qquad i = 1, \ldots, \bar{m} \qquad (12.64)$$

where $\mathbf{M}$ is a $m \times \bar{m}$ block-diagonal matrix of $\bar{m}$ vectors with all unit components, and $\bar{\xi}$ is the vector of the $\bar{m} < m$ independent design variables. A relation like expression (12.63), introduced in expressions (12.34) and (12.35) and (12.42) and (12.43), does not alter the essential structure of the problem, and is often possible or desirable for technological or aesthetic reasons (shape regularity for easier fabrication, symmetry, etc.).

(b) A 'nucleus' of the optimal structure, i.e. a part of $\mathbf{R}_a$ where material is likely to be needed, can be often conjectured, and, accordingly, the following transformation can be used in either program:

$$[\xi]^\mathrm{T} = \{[\mathbf{U}]^\mathrm{T} : [\xi']^\mathrm{T}\}, \qquad \xi'^i = (0, 1), \qquad i = 1, \ldots, m' \qquad (12.65)$$

where $\mathbf{U}$ collects as components the units preassigned to the elements of the nucleus, $\xi'$ refers to the other $m' < m$.

The above conjecture of a nucleus may rest, instead of on intuition, on the relatively easy solution of the 'associated continuous' linear-programming problem. This is obtained from equations (12.55) and (12.56) by relaxing the zero–one constraints into continued inequalities $0 \leqslant \xi^i \leqslant 1$ ($i = 1, \ldots, m$); physically it would correspond to a minimum-weight design based on the assumption that material can be arbitrarily rarefied and that its strength varies proportionally to its density. Techniques have been established for dealing efficiently with bounded variables in linear programming[17].

A rather obvious procedure for reducing $y$, and, hence, the number of plastic conformity constraints in the primal and of $\lambda$ variables in the dual, can be outlined as follows.

(1) Start from a tentative design $\xi_T$ and a relevant plausible stress state $\mathbf{Q}_T$ (e.g. a $\xi_T$ can be derived by experience and intuition, $\mathbf{Q}_T$ can be derived by an elastic analysis of $\xi_T$; alternatively, both $\xi_T$ and $\mathbf{Q}_T$ can be suggested by the solution of the above associated continuous linear program).

(2) Choosing a tolerance $t^i$, e.g. $t^i$ = half the mean value of the constants $K_j^i$, perform the test

$$[\mathbf{N}_j^i]^T\mathbf{Q}_T^i - K_j^i\xi_T^i \leqslant -t^i, \qquad i = 1,\ldots,m, \quad j = 1,\ldots,y. \qquad (12.66)$$

(3) Formulate and solve a 'reduced' program by neglecting all the 'potentially inactive' conformity constraints, with indices $i, j$ for which expression (12.66) was fulfilled.

(4) Check whether all these constraints are complied with by the solution $\xi_R, \mathbf{Q}_R$ to the reduced program. If they are, stop; otherwise go to (2), and replace $\xi_T, \mathbf{Q}_T$ by $\xi_R, \mathbf{Q}_R$, and continue.

Finally, it is worth stressing that two-dimensional design, implicitly covered by the present study, obviously leads to problems of drastically smaller size (triangle elements with three components, as in Figure 12.1, instead of tetrahedra with six components). Practical examples of plane problems may concern the optimal cut of a given steel sheet for a support (plane stress) and the optimal profile of a gravity dam (plane strain); in both, allowance can be made for any geometric requirement in view of surrounding or embedded installations (the available area $R_a$ need not be simply connected).

## 12.6 Closure

The main conclusions can be summarized as follows.

(a) A minimax characterization of the optimal structure(s) can be derived from the kinematic theorem of limit analysis and proves useful when the design problem is suitably discretized.

(b) Finite-element models and piecewise linearization of yield surfaces reduce the shape optimization in three and two dimensions to tractable, although large, linear mixed zero–one programming problems.

(c) Available numerical techniques, combined with artifices which reduce the problem size, and possibly adjusted to exploit the special mathematical structure, seem to be capable of supplying solutions of practical interest in many cases.

(d) Duality theory in integer programming provides some theoretical results in a more systematic and straightforward way than physical arguments (as employed in Sections 12.2 and 12.3).

Natural developments of this study are: computational experiments, the wider use of mathematical-programming theory for a better insight into the mechanical problem, and an extension of the present approach to shape optimization under different behaviour constraints with possible recourse to non-linear integer programming.

## References

1. D. C. Drucker, 'On the postulate of stability in the mechanics of continua', *Journ. Mécanique*, **3**, 235–249 (1964).
2. A. G. M. Michell, 'The limits of economy of material in framed structures', *Phil. Mag.* (*Series 6*), **8**, 589–597 (1904).
3. C. Y. Sheu and W. Prager, 'Recent developments in optimal structural design', *Appl. Mech. Rev.*, **21**, No. 10, 985–992 (1968).
4. D. C. Drucker and R. T. Shield, 'Design for minimum weight', in *Proc. 9th Intern. Congr. Appl. Mech.*; *Brussels, 1957*, Vol. 5, pp. 212–222.
5. D. C. Drucker and R. T. Shield, 'Bounds on minimum weight design', *Quart. Appl. Math.*, **15**, 269–281 (1957).
6. Z. Mróz, 'Limit analysis of plastic structures subject to boundary variations', *Arch. Mech. Stos.*, **1**, No. 15, 63–76 (1963).
7. W. Prager, 'Optimality criteria derived from classical extremum principles', in *An Introduction to Structural Optimization* (ed. M. Z. Cohen), Sol. Mech. Div., University of Waterloo, Waterloo, Ontario, 1969, pp. 165–178.
8. D. C. Drucker, W. Prager and H. J. Greenberg, 'Extended limit design theorems for continuous media', *Quart. Appl. Math.*, **9**, 381–389 (1952).
9. W. Prager, *An Introduction to Plasticity*, Addison-Wesley, Reading, Mass., 1959.
10. J. H. Argyris, 'Continua and discontinua', in *Proc. of Air Force Conf. on Matrix Methods in Structural Mechanics*, AFFDL-TR-66-80, Oct. 1965.
11. O. C. Zienkiewicz, *The Finite Element Method in Engineering Sciences*, McGraw-Hill, New York, 1971.
12. G. Hadley, *Linear Programming*, Addison-Wesley, Reading, Mass., 1962.
13. G. Maier and A. Zavelani-Rossi, 'A finite element approach to optimal design of plastic discs', Report, 1st. Sci. Tecn. Costr., Politecnico, Milano, May 1970.
14. E. Balas, 'Discrete-programming by the filter method', *Operations Research*, **15**, 915–957 (1967).
15. E. Balas, 'Implicit enumeration techniques for mixed-integer zero–one programs', Report, Grad. School Ind. Admin., Carnegie-Mellon Univ., Pittsburgh, Feb. 1968.
16. E. Balas, 'Minimax and duality for linear and nonlinear mixed-integer programming', in *Integer and Nonlinear Programming* (ed. J. Abadie), North-Holland, 1970.
17. H. M. Wagner, 'Linear programming with relative bounded variables', *Proc. Conf. on Applications of Optimisation Methods for Large-Scale Resource-Allocation Problems, Elsinore, 1971* (to be published).

*Chapter 13*

# Design for Reliability—
# Concepts and Applications

*Fred Moses*

## 13.1 Introduction

The aim of structural optimization is to achieve lighter or lower-cost structures that satisfy structural requirements and safety. In general this has been interpreted in the following manner. The load or load conditions are given, and a set of design limitations such as stress, displacement or stability are imposed. The values associated with load and strength terms are treated in a non-statistical deterministic fashion, and it is usually implied that specification or design codes provide adequate safety factor values to cover any likelihood of failure. Most optimization studies to date report on how a structure should be proportioned for the given load and strength terms. However, it is clear that the implied safety factor has a much greater effect on the cost or weight than any of the subsequent results of mathematical-programming techniques. It can be said that, given a code or a set of design specifications, structural optimization gives the 'best' results that can then be obtained.

The deterministic design methodology suggested above represents a narrow view of design optimization, and has implications that need to be examined. In particular, the viewpoint will be discussed that optimization should be considered in the context of safety which describes load and strength terms probabilistically. The failure probability should be calculated, or at least evaluated, in some approximate manner that is consistent from structure to structure, and should be used either as a constraint in the optimization process or as part of the objective function. There are several advantages, both real and philosophical, in including failure probability or its converse, reliability, in an optimization process.

It should be noted that this approach is in its early stage of development, and only future efforts will determine whether reliability can be directly incorporated in a specific structural design or else form the foundation for code safety factors. Such factors will need to relate the design of the elements and the whole system in an optimal way. It should be observed that:

(1) The safety factors in most specifications have been developed in an evolutionary manner over relatively long periods of time for structures that

241

have not been optimally designed. The introduction of automated design procedures usually has the effect of causing many more constraints to be active in the design than in an unoptimized design. Thus optimized designs will usually have higher failure probabilities than unoptimized structures. The reason for this is that the specification safety factors usually refer to the design of elements, such as beams and columns, and do not allow for a simultaneous occurrence of many failure modes in an optimized structure designed to its limits. The use of reliability failure analysis in the design optimization procedure should both allow more rational safe designs and permit the continuing evolution of code safety factors in a consistent unbiased manner by being based on structures that are both optimized and not optimized.

(2) Structures are being designed to resist more 'extreme' types of environments, such as earthquake, wind, tornado and ocean waves, than considered previously. Such loads can only rationally be considered in a probabilistic sense. Furthermore, developments and application of random vibration analysis are becoming more accurate for such loadings, and the results can be incorporated directly into a reliability analysis for a design procedure. The same is true for material strength properties such as yield, fatigue life, ultimate strength and creep, which are being described probabilistically and can also be incorporated in reliability analysis. The luxury of using the highest recorded value or worst possible load and strength combination cannot be used with the loadings mentioned above or with materials such as composites, glass fibre or even concrete.

(3) It is being increasingly recognized that structural failure is usually a complex phenomenological process involving various stages of failure, including initial yielding, large displacements and cracking, local instability and, finally, collapse. The behaviour differences between elastic design and limit design thus become academic, as all levels of failure should be considered in a design process. Optimization then becomes a process of properly proportioning the structure against all levels of failure. A rational procedure in allocating the structural cost is to consider the probability of the occurrence of, and the likely damage cost associated with, each failure level.

Many problems, of course, exist in the implementation of a reliability-based optimization procedure. Most of the reliability studies under way in various parts of the world are aimed at incorporating probabilistic methods in choosing proper safety-factor values for code specifications. As indicated above, the codes deal mainly with the design of elements. This chapter surveys reliability analysis as it principally relates to systems of elements, as in towers and frames, and how it can be incorporated into optimization methods. In particular, reliability models will be shown for failure by yielding and collapse, and optimization methods will be illustrated for both. Further, an example of reliability-based optimization of complex elements, such as reinforced-concrete beams, in which strength properties are random variables, will be described.

## 13.2 Fundamental reliability analysis

An important factor in the reliability analysis of structures (i.e. the calculation of the failure probability) is the level of failure or type of failure mode one is examining. Thus reliability analysis may be developed for predicting yielding, large displacements, collapse, instability, dynamic inserviceability, etc. In all these cases an important feature of reliability analysis may be observed from what is usually called the fundamental case. It consists of a single member subject to one load. All of the load variability is lumped into one load term $P$, and all the strength variability is lumped into one term $R$ (resistance).

This fundamental structural-reliability model has many of the features that distinguish structural reliability from other reliability problems in electrical networks and systems; namely, that both the system and the environment are random variables. The failure probability $P_f$ (reliability $= 1 - P_f$) is the probability that the load variable exceeds the strength and may be evaluated from either of the two integrals[1,2]:

$$P_f = \int_0^\infty [F_R(t)] f_P(t) \, dt = 1 - \int_0^\infty [F_P(t)] f_R(t) \, dt \qquad (13.1)$$

where $F(t)$ is a probability distribution, and $f(t)$ is a density or frequency distribution.

In the special case where both $R$ and $P$ are normally distributed we can define a random variable $Z$, which is the reserve strength:

$$Z = R - P. \qquad (13.2)$$

Since $R$ and $P$ are normal, $Z$ is also normal, with mean

$$\bar{Z} = \bar{R} - \bar{P} \qquad (13.3a)$$

and standard deviation

$$\sigma_Z^2 = \sigma_R^2 + \sigma_P^2. \qquad (13.3b)$$

By definition

$$P_f = \text{probability that } Z < 0. \qquad (13.4)$$

244                    *Optimum Structural Design*

Substitution into the standard normal distribution form gives

$$P_f = \Phi\left[\frac{\bar{R} - \bar{P}}{(\sigma_R^2 + \sigma_P^2)^{\frac{1}{2}}}\right] \tag{13.5}$$

where $\Phi$ is available in tabular form.

Defining the central safety factor $n = \bar{R}/\bar{P}$ gives

$$P_f = \Phi\left[\frac{n - 1}{(n^2\gamma_R^2 + \gamma_P^2)^{1/2}}\right] \tag{13.6}$$

where $\gamma_R = \sigma_R/\bar{R}$ is the coefficient of variation (c.v.) of resistance, and $\gamma_P = \sigma_P/\bar{P}$ is the c.v. of load.

Equation (13.6) is illustrated in Figure 13.1 for a particular set of parameters[3]. Except for normal and log–normal frequency functions, the fundamental case $P_f$ must, in general, be evaluated numerically with a computer,

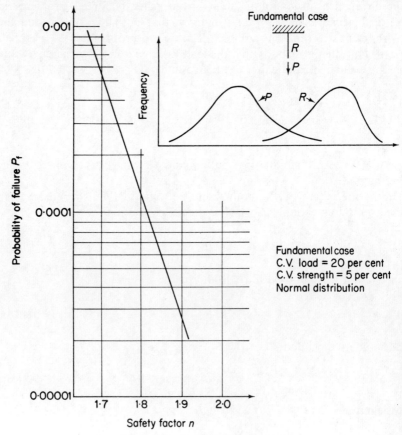

Figure 13.1   $P_f$ against safety factor $n$

as has been done by Freudenthal and others for various parameters[1,2]. A most comprehensive survey of these results has been given recently by J. Ferry Borges[4], who shows curves similar to Figure 13.1 of safety factor against specified failure probability for different cases, including:

(a) Coefficients of variation of both $R$ and $P$.
(b) Frequency distributions for $R$ and $P$, such as normal, log–normal, Weibull, extremal, etc.
(c) Definition of the safety factor in terms of mean values (central safety factor defined above) and nominal values (conventional safety factors) defined by their characteristics values (0·05 and 0·005 fractiles).

Such curves are useful in clarifying assumptions on frequency distributions and indicating the sensitivity of $P_f$ to input statistical information. To extend the fundamental case, consider situations where the element strength (or load) is made up of a function of random variables. Examples include reinforced-concrete beams and columns, composite material and cold-formed steel elements. Let

$$y = f(x_1, x_2, \ldots, x_n) \tag{13.7}$$

where $y$ is the fundamental strength term, and $x_1, \ldots, x_n$ are independent element random variables. A direct method of evaluating the distribution of $y$ is by the convolution method, but this requires multiple integrals. An approximation technique for the distribution can be made with a Taylor series, preferably expanded about the mean values of $x$. Thus

$$y = f(\bar{x}_1, \bar{x}_2, \ldots, \bar{x}_n) + \sum_{i=1}^{n} \frac{\partial f}{\partial x_i}\bigg|_{x=\bar{x}} (x_i - \bar{x}_i) \tag{13.8a}$$

where second-order terms are dropped. The statistical moments of $y$ give a mean value

$$\bar{y} = f(\bar{x}_1, \bar{x}_2, \ldots, \bar{x}_n) \tag{13.8b}$$

(i.e. the mean of a function equals function of the means) and standard deviation

$$\sigma_y^2 = \sum_{i=1}^{n} \left( \frac{\partial f}{\partial x_i}\bigg|_{x=\bar{x}} \right)^2 \sigma_{x_i}^2. \tag{13.8c}$$

Therefore, as an approximation, the mean and variance of a function can be easily found from the mean and variance of the independent variables. An example of this application will be given in Section 13.5 on the optimization of reinforced-concrete beams. Although the Taylor-series approximation only gives the coefficient of variation, studies of the fundamental case have indicated this to be the most important parameter affecting $P_f$, while the exact form of the frequency distribution is of less importance.

246 Optimum Structural Design

In general, the determination of a frequency distribution with a high degree of statistical confidence requires a large amount of data (usually a prohibitive amount for structural problems). Thus assumptions on frequency distributions have to be made using probabilistic models of behaviour or other insights such as, for example, the central-limit theorem for a sum of random variables approaching a normal distribution. Accuracy in the reliability analysis must always be a relative term, consistent with the sensitivity of overall costs to failure probabilities. Also, other types of failure not normally covered by structural codes, such as fire, blast, construction blunders, improper supervision, ignoring critical-load conditions, etc., are usually responsible for most damages and are also associated with larger degrees of uncertaining in predicting occurrences. The structural code and the costs associated with structural protection must be balanced in an optimization procedure consistent with the overall data available.

## 13.3  Reliability of structural systems

The fundamental case is only of limited direct application in most structures because of multiple-load combinations of both a static and a dynamic nature, and also various alternative failures of members and member combinations as already indicated. This section will survey some applications of the reliability of systems that are used as tools in the reliability-based optimization of structures.

### 13.3.1  Yield, 'weakest-link', models

An important factor in controlling structural deflections and serviceability is the prediction of the onset of yielding as determined from elastic analysis [see Figure 13.2(a)].

*One member, m load conditions*

$$P_f = 1 - \int_0^\infty \left[ \prod_{j=1}^m F_{P_j}(t) \right] f_R(t) \, dt. \tag{13.9}$$

Equation (13.9) is useful if a load distribution exists, say, on a maximum annual basis and failure probability is needed for $m$ years.

*n members, one load*  See Figure 13.2(b)[5].

$$P_f = 1 - \int_0^\infty \left\{ \prod_{i=1}^n [1 - F_{R_i}(a_i t)] \right\} f_P(t) \, dt \tag{13.10}$$

where $a_i$ relates the force or stress level in member $i$ to the load $P = t$. The analysis in equation (13.10) is for a 'weakest-link' system, since failure is defined to occur if any single element of the system reaches its critical

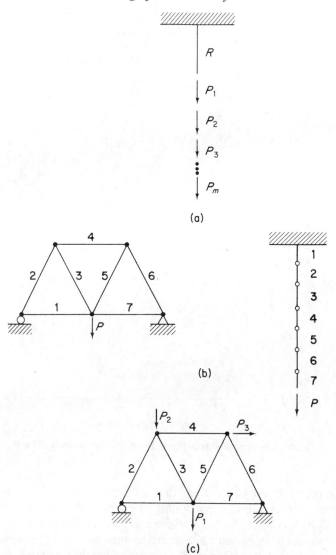

**Figure 13.2** Examples of structural system reliability models: (a) one member, $m$ loads; (b) $n$ members, one load; (c) $n$ members, $m$ loads

capacity. Such a model is useful for predicting yielding and associated deflections under a given load condition. Some results of 'weakest-link' analysis[6] are in Figure 13.3(a), showing the variation of central safety factor with $P_f$. Figure 13.3(b) shows the effect of the number of element failure modes for various assumed coefficients of variation and failure probability.

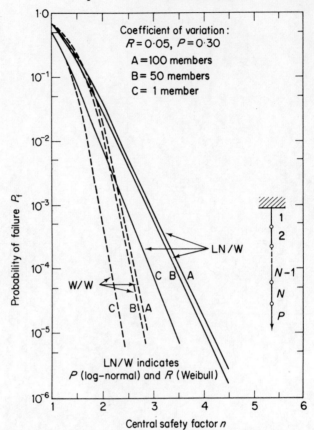

**Figure 13.3a**   Probability of failure $P_f$ against central safety factor $n$

It is assumed in these calculations that all elements have the same size (or safety factor). Optimum proportioning of elements for specified $P_f$ will be discussed in Section 13.5. The curves in Figure 13.3(b) are normalized to show how much an element size must be increased for the same overall failure probability when it is part of a system rather than isolated. $P_f = 10^{-4}$ in all cases and ratio of $n$ divided by safety factor is required for the one-member case.

If load variability (c.v.) greatly exceeds strength variability, as in dynamic-load cases, the $P_f$ of a system is about equal to the $P_f$ of the most critical member. That is, if any element fails, all the others will also fail, since a high load level has been reached. If strength variability dominates, as in brittle elements, the $P_f$ of the system is approximately the sum of the $P_f$ values of the elements[5,6].

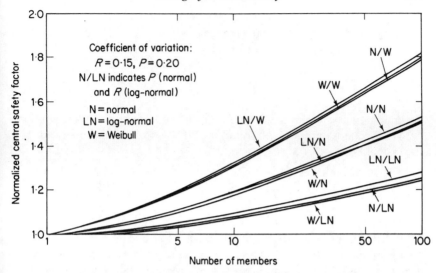

**Figure 13.3b**   Normalized central safety factor against number of members

*n members, m loads*   An exact analysis for this case requires the evaluation of multiple integrals. As an approximation, each load condition could be analysed separately, thereby disregarding the statistical correlation of failure among load conditions. The $P_f$ for the system would then be the sum of the $P_f$ for each load condition evaluated as in equation (13.10). Another approximation is to treat each element under every load condition as a fundamental reliability case and evaluate $P_{f_{ij}}$, the failure probability of the *i*th element under the *j*th load condition. As an upper bound[7]

$$P_{f_{\text{system}}} \simeq \sum_{i}^{n} \sum_{j}^{m} P_{f_{ij}}. \qquad (13.11)$$

*13.3.2   Collapse models ('fail safe')*

For most ductile and semiductile elements, ultimate failure occurs only when several elements reach their maximum capacity simultaneously. Examples include the basic mechanisms in frame collapse, the instability of multicolumn building frames, and the collapse of structures under dynamic excitation. In such cases a general expression for reserve strength can be established[8]:

$$Z_j = \sum_{i=1}^{R} a_{ji}M_i - \sum_{k=1}^{L} b_{jk}P_k \qquad j = 1, \ldots, N \qquad (13.12)$$

where $Z_j$ is the reserve strength of the *j*th mode, $M_i$ is the *i*th strength term in the *j*th collapse mode, and $P_k$ is the *k*th load term causing the *j*th collapse mode. The random variables are the $M_i$ and $P_k$ terms, which may in themselves be functions of other independent random variables, as in equation (13.7). The reliability analysis for the mode is to find $P_{f_j}$, the probability that mode $j$ collapses, which is

$$P_{f_j} = \text{probability that } Z_j < 0. \qquad (13.13)$$

To handle the system effect, Stevenson[9] studied frames with up to fifty-one potential collapse modes, and found that it was reasonable to let the $P_f$ of the system be

$$P_f = \sum_{j=1}^{N} P_{f_j}. \qquad (13.14)$$

Exact evaluation of each $P_{f_j}$ requires multiple integrations, except in the case where all the variables are normally distributed. Approximations based on a Pearson distribution using statistical moments have been presented by Stevenson and Moses[8] and require only a single numerical integration analogous to the fundamental reliability example presented above.

### 13.3.3   Brittle members

In redundant brittle structures, members that reach their capacity and fail carry little or no load thereafter. This greatly complicates the reliability analysis, and only a few analyses have been given[10]. Several factors suggest, however, that 'weakest-link' analysis would be applicable to brittle elements. Unless strength variability is large, failure of one element will usually 'trigger' consecutive failures of other members following load redistribution. Also, redundant structures usually have some elements, such as foundations or a weld point, which are statically determinate and thus also are part of a 'weakest-link' model. Only in brittle yarn-type systems is the 'fail-safe' probability significantly larger than 'weakest-link' analysis[11].

### 13.3.4   Dynamic analysis

Most studies of structural response to random load inputs have been based on linear elastic behaviour. A typical result found by Racicot is shown in Figure 13.4 for wind-excited structural response[12]. The failure probability is annual and is based on integrating responses over a distribution of wind spectra, maximum annual storms, durations and random-vibration critical-response analysis[13]. Other examples have been presented for wind, earthquake, highway-bridge truck loading, ocean wave, tornado and high-altitude turbulence[14-17]. An analysis for $P_f$ against safety factor can be treated similarly to a fundamental reliability case and used in subsequent optimization work. Veletsos[18] has recently extended the random-vibration response to multidegree-of-freedom inelastic systems for earthquake-type loading,

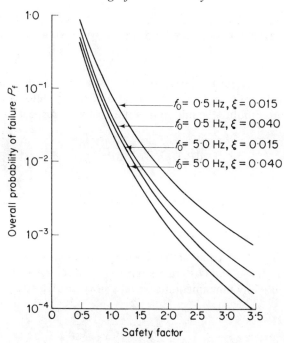

**Figure 13.4** $P_f$ against safety factor in dynamic wind load—results indicate the effect of damping and natural frequency (Reference 13)

and has found response characteristics that are similar for elastic and inelastic systems, particularly for low natural frequencies (less than 1·0 Hz).

### 13.3.5 Approximate reliability methods

Because of the computational aspects of evaluating failure probabilities for complex systems under random loading, various approximate techniques have been offered for code design applications[7,19–21]. Also, the lack of sufficient empirical data raises questions on the accuracy of $P_f$ predictions. Most of the work in this area centres on the development of design specifications utilizing fundamental probabilistic concepts and the result of the various models described above. Since this work is in its early stages, further discussion is beyond the scope of this chapter. The implications of this approach and comparisons with alternative formulations have been examined by Mau[22].

## 13.4 Reliability-based optimization

Introducing reliability or failure probability into the structural design process poses the question of what constitutes optimality. Obviously both initial economy and overall safety must be considered. Several alternative formulations exist, including the following, in which total cost is minimized:

$$C = C_i + P_f C_f \tag{13.15}$$

where $C$ is the total cost, $C_i$ is the initial cost and $C_f$ is the cost of failure. An optimum $P_f$ for the system and its associated safety factors or design sizes must be found. Finding the minimum total cost with respect to $P_f$ gives[23]

$$\frac{dC_i}{dP_f} + C_f + P_f \frac{dC_f}{dP_f} = 0 \tag{13.16a}$$

or

$$\frac{dC_i}{dP_f} = -C_f \quad \text{at the optimum} \tag{13.16b}$$

as $C_f$ does not depend on $P_f$.

Letting failure cost equal initial cost plus damage costs ($C_d$), and also assuming that the initial cost varies linearly with $\log P_f$, which is approximately true in many fundamental-case reliability results, gives

$$P_{f_{\text{optimum}}} = \frac{c}{C_i + C_d} \simeq \frac{c}{C_d} \tag{13.16c}$$

noting that $\frac{dC_i}{dP_f} = -c/P_f$, where $c$ is a constant from reliability analysis.

Equation (13.16c) assumes that the damage cost greatly exceeds the initial cost, as is often the case in, say, bridges, aircraft, dams, etc.

A more conventional approach to reliability optimization, analogous to deterministic design, is to (a) make $P_f$ the objective to be minimized, with structural cost specified as a constraint, or (b) set $P_f$ as the constraint, with initial cost or weight to be minimized. The latter approach is used in several subsequent examples.

In mathematical-programming form, the optimization is a constrained minimization of the form:

Minimize the weight (or cost)

$$W = W(A_k) \tag{13.17a}$$

subject to

$$P_f(A_k) \leqslant P_{f_{\text{allowable}}}. \tag{13.17b}$$

Fabrication or other side constraints can also be imposed of the form $g_j(A_k) \leqslant 0$. A convenient technique used for this problem has been the usable feasible-gradient method (see Chapter 8). Its advantage is that the calculations for $P_f(A_k)$ are time-consuming integrations that are similar to the fundamental reliability case. Hence the number of design points examined should be made as small as possible. This may preclude a penalty-function method, which, in my experience, requires more design points to be examined than with feasible-direction methods.

## 13.5 Examples

### 13.5.1 'Weakest-link' examples

The optimization problem, namely the minimum-weight proportioning of members of a 'weakest-link' system, has been considered in several investigations, in particular using the approximate $P_f$ expression given in equation (13.11). Hilton[24] used a Lagrange-multiplier technique to minimize the weight, while Kalaba[25] showed that dynamic programming was applicable and more efficient (one load condition). Switzky, in a further elaboration, showed that the following result might be applicable at the optimum[26]:

$$\frac{\text{weight (member } i)}{\text{total weight}} = \frac{P_{f_{(\text{member } i)}}}{P_{f_{\text{allowable}}}}. \tag{13.18}$$

i.e. heavier members should have lower safety factors than lighter members. Weight savings are reported, compared with a direct design procedure which assumes equal safety factors for all elements, so that

$$P_{f_i} = \frac{P_{f_{\text{allowable}}}}{n} \tag{13.19}$$

where $n$ is the number of elements.

Moses and Kinser[27] have shown the effect of computing $P_f$ by the exact expression in equation (13.10), which includes the effect of statistical correlation between failure modes. This is particularly important when load variability greatly exceeds strength variability, as in many civil and vehicle-design cases. In effect, the approximation in equation (13.11) adds to the failure probability by summing all the element failure probabilities, although many of the failure events occur with the simultaneous failure of several members, particularly if the load variance exceeds the strength variance. Table 13.1 shows the effect of the statistical correlation in a simple example with ten members of unequal mean loads. Results show the equal safety factor design optimization based on equation (13.11) (neglecting correlation) and optimization including correlation and using a gradient method. The weight saved by the more exact procedure depends on the ratio of load to strength variance, the number of members, the allowable $P_f$ and the frequency distribution. Studies with equal mean loads in the elements indicate a weight reduction of 7 per cent for fifty-element systems and normal distributions of load $\gamma_P = 10$ per cent and strength $\gamma_R = 1$ per cent[27]. The extreme case is when $\gamma_R$ approaches zero and the approximate $P_f$ expression requires an increased

Table 13.1   Optimum design using exact failure-probability expression including correlation—
'weakest-link' structures[a]

| Member | Member load | Equal safety factors | | Optimum design[b] neglecting correlation | | Optimum design[c] including correlation | |
|---|---|---|---|---|---|---|---|
| | | Area (in²) | $P_{f_i} \times 10^{-4}$ | Area (in²) | $P_{f_i} \times 10^{-4}$ | Area (in²) | $P_{f_i} \times 10^{-4}$ |
| 1 | 0·1P | 0·274 | 1·0 | 0·287 | 0·193 | 0·297 | 0·0519 |
| 2 | 0·2P | 0·547 | 1·0 | 0·562 | 0·377 | 0·554 | 0·604 |
| 3 | 0·3P | 0·817 | 1·0 | 0·833 | 0·561 | 0·818 | 0·958 |
| 4 | 0·4P | 1·09 | 1·0 | 1·101 | 0·735 | 1·09 | 0·991 |
| 5 | 0·5P | 1·37 | 1·0 | 1·367 | 0·917 | 1·35 | 1·23 |
| 6 | 0·6P | 1·64 | 1·0 | 1·630 | 1·095 | 1·61 | 1·61 |
| 7 | 0·7P | 1·92 | 1·0 | 1·893 | 1·271 | 1·86 | 2·08 |
| 8 | 0·8P | 2·19 | 1·0 | 2·153 | 1·45 | 2·11 | 2·65 |
| 9 | 0·9P | 2·46 | 1·0 | 2·413 | 1·63 | 2·35 | 3·25 |
| 10 | 1·0P | 2·74 | 1·0 | 2·672 | 1·79 | 2·59 | 3·91 |
| Weight | | 255·6 | | 253·2 | | 248·6 | |

[a] Weight based on 60 in member and density of 0·283 lb/in³. $P_{f_{allowable}} = 0.001$ for system; mean load $P = 60$ kip; mean strength $= 40$ kip/in²; normal distribution $\gamma_{load} = 20$ per cent, $\gamma_{strength} = 5$ per cent.
[b] Optimum based on equation (13.18).
[c] Optimum based on $P_f$ analysis using equation (13.10) and mathematical-programming solution of equations (13.17).

Table 13.2   Optimum design—'weakest-link' structure including proof loading

| Member | Member load | Optimum design, no proof-loading area[a] | Optimum design[b] | |
|---|---|---|---|---|
| | | | Area ($\gamma = 10^{-6}$)[c] | Area ($\gamma = 10^{-4}$) |
| 1 | 0·1P | 0·287 | 0·257 | 0·283 |
| 2 | 0·2P | 0·562 | 0·498 | 0·550 |
| 3 | 0·3P | 0·833 | 0·734 | 0·812 |
| 4 | 0·4P | 1·101 | 0·966 | 1·060 |
| 5 | 0·5P | 1·367 | 1·196 | 1·322 |
| 6 | 0·6P | 1·630 | 1·424 | 1·573 |
| 7 | 0·7P | 1·893 | 1·65 | 1·821 |
| 8 | 0·8P | 2·153 | 1·875 | 2·068 |
| 9 | 0·9P | 2·413 | 2·098 | 2·313 |
| 10 | 1·0P | 2·672 | 2·320 | 2·556 |
| Weight | | 253·2 | 221·0 | 243·9 |
| $P_f$ [d] | | 10⁻³ | 0·613 × 10⁻³ | 0·625 × 10⁻³ |

[a] See Table 13.1.
[b] From Reference 28.
[c] $\gamma$ is the ratio of cost of element to cost of failure. Reference 28 also shows optimum levels of proof-load testing.
[d] $P_f$ completely based on neglecting correlation in all cases.

safety factor (greater weight) for high number of elements, while the exact expression has a safety factor that does not depend on the number of elements. This is important in the design of tall frames and transmission, roof and offshore trusses.

Shinozuka[28] extended the 'weakest-link' design cited above by including proof-load testing in the model. Thus the integral on each strength distribution gives a lower bound (truncation), which becomes one of the design variables. The cost of testing and discarding weak samples was also included. Optimum design results for the example of Table 13.1 are present in Table 13.2. Table 13.2 neglects, however, the correlation effect, which would allow further weight reduction in the optimum shown.

### 13.5.2   Collapse ('fail-safe') systems

A recent study[5,9] illustrated the optimum proportioning of elements of redundant structures for collapse failure using the reliability analysis of equation (13.13). The specific application was limit analysis of frames with up to six independent design variables and as many as fifty-one collapse modes. The usable feasible-direction method of Zoutendijk[22] was used for

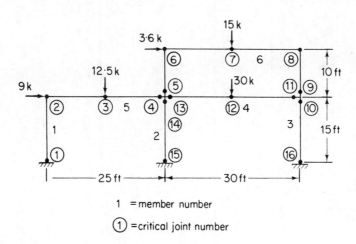

1  = member number

① =critical joint number

**Figure 13.5**   Two-storey two-bay frame

optimum proportioning of the elements. The results show a complex interplay of members in collapse modes and the difficulty of estimating in advance which collapse modes will dominate in contributing to the overall $P_f$ of the system and to which elements additional weight should be allotted for maximum reduction of $P_f$. The design problem is more complex than the

# Optimum Structural Design

Table 13.3 Optimum-design results of two-storey two-bay frame shown in Figure 13.5

| Example number | Optimum moment capacities kip ft | | | | | | C.V. | | Allowable $P_f$ | Weight function | Frequency distribution |
|---|---|---|---|---|---|---|---|---|---|---|---|
| | $M_1$ | $M_2$ | $M_3$ | $M_4$ | $M_5$ | $M_6$ | Moment capacity | Load | | | |
| 1 | 29.2 | 95.8 | 84.4 | 175.0 | 73.2 | 74.4 | 0.10 | 0.20 | 7.78(−2)$^a$ | 312.47 | Normal |
| 2 | 27.8 | 96.3 | 84.4 | 173.8 | 72.0 | 77.9 | 0.10 | 0.20 | 7.80(−2) | 312.89 | Log normal |
| Monte Carlo value of $P_f$ (9500 trials) | | | | | | | | | 7.59(−2) | | |
| 3 | 28.0 | 78.7 | 71.0 | 170.9 | 69.4 | 74.9 | 0.20 | 0.10 | 7.72(−2) | 297.26 | Normal |
| 4 | 27.3 | 78.3 | 71.3 | 166.4 | 65.1 | 74.9 | 0.20 | 0.10 | 7.16(−2) | 293.53 | Log normal |
| 5 | 29.1 | 87.8 | 72.3 | 170.3 | 68.0 | 74.1 | 0.15 | 0.15 | 7.52(−2) | 300.56 | Normal |
| Monte Carlo value of $P_f$ (7000 trials) | | | | | | | | | 7.50(−2) | | |

$^a$ Exponents of failure probability are shown in parentheses ($m$) and should be read as $10^m$.

'weakest-link' system discussed above, with no simplified result analogous to equation (13.18), and it is difficult to see how the design process can be performed without a mathematical-programming method.

A two-storey two-bay frame with six design variables is illustrated in Figure 13.5. A deterministic collapse analysis using linear programming would show at least six simultaneous active collapse mechanisms at the optimum. Table 13.3 shows some examples of optima. As may be seen[5], only one mode dominates in the $P_f$ analysis. A further conclusion was that the central-safety-factor checks alone, as used in conventional deterministic optimization methods, were not good indications of failure probability, and that some collapse modes in the same structure with higher safety factors also have higher failure probabilities.

### 13.5.3 Element optimization

An example of reinforced-concrete beam design will be used as an illustration[29]. For simplicity, the load will be assumed to be primarily dead weight and not be included as a random variable. $P_f$ will be evaluated using the Taylor-series method of calculating variances given in Section 13.3. The ultimate bending moment capacity (ACI Code) for under-reinforced-concrete beams is

$$M_u = A_s f_y \left( d - 0.59 \frac{A_s f_y}{b f_c} \right) \tag{13.20}$$

where $M_u$ is the design moment capacity (taken here as fixed), $A_s$ is the steel area, $d$ is the depth, $b$ is the width, $f_y$ is the steel bar strength and $f_c$ is the concrete strength. The design variables, which are also random variables, are taken to be $d, b, f_c, f_y$ and $A_s$. These parameters define the cross-section of the beam, which will be uniform. Thus strength, dimensions and fabrication variabilities are included in the design. The variable (mean values) must be chosen to minimize cost and provide a mean moment capacity that exceeds the applied bending moment ($M^*$) by some specified number of standard deviations.

Equation (13.8c) gives the standard deviation of the ultimate moment capacity in terms of the mean design variables:

$$\sigma_{M_u}^2 = \left( f_y d - 1.18 \frac{A_s f_y^2}{b f_c} \right)^2 A_s^2 \gamma_{A_s}^2 + A_s^2 f_y^2 d^2 \gamma_d^2$$

$$+ \left( A_s d - 1.18 \frac{A_s^2 f_y}{b f_c} \right)^2 f_y^2 \gamma_{f_y}^2 + \frac{A_s^4 f_y^4}{b^2 f_c^4} (0.59)^2 \gamma_b^2$$

$$+ (0.59)^2 \frac{A_s^4 f_y^4}{b^2 f_c^2} \gamma_{f_c}^2. \tag{13.21a}$$

where all the design variables in equation (13.21a) are mean values and $\gamma$ denotes the respective coefficients of variation.

The constraint on ultimate moment capacity is formulated so that the required moment capacity is exceeded by $k$ standard deviations (usually $k$ was chosen as 3·0). Therefore, in terms of the design variables,

$$A_s f_y \left( d - 0.59 \frac{A_s f_y}{b f_c} \right) - (M^* + k\sigma_{M_u}) \geqslant 0. \tag{13.21b}$$

Other imposed constraints needed for the problem are:

$$\left. \begin{array}{l} U_d - d \geqslant 0 \\[2mm] U_f - f_c \geqslant 0 \\[2mm] U_{f_y} - f_y \geqslant 0 \end{array} \right\} \text{ upper bound limits} \tag{13.21c}$$

$$\frac{3}{4}(0.85)k_1 \frac{f_c}{f_y} bd \frac{87}{87 + f_y} - A_s \geqslant 0. \tag{13.21d}$$

Equation (13.21d) contains the balanced steel requirement for under-reinforced beams, where

$$k_1 = 0.85, \qquad f_c < 4$$

$$= 0.85 - 0.05(f_c - 4), \qquad f_c > 4.$$

Based on some available cost information[29], the total cost includes concrete, steel and forming and finishing costs as follows:

$$f(d, b, f_c, f_y, A_s) = C_{\text{concrete}} + C_{\text{steel}} + C_{\text{forming}}$$

where

$$C_{\text{concrete}} = (1.006 + 0.031 f_c + 0.005 f_c^2)bd/144$$

$$C_{\text{steel}} = 78.8/144.0\, A_s \tag{13.22}$$

$$C_{\text{forming}} = 1.76(b + d)/12.0.$$

All terms are dollars per foot of beam length.

The design problem is to find $b, d, A_s, f_c$ and $f_y$ to minimize equation (13.22) subject to equations (13.21). A penalty-function (Sumt) formulation of optimization was used. It was found that the Powell search[30] method gave efficient results. A gradient technique was also tried, but did not give such a good convergence time.

Some examples of results are presented in Figures 13.6–13.8, showing the variation of the optimum-cost beam with design moment capacity ($M_u^*$), $k$ (standard deviations above mean required moment demand) and coefficient of variation of concrete strength (which usually has the highest variance).

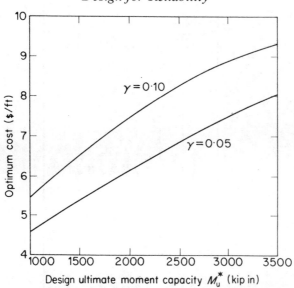

**Figure 13.6** Optimum-cost beams against design moment capacity, $k = 3.0$

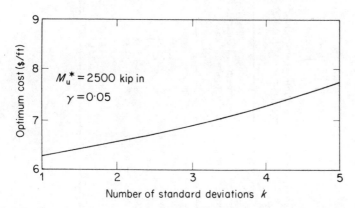

**Figure 13.7** Optimum cost against safety level—$k$ standard deviations above mean

In Figures 13.6 and 13.7, $U_d = 30$ in, $U_f = 5.0$ kip/in$^2$ and $U_{f_y} = 55.0$ kip/in$^2$. Among the variables, as expected, the steel-bar strength $f_y$ always equals the maximum value, and could, in effect, be removed as a design variable unless cost data different from that of equation (13.22) were available.

A further level of optimization can be obtained by looking at Figure 13.7 and allowing the measure of safety $k$ to become a design variable. To do this, the cost of failure would have to be included and the likelihood of failure expressed in terms of $k$. Assuming a normal distribution for $M_u$ would simplify

Table 13.4 Effect of costs on optimum values. Reliability constraint: $M_u M_u^* + 3m_u$ ($M_u^* = 2500 \cdot 0$); $\gamma_d = \gamma_{f_c} = \gamma_{f_y} = \gamma_{A_s} = \gamma_b = 0 \cdot 05$; upper bounds on: $d = 30 \cdot 0, f_c = 5 \cdot 0, f_y = 55 \cdot 0$

| Increase in cost | | | Optimum objective | Percentage increase in optimum cost compared to normal case (%) | Optimum $X_i$ | | | | |
| Forming (%) | Steel (%) | Concrete (%) | | | $d$ | $f_c$ | $f_y$ | $A_s$ | $b$ |
|---|---|---|---|---|---|---|---|---|---|
| Normal | Normal | Normal | 6·8851 | 0·00 | 25·60 | 4·98 | 54·89 | 2·89 | 4·13 |
| −25 | 0·00 | 0·00 | 5·7871 | −15·93 | 27·96 | 4·97 | 54·97 | 2·62 | 3·55 |
| +25 | 0·00 | 0·00 | 7·9400 | +15·31 | 22·97 | 4·95 | 54·97 | 3·23 | 5·05 |
| 0·00 | −25 | 0·00 | 6·4513 | −6·29 | 23·89 | 4·98 | 54·97 | 3·11 | 4·65 |
| 0·00 | +25 | 0·00 | 7·2568 | +5·39 | 26·48 | 4·97 | 54·97 | 2·80 | 3·81 |
| 0·00 | 0·00 | −25 | 6·6638 | −3·22 | 24·80 | 4·98 | 54·97 | 2·96 | 4·50 |
| 0·00 | 0·00 | +25 | 7·1022 | +3·15 | 25·15 | 4·98 | 54·97 | 2·95 | 4·21 |
| −25 | −25 | −25 | 5·1537 | −25·12 | 24·57 | 4·97 | 54·97 | 3·02 | 4·42 |
| +25 | +25 | +25 | 8·5800 | +24·63 | 25·81 | 4·98 | 54·97 | 2·88 | 3·99 |

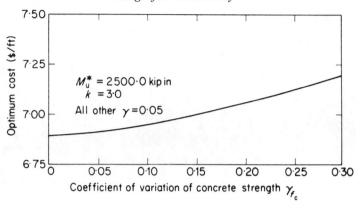

**Figure 13.8**   Optimum cost against coefficient of variation of concrete strength

the approach, so that the variation of $P_f$ with $k$ could be found from normal-distribution tables.

An interesting aspect of this cost optimization can be seen in Table 13.4. $f_c$ and $f_y$ are insensitive to variations of 25 per cent in cost input; however, $A_s$, $b$ and $d$ change by as much as 12 per cent, 25 per cent and 20 per cent, respectively, owing to cost variations. The concrete cost term might have been more sensitive if a different expression for increase in concrete cost with strength were used. The cost data is about ten years old, but only relative cost information and not its absolute value is needed in minimum-cost design. Thus inflationary effects over time are not important, so long as they affect all material and labour equally.

Other optimum-reliability-based design examples have been presented for composite materials, shells and pressure vessels[31,32]. In all reliability optimization it is necessary to subject the results to sensitivity analysis to determine the influence of input statistical parameters, including distribution functions and coefficients of variation, on optimum design variables and cost.

### 13.6   Stochastic programming (by C. Gavarini—a discussion)

Reliability-based limit analysis can be put into the frame of mathematical stochastic programming, with theoretical advantages both for the analysis itself and for the optimum design when uncertainty of data exists.

First, let us consider limit analysis with the deterministic approach. Under some hypotheses it can be presented as a linear-programming problem; introducing appropriate generalized variables, like hyperstatic unknowns, the program can include only inequalities; so, with matrix notation, we can write the two dual programs (LP1 = static approach; LP2 = dual =

kinematic approach):

$$LP1\begin{cases} \mathbf{Ax} + \mathbf{a}\lambda \leqslant \mathbf{b} \\ \lambda = \max \end{cases} \tag{13.23a}$$

$$LP2\begin{cases} \mathbf{A}^T\mathbf{v} = 0 \\ \mathbf{a}^T\mathbf{v} = 1 \\ \mathbf{v} \geqslant \mathbf{O} \\ \mathbf{b}^T\mathbf{v} = \min. \end{cases} \tag{13.23b}$$

$\lambda$ is the load multiplier, $\mathbf{x}$ is the vector of the hyperstatic unknowns and $\mathbf{v}$ is the vector of the dual variables governing the mechanism. Let us suppose now that strength has a random character; then vector $\mathbf{b}$ depends on some random variables, vector $\mathbf{t}$:

$$\mathbf{b} = \mathbf{Tt}. \tag{13.24}$$

The programs LP1 and LP2 become stochastic, and stochastic programming theory gives us the opportunity to compute the *distribution function* of the collapse multiplier[33,34]; obviously we need a knowledge of the distribution functions for the components of $\mathbf{t}$. We thereby obtain the statistical response of the structure, i.e. its statistical strength in terms of a load multiplier. Further, if the external load also has a random character, with a given distribution function, we can then compute the failure probability with equation (13.1).

Let us now consider a more general approach, including more stochastic loads and not necessarily restricted to plastic analysis. For the sake of simplicity, however, we will limit ourselves to plastic analysis. If $\mathbf{s}$ is the vector of generalized stresses we can write

$$\mathbf{s} = \mathbf{A}_1\mathbf{t}_1 + \mathbf{Bx} \tag{13.25}$$

where $\mathbf{t}_1$ is a vector of load multipliers and $\mathbf{x}$ has the same meaning as before. In terms of stresses, the yield constraints can be written

$$\mathbf{Ds} \leqslant \mathbf{Tt}_2 \tag{13.26}$$

where $\mathbf{t}_2$ is now the vector of parameters governing the strength. Substitution of equation (13.25) into expression (13.26) gives:

$$\mathbf{Cx} \leqslant \mathbf{At} \tag{13.27}$$

where

$$\mathbf{t} = \begin{bmatrix} \mathbf{t}_1 \\ \cdots \\ \mathbf{t}_2 \end{bmatrix}, \quad \mathbf{A} = [-\mathbf{DA}_1 \vdots \mathbf{T}], \quad \mathbf{C} = \mathbf{DB}. \tag{13.28a,b,c}$$

We now consider the space S of t components and give the following definitions. In that space there will be a *region*, either limited or unlimited, in which
the system of inequalities [expression (13.27)] has solutions; no solutions are
possible outside the region. We call the region the *safety domain* D of the
structure in the parameter space[35]. In the limit plastic case considered, the
domain is a polyhedron (plastic domain) and each face of it corresponds to a
collapse mode. The *parametric analysis* of system (13.27) (in physical terms,
the deterministic parametric analysis of limit structural behaviour) gives
us the frontier of the safety domain. The analysis is performed with the
techniques of so-called *parametric programming*[36].

Let us now suppose that the safety domain is known. The whole limit
behaviour of the structure is indicated by this domain, and we can now
forget which was the structure. If we know the joint density probability
function of the parameters, we need only compute the following integral to
obtain the probability of collapse[37]:

$$P_f = \int_{S-D} f_t(\mathbf{t}) \, d\mathbf{t}. \tag{13.29}$$

We can now undertake the optimum design problem, in particular with
the criterion of minimum weight. If $\mathbf{x}$ is the vector of design variables, the
matrix $\mathbf{D}$ becomes a function, generally non-linear, of $\mathbf{y}$—$\mathbf{D(y)}$—and we can
put inequality (13.26) in the form

$$\mathbf{D(y)s} \leqslant \mathbf{T_2 y} \tag{13.30}$$

where $\mathbf{T_2}$ is the diagonal matrix associated with $\mathbf{t_2}$. System (13.27) becomes:

$$\mathbf{D(y)Bx} \leqslant \mathbf{T_2 y} - \mathbf{D(y)A_1 t_1}. \tag{13.31}$$

Obviously the components of $\mathbf{y}$ must be positive and the weight will be a
linear function of $\mathbf{y}$:

$$W = \mathbf{c^T y}. \tag{13.32}$$

We can now formulate the minimum-weight limit design under uncertainty
in the following way[38]:

$$\left. \begin{array}{c} W = \mathbf{cy} = \min \\ \text{under} \\ \mathbf{y} \geqslant 0 \end{array} \right\} \tag{13.33}$$

$$P_f = P_f(\mathbf{y}) = \int_{S-D(\mathbf{y})} f_{t_1, T_2}(\mathbf{t}) \, d\mathbf{t} \leqslant \bar{P}_f$$

with $D(\mathbf{y}) = \{\mathbf{t_1}, \mathbf{T_2} | \mathbf{D(y)Bx} \leqslant \mathbf{T_2 y} - \mathbf{D(y)A_1 t_1}\}$.

The optimum can be reached by one of the known techniques for optimization with non-linear constraints, paying attention to the fact that the main

constraint (the one on probability) does not have an analytical explicit form. A computer program has been coded to solve problem (13.33), and some numerical results have been obtained for a four-bar truss and a six-bar truss, with two hyperstatic unknowns and two random variables[38]. Depending on the complication of the problem, the computing times are rather long (14 min of UNIVAC 1108). However, the code is an experimental one and could be improved. Two points are to be emphasized. First, the code does not need the parametric analysis at each step of the optimization procedure. Secondly, the complication of the computations does not depend essentially on the structure complexity, i.e. on the number of design variables, hyperstatic unknowns and inequalities. Rather, it depends mainly on the number of random variables involved, i.e. on the size of the space in which we must determine the safety domain.

## References

1. A. M. Freudenthal, J. M. Garrelts and M. Shinozuka, 'The analysis of structural safety', *J. Struct. Div. ASCE*, **92**, No. ST1, (1966).
2. A. M. Freudenthal, 'Safety and probability of structural failure', *Trans. ASCE*, **121** (1956).
3. F. Moses, 'Approaches to structural reliability and optimization', in *An Introduction to Structural Optimization* (ed. M. Z. Cohen), S. M. Study No. 1, University of Waterloo Press, 1969.
4. J. Ferry Borges and M. Costanheta, *Structural Safety*, 2nd ed., National Civil Engrg. Lab., Lisbon, Portugal, 1971.
5. F. Moses and J. D. Stevenson, 'Reliability based structural design', *J. Struct. Div. ASCE*, **96**, No. ST2 (1970).
6. Y. S. Shin, 'Studies on reliability based analysis of structures', *Ph.D. Thesis*, Case Western Reserve University, August 1971.
7. C. A. Cornell, 'Bounds on reliability of structural system', *J. Struct. Div. ASCE*, **93**, No. ST1 (1967).
8. J. D. Stevenson and F. Moses, 'Reliability analysis of frame structures', *J. Struct. Div. ASCE*, **96**, No. ST11 (1970).
9. J. D. Stevenson, 'Reliability analysis and optimum design of structural systems with applications to rigid frames', Case Western Reserve University, Solid Mech., Struct. and Mech. Design Div. Report 14, Nov. 1967.
10. M. Shinozuka and H. Itagaki, 'On the reliability of redundant structures', in *Proc. of 5th Reliability and Maintainability Conf.*, *N.Y.*, *July, 1966*.
11. D. W. Gucer and J. Gurland, 'Comparison of the statistics of two fracture modes', *J. Mech. Phys. Solids*, **10**, 365–373 (1962).
12. F. Moses, 'Sensitivity studies in structural reliability' in *Structural Reliability and Codified Design* (ed. N. C. Lind), S. M. Study No. 3, University of Waterloo Press, 1970.
13. R. L. Racicot and F. Moses, 'A first-passage approximation in random vibration', *J. Appl. Mech.*, *Trans. ASME*, **93** (1971).
14. A. G. Davenport, 'Gust loading factors', *J. Struct. Div. ASCE*, **93**, No. ST3 (1967).
15. C. A. Cornell, 'Analysis of uncertainty in ground motion and structural response due to earthquake', MIT Dept. of Civil Engineering Report, April 1969.

16. R. Garson and F. Moses, 'Reliability design for bridge fatigue under random loading', paper presented at ASCE Specialty Conf. on Safety and Reliability of Metal Structures, Pittsburgh, Nov. 1972.
17. A. K. Malhotra and J. Penzien, 'Nondeterministic analysis of offshore structures', *J. Eng. Mech. Div. ASCE*, **96**, No. EM 6 (1970).
18. A. S. Veletsos and P. Vann, 'Response of ground-excited elastoplastic systems', *J. Struct. Div. ASCE*, **97**, No. ST4 (1971).
19. N. C. Lind, 'Consistent partial safety factors', *J. Struct. Div. ASCE*, **97**, No. ST6 (1971).
20. C. A. Cornell, 'A probability based structural code', *ACI Journal*, **66**, No. 12 (1969).
21. J. R. Benjamin and N. C. Lind, 'A probabilistic basis for a deterministic code', *ACI Journal*, **66**, No. 11 (1969).
22. S. T. Mau, 'Optimum design of structures with a minimum expected cost criterion', Cornell Univ. Dept. of Structural Engrg. Report No. 340, 1971.
23. A. M. Freudenthal, 'Critical appraisal of safety criteria and their basic concepts', Preliminary Report, 8th Congress Int. Assoc. of Bridge and Struct. Engrs., New York, 1968.
24. H. H. Hilton and M. Feigen, 'Minimum weight analysis based on structural reliability', *J. Aero. Sci.*, **27**, No. 9, (1960).
25. R. Kalaba, *Introduction to Dynamic Programming*, Academic Press, New York, 1962.
26. H. Switzky, 'Minimum weight design with structural reliability', in *Proc. AIAA/ASME 5th Struct. and Materials Conf.*, *1964*, pp. 315–322.
27. F. Moses and D. E. Kinser, 'Optimum structural design with failure probability constraints', *AIAA J.*, **5**, (1967).
28. M. Shinozuka and J. N. Yang, 'Optimum structural design based on reliability and proof load test', in *Annals of Assurance Science, Proc. of 8th Reliability and Maintainability Conf.*, Vol. 8, 1967.
29. S. S. Rao, 'Optimum-cost design of concrete beams with a reliability based constraint', Case Western Reserve University Solid Mech., Struct., and Mech. Design Div. Report, 1970.
30. M. J. D. Powell, 'An efficient method for finding the minimum of a function of several variables without calculating the derivatives', *Computer J.*, **7**, No. 3 (1964).
31. F. W. Diederich, W. C. Broding, A. J. Hanawalt and R. Sirull, 'Reliability as a thermostructural design criterion', in *Proc. 6th Symp. on Ballistic Missiles and Space Technology*, Vol. 1, Aug. 1962.
32. M. Shinozuka, J. N. Yang and E. Heer, 'Optimum structural design based on reliability analysis', in *Proc. 8th Int. Symp. on Space Science and Tech., Tokyo, Japan, Aug. 1969.*
33. C. Gavarini, 'Concezione probabilistica del calcolo a rottura', *Giornale del Genio Civile*, agosto (1969).
34. C. Gavarini and D. Veneziano, 'Calcolo a rottura e programmaxione stochastica—Problemi con una variable casuale', *Giornale del Genio Civile*, No. 4 (1970).
35. C. Gavarini and D. Veneziano, 'On the safety domain of structures—1', in *Convegno nazionale di Meccanica teorica ed applicata, Udine, giugno 1971*.
36. D. Veneziano, 'Calcolo a rottura probabilistico con piu parametri casuali', *Giornale del Genio Civile*, No. 2 (1971).
37. C. Gavarini and D. Veneziano, 'Sulla teoria probabilistica degli stati limite delle strutture', *Giornale del Genio Civile*, No. 11–12 (1970).
38. C. Gavarini and D. Veneziano, 'Minimum weight limit design under uncertainty', Preprint, December 1971.

*Chapter 14*

# Optimum Design of Multistorey Rigid Frames

M. R. Horne and L. J. Morris

## 14.1 Introduction

Designs for the majority of multistorey frames have traditionally been based on a sufficient number of simplifying assumptions to enable the design to be defined directly by the specified loading. Wind and any other lateral loads are taken by bracing, thus enabling the beams and columns to be designed for vertical loading only. By assuming certain support conditions for the beams and 'effective lengths' for the columns, the various member sizes can be determined directly, and hence the interaction effects between stiffness ratios and induced stresses bypassed.

On the other hand, when bracing is absent and horizontal loads are taken by rigid frame action (i.e. in sway frames) there is a wider range of possibilities open to the designer by which the relative proportions of the beam and column members can be obtained. When the fundamental criterion is that of maximum induced elastic stress, no start can be made to the design process without first assuming stiffness ratios, which subsequently have to be revised and the analysis, if necessary, repeated. While the elastic procedure has the advantage that strength, deflection and maximum-stress criteria follow from the same elastic analysis, the strength criterion (based on maximum local stresses) is not particularly consistent, and it is more economical to use plastic or ultimate-strength criteria to establish overall safety.

One is, however, then faced with designing the structure simultaneously on two sets of requirements—safety, based on ultimate strength (plastic or elastic–plastic behaviour) and limiting deflections, and maximum-stress requirements at working loads (elastic behaviour).

It is theoretically possible to satisfy all the design criteria simultaneously. Such a solution has been achieved[1], but only for very small structures, and even then the result is a design method into which additional design criteria can only be fitted with difficulty. The engineer will therefore prefer to use his engineering knowledge in a more adaptable procedure by exploiting the fact that certain criteria are known to be of overriding importance. When such criteria are used as the prime factors in an iterative procedure, rapid

267

convergence to a fully rational design satisfying all the necessary criteria (including that of minimum weight) can be achieved.

The most universally successful single criterion in proportioning the members of rigid frames is that of ultimate strength by simple plastic theory. Computationally, this has the enormous advantage that minimum-weight design can then readily be introduced as a linear problem, thereby establishing a fully determined set of desired plastic moments from which the full iterative design procedure can start. On the other hand, initial proportioning based on limiting deflections or on maximum stresses is highly non-linear, making these criteria an unsuitable starting point computationally.

The primary purpose of a study carried out at Manchester has been to establish iterative design sequences based, in the first instance, on the linear optimization of the plastic strength of a frame, but including allowances for the effect of change of geometry in the actual elastic–plastic structure. Various important practical design considerations have also been taken into account at this stage of the process. The strength design is then systematically amended in an iterative process, still using optimum techniques, to satisfy other required conditions (mainly limitations on deflections) at working loads. A certain amount of engineering intuition has been built into the design process to effect a rapid, but automatic, convergence of the design procedure.

It has been possible to establish an automated design system[2] that is economically viable in comparison with more conventional design methods, manual or otherwise, and produces safe structures within the limitations of the design specification. The various computer techniques are described, and an assessment of the computer-aided design system is made.

## 14.2  Optimum-weight design based on strength

The major step in design is to assign suitable sections to all the members of the frame so that the applied load system or systems can be carried at the specified load factors. The starting point is to obtain an optimum design according to rigid plastic theory, but the resulting design has then to be modified for a number of factors not allowed for in the basic design process. The factors not allowed for are those which introduce non-linearity, so that it is possible to use linear programming for the basic design procedure.

As a result of investigations carried out by Toakley[3] into the suitability of different linear-programming techniques, the application of the dual algorithm to the primal problem has been selected. When an optimum design is being considered, the main purpose is to minimize a function that expresses the total weight or cost of a frame. Though such a function represents an exact solution of the problem, it cannot be used in that form. This is because the constraints provided by the moment equilibrium consideration relate the plastic moment capacities of the individual members and not their weights. However, by assuming a linear relation between the plastic moment capacity $M_p$ and the weight $W_i$ of a member $i$, we have

$$W_i = (a_i + b_i M_{p_i})L_i \tag{14.1}$$

and hence a total weight function for a structure with $n$ members is

$$W = \sum_{i=1}^{n} W_i. \tag{14.2}$$

Thus

$$W = \sum_{i=1}^{n} a_i L_i + \sum_{i=1}^{n} b_i M_{p_i} L_i. \tag{14.3}$$

Since the first term and $b_i$ are constant and do not influence a minimization, the 'effective' weight function $f$ becomes (without change of symbolism):

$$W = \sum_{i=1}^{n} M_{p_i} L_i = \sum_{j=1}^{g} M_{p_j} L_j \tag{14.4}$$

where now $M_{p_j}$ is the group plastic moment (scalar quantity), $L_j$ is the total length of all members in group $j$, and $g$ represents the total number of groups in a structure. The concept of grouping is adopted so that members in any particular group are automatically required to have the same moment capacity and hence the same properties, thereby permitting the important practical advantage of repetitive sections where this appears desirable to the engineer.

In dealing with the optimization problem, a 'force-method' approach has been adopted, and the first step is to make the rigid frame statically determinate. This is achieved by cutting sufficient members within the structure so as to create rooted 'trees' and applying arbitrary sets of re-dundants $x_q$ to each cut. This enables the moments, and hence a set of constraints, which satisfy both the equilibrium and the yield criteria of the plastic theory to be established, i.e.

$$M_{p_j} \leqslant \sum_{q=1}^{e} a_{r,q} x_q + m_{0r} \leqslant +M_{p_j}. \tag{14.5}$$

These inequalities express the fact that the admissible moments nowhere exceed the plastic moment capacities of the members. Strictly, these moment constraints must apply at any point in the structure, but, in practice, it is necessary to confine their application to a finite number of possible hinge positions. The coefficients $a_{r,q}$ depend only on the geometry of the structure and $m_{0r}$ represents the moment at point $i$ on the frame due to external loading, with all redundant forces zero, the total number of redundants being $e$. The redundants $x_q$ can either be positive or negative, and, since the dual simplex algorithm operates only with non-negative variables, each redundant is made equal to the difference between a non-negative variable and a positive integer which has a numerical value larger than the expected value of the redundant.

The restriction in the number of locations for which the inequalities (14.5) are included is achieved by confining attention to the ends of members and to the position of maximum sagging moment in loaded beams. Loading can consist of point loads applied to any joint, and beam loading, which can be point loads and/or uniformly distributed loading. Moreover, in most of the hinges, the sense in which a plastic hinge would form is evident from the general loading pattern, and only one of the two inequalities (14.5) need be imposed. Though only one of the constraints is applied, the design program must, however, contain a check to ensure that the correct constraint is used.

It is a practical construction requirement that column members should not show reverse taper, and to ensure this additional constraints of the form

$$-M_{p_j} + M_{p_k} \leqslant 0 \qquad (14.6)$$

are imposed, where a column length in group $j$ lies immediately below a column length in group $k$.

The preceding linear optimization technique produces the required plastic moments, together with the values of the redundant forces at each cut. The following factors, all introducing non-linear or discontinuous features, have not so far been taken into account:

(1) The reduction of plastic moment due to axial load.
(2) The availability, in practice, of only a finite number of discrete hot-rolled sections, all of different shape, thus destroying any simple relationship between weight and plastic moment capacity.
(3) The effect of elastic and plastic deformations of the structure, causing finite changes of geometry and thus modifying the equilibrium conditions. This has been called the $P$–$\Delta$ effect because the changes in moments may be thought of as due to a typical axial load $P$ changing its line of action by an amount $\Delta$ where $\Delta$ is the deflection of a node of the structure perpendicular to the direction of $P$.
(4) The effect of elastic flexibility, combined with the deterioration of stiffness, due to plasticity, introducing problems of instability which may be classified into three categories:
   (a) Local instability within the cross-section of a member, leading to plate buckling of a web or of an outstanding flange.
   (b) Instability of a member, for example the buckling of a column or the lateral torsional buckling of a beam or column.
   (c) Instability of the structure involving relative deflections of the ends of members perpendicular to the directions of the members. This is

referred to as frame instability and may be discussed in terms of 'deteriorated critical loads', a term introduced by Wood[4].

As well as these factors, it is necessary to discuss the loads and load combinations for which the structure is to be designed. In the following description, loads are discussed first, and then methods of modifying the linear optimizing design procedure to allow for these non-linear factors are considered.

## 14.3 Loading

Usually a rigid multistorey frame has to be developed for the following loading conditions:

(a) Dead loads + imposed vertical loads, with load factor $\lambda_1$.
(b) Dead loads + imposed vertical loads + wind loads, with load factor $\lambda_2$.

The design procedure would, of course, allow for any desired values of $\lambda_1$ and $\lambda_2$. It has become customary in British practice to assume that $\lambda_1 = 1.75$ and $\lambda_2 = 1.40$, and these values have been adopted in the program.

In the initial versions of the system, the two loading conditions were manipulated simultaneously, leading to a storage requirement of about four times that for one loading case, since each loading case necessitates a complete set of redundants and inequalities. In some unsymmetrical structures, the number of loading cases is increased to three, since the effect of wind from two directions has to be considered. Even more loading cases may be involved, and it is desirable to avoid the possible escalation of storage demands and computer time.

Fortunately, investigation has shown that, provided that the results of the design for the first loading case are used as lower bounds for the second loading case, and so on, the change in the weight function is marginal when the loading conditions are manipulated independently and sequentially instead of simultaneously. The application of lower bounds from the preceding load case ensures that each load condition is satisfied. Though the final result does not necessarily represent the absolute minimum weight/cost design, it has been found to produce a very close approximation. Such a design is safe with respect to the stipulated load factors and all the specified load conditions, and the design therefore satisfies all the required strength criteria. The lower bounds are inserted into the optimization array as additional constraints and are imposed on the group moments.

## 14.4 Effect of axial load on plastic moments

The effect of axial thrust on a symmetrical section is to reduce a plastic moment capacity $M_p$ to some value $\alpha M_p$, where $\alpha$ depends on the axial thrust and the shape of the cross-section. For a given section, $\alpha$ is a non-linear function of the axial thrust, and is, in practice, not far from unity for all the beams and for the upper columns of a multistorey frame. The procedure adopted in allowing for axial thrusts is therefore to modify the inequalities (14.5) to the form

$$-\alpha_r M_{p_j} \leqslant \sum_{q=1}^{e} a_{r,q} x_q + m_{0r} \leqslant \alpha_r M_{p_j} \tag{14.7}$$

where $\alpha_r$ is the estimated modifying factor due to axial thrust at the point $r$ for the member with full plastic moment capacity $M_{p_j}$. The values of the $\alpha_r$ are, at the first design iteration, assumed to be either unity or some other value based on experience. The first design cycle results (see Section 14.5) in the tentative choice of a cross-section at each potential hinge point in the frame, together with a value for the axial thrust. This permits the calculation of a more accurate value of $\alpha_r$ for use in the second design cycle. This procedure does not appear to introduce any problems of non-convergence, and it is found that, if a third design cycle is required for other reasons, a second review of $\alpha_r$ values leads to negligible changes.

**14.5 Selection of discrete sections**

The use of the linearized relationship between weight and plastic moment [equation (14.1)] has to be followed by the selection of a suitable section from the handbook list of available sections. A safe design results if a section is chosen giving a plastic moment (reduced for the effect of axial load) not less than that which emerges from the linear optimization design routine. Manifestly, by this process, the overall design will then be stronger than required, and it may ideally be possible to reduce some of the sections to give a reduced plastic moment less than that emerging from the linear analysis. In practice, because the linear optimization process ignores various factors all tending to overestimate the carrying capacity of the structure as designed, it is found advantageous to accept the degree of overdesign inherent in selecting the discrete sections in the manner described.

The part of the computer program that selects discrete sections must choose the most economical section for the given axial thrust and reduced plastic moment, i.e. it must choose the appropriate 'economy' section. Previous researchers have defined an economy section as being one which, for a certain range of plastic moment capacity (without reference to reduction due to axial thrust), has a weight equal to or less than that of any other available section. However, the criteria for a column should be with respect to its performance under varying degrees of axial load, i.e. the actual plastic moment capacity for a given axial thrust. Accordingly, the reduced plastic moment capacity has been determined for different values of axial load for the complete ranges of universal beams and columns for the two grades of steel 43 and 50 to BS 4360, taking any variation of yield stress into account. The yield stress for any grade of steel is a function of the flange thickness of the design section. In addition, because of local instability problems, certain sections (depending on the grade of steel being used) are not suitable for plastic action[5] and are not included in the final selection of the economy sections. The various sections have first been compared purely on a weight basis, and then from a cost aspect, so that finally the true 'cost economy' sections are listed. The resulting economy sections are found to be almost identical to those selected on the basis of weight and unreduced plastic moment.

Usually the program selects universal column sections for column members and universal beam sections for beam members. Where depth restrictions do not apply and member instability problems are unlikely to be severely critical, universal beam sections may be used with advantage for column members, and an option is included in the program whereby the engineer may stipulate universal beam sections for certain columns if he desires. Conversely, where depth restrictions are stringent, a facility exists for re- stricting the depth of any particular member, and this can result in the choice, within the program, of a non-economy section for that member.

As mentioned previously, the choice of a discrete section for each member enables a revision to be made of the $\alpha_r$ values that allow for the effect of axial thrust on the plastic moment. The iterative cycle of design by linear pro- gramming, followed by the automatic selection of sections, is continued until the sizes from successive design cycles are identical.

The above process does not necessarily lead to the optimum design which would result if the discrete nature of available sections were fully allowed for. That particular problem has been solved by Toakley[3] using integer pro- gramming. Unfortunately, although a solution is attainable, the computing time is high, and comparisons between results obtained from the present procedure and that of Toakley show little or no improvement in the weight of the final design. It even appears that in certain cases the present procedure can give results better than that of Toakley, indicating that Toakley's algorithm does not always produce the minimum-weight frame.

An important feature of this basic design method is that it does not enforce any particular pattern of hinge development in the optimum structure, nor is there any requirement that complete mechanisms be developed. The convergence of the method is rapid, the number of analyses and design cycles required for the majority of the frame (so far as has been examined) being no more than three. This should be viewed in conjunction with the fact that the last iteration is a check on the preceding cycle.

### 14.6　Effects of change of geometry and instability

As already mentioned, local instability is avoided by restricting the sections that can be selected, depending on the grade of steel used. Member instability does not usually rule design in multistorey sway frames, the only circumstances in which it may do so being in the columns in the upper storeys or in any columns for which universal beam sections are employed. Though it is uncommon for lateral instability to govern the design of columns, a routine, based on an analysis by Horne[6], is used by which the columns in the upper storeys are checked for member instability. It is assumed that the beam members are adequately restrained by the floors, so that no lateral torsional buckling of the beams can occur.

The problem of overall frame instability is intimately bound up with the effect of change of geometry. The effect of change of geometry by itself on the plastic-collapse load may be studied by considering what happens to the equilibrium conditions in a rigid–plastic collapse mechanism. Figure 14.1

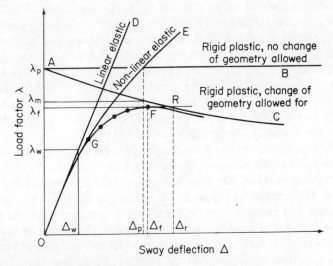

**Figure 14.1**　Typical load–deflection curves

represents a plot of load factor $\lambda$ against a typical sway deflection $\Delta$ of a multistorey frame, and OAB represents the rigid–plastic behaviour of a frame, ignoring change of geometry or $P$–$\Delta$ effects. $\lambda_p$ is the simple plastic collapse load. When $P$–$\Delta$ effects are allowed for, the drooping characteristic AC is obtained. OD represents linear elastic behaviour, and OE represents non-linear elastic behaviour, i.e. elastic behaviour allowing for change of geometry effects. Actual elastic–plastic behaviour OGF involves the successive formation of plastic hinges, thereby causing a departure from the elastic curve OE, but, usually before sufficient plastic hinges necessary for a mechanism are formed, a stage is reached at F at which the structure becomes unstable. At load factor $\lambda_f$, corresponding to point F, the deteriorated critical load is reduced to the current load level. Hence the failure load factor $\lambda_f$ depends both on change of geometry ($P$–$\Delta$ effect) and on instability brought on by reduction of stiffness due to plasticity.

The only accurate means of determining $\lambda_f$ is by an elastic–plastic analysis, and Majid and Anderson[7] have derived a design method based on complete elastic–plastic analyses in which certain design criteria are satisfied on an iterative basis. Unfortunately, a large amount of computing time is involved, despite the fact that no strictly minimum-weight procedure is included in the process.

Since the difference between $\lambda_f$ and $\lambda_p$ is, although important, not usually large, it was decided to seek some approximate means of determining the reduction $\lambda_p - \lambda_f$ in collapse load.

Majid[8] proposed such a method for evaluating the loads at failure of plane frames. This method assumes that the deflection $\Delta_p$ given by the intersection of the rigid–plastic line (ignoring change of geometry) and the non-linear elastic curve (see Figure 14.1) is approximately equal to that which occurs at collapse, as predicted by an elastic–plastic analysis ($\Delta_f$). The load factor at failure ($\lambda_m$) is then determined by using the $\Delta_p$ deflections in a rigid–plastic analysis. Though this may be true for small frames, it does not necessarily apply to large frames.

An alternative possibility is to assess the sway deflection $\Delta_r$ that corresponds to point R on the drooping characteristic AC at the actual failure load factor $\lambda_f$ (see Figure 14.1). $\Delta_r$ is assessed by assuming that $\Delta_r/\Delta_w$ (where $\Delta_w$ is the linear elastic deflection at working load) is some simple function of the number of storeys. This assumption has been explored for a number of typical frames, with the results shown in Table 14.1. While there is some scattering of results, the simple relationship given in Table 14.1 is found to lead to sufficiently close estimates of $\lambda_p - \lambda_f$.

The load factor at point R on the drooping characteristic AC in Figure 14.1 can be obtained by a rigid–plastic analysis if the initial geometry of the frame is suitably modified. Strictly speaking, the geometry should be that of the rigid–plastic mechanism, but it is assumed that, provided that the general magnitude of the deformations is correct, the precise pattern is immaterial. The estimated value of the function $\Delta_r/\Delta_w$ is therefore used as a common multiplier applied to the linear elastic working load deflections $\Delta_w$ of the first trial design. When the second design cycle is applied, the geometry of the frame is adjusted by the amount of these factored deflections, thus including in the optimum design process a suitable allowance for $P$–$\Delta$ and frame

Table 14.1   Deflection parameters for different frames

| Storeys | Bays | Load factor (nominal 1·40) | Deflection parameter $\Delta_r/\Delta_w$ |
|---|---|---|---|
| 2 | 1 | 1·70 | 8 |
| 4 | 1 | 1·48 | 4 |
| 6 | 1 | 1·61 | 5 |
| 8 | 1 | 1·49 | 4 |
| 8 | 2 | 1·44 | 3 |

instability effects. If desired, estimated change of geometry deflections may be introduced into the first design iteration—this being done most simply by the application of a 'sway index', i.e. a sway deflection equal to the sway index multiplied by the height of each storey. The program includes an input facility for this procedure.

### 14.7   Control of sway deflections

The procedures so far discussed lead to a minimum-weight-strength design that includes allowances for instability and change of geometry. Limitations on sway deflections at working load are, however, likely to be controlling factors on the design for some frames. The tendency for deflection criteria to predominate is necessarily increased by the introduction of more refined design methods based on strength and the introduction of higher-strength steels. It is therefore unrealistic to introduce advanced minimum-weight-ultimate-strength design without at the same time introducing some form of minimum-weight design for modifications which may be required to satisfy constraints on deflections. Unfortunately, when dealing with deflection constraints on rigid frames, there is the difficulty that the reciprocal of the moment of inertia is involved. The problem of controlling sway deflections thus becomes markedly non-linear.

The minimum-weight-ultimate-strength design is used as a starting point. If certain deflection constraints are found to be violated, the member sizes are increased to satisfy the constraints and the minimum weight/sway problem is resolved by reducing it to a linear one by piecewise linearization.

The solution that has been adopted is to establish an approximate relationship between the stiffnesses of the members, and hence their moments of inertia for each storey level[2]. Using these derived expressions, the relative sway deflections can be calculated and compared with the allowable deflection limits. If they are excessive, the sway control is executed until the limiting conditions are satisfied. The outline of the method is as follows.

By using the 'portal method' of analysis[9] for the condition when the frame is subjected to wind loading only, a statically determinate analysis is possible by making certain assumptions. These are that there is a point of contra-flexure at the mid-length of each member, and that the wind shear $q_k$ at any given storey level $k$ is proportioned between the columns with respect to the connecting beam spans. With these assumptions, and the use of energy methods, the sway deflection in any one storey can be assessed by taking

$$\Delta_k = \frac{1}{E} \int \int \frac{m}{I} \frac{\partial m}{\partial q_k} ds \qquad (14.8)$$

where $m$ is the moment due to the proportional wind shear and $I$ is the moment of inertia. The resulting expressions contain a number of terms that can be regrouped to obtain coefficients of the reciprocals of the moment of inertia for the various groups such that

$$\Delta_k = \sum_{j=1}^{g} a_{k,j} \frac{1}{I_j}. \qquad (14.9)$$

The moment of inertia for the various structural shapes, as listed in the handbooks, can be reduced to approximate continuous functions involving the square of the cross-sectional area. This means that the functions are non-linear, but they can be linearized by making use of polygonal approximations. The maximum deviation of the straight-line approximations from the true value is constrained to be less than 1 per cent.

To illustrate the application of sway control, a twelve-storey one-bay frame has first been designed by the plastic strength criterion only and then redesigned with differing deflection controls. The results obtained are given in Figure 14.2 and show that, as the sway control increased in severity, the frame became more stiff, and consequently the frame weight increased. Nakamura and Litle[10] obtain a similar curve for a 24 × 3 frame, which is

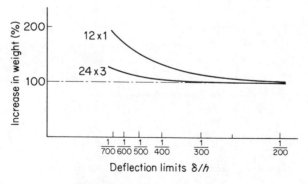

**Figure 14.2** Effect of sway control

278     *Optimum Structural Design*

inherently stiffer owing to the three bays. It would appear that a deflection of 1/400 of the overall height represents the practical limit for this type of structural framing. Beyond this value other structural systems may prove more economic. It should be remembered that these are bare frame analyses, and that the cladded structures would probably only deflect by a proportion of the values given.

The design system has an option by which a frame can be checked by the sway-control routine, without prior reference to the plastic-design routine. Although initial sizes for the frame are required, these sizes may be obtained outside the design system. Thus the engineer who has executed a conventional elastic design has here a means of modifying his design to satisfy deflection constraints so that the increase of weight is a minimum.

### 14.8 Application of design system

To illustrate the design system, an eight-storey two-bay frame, details of which are given in Figure 14.3, is designed under different conditions. The example originated from BCSA Publication 16[11], where it had been designed by traditional elastic methods, i.e. the frame was first designed according to

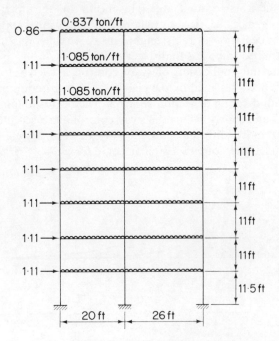

**Figure 14.3**   Details of design frame

the simple method outlined in BS 449 and then the member sizes amended where necessary to accommodate the wind moments. The resulting frame weight was 15·12 ton, using grade 43 steel (16 ton/in$^2$). Details of the design sizes are given in Table 14.2. This particular method constitutes the main technique used in design offices to produce a preliminary design for medium-rise structures.

Table 14.2  Member sizes of design frames

| Group number | Design case | | | | | | |
|---|---|---|---|---|---|---|---|
| | Elastic | Anderson | 1 | 2 | 3 | 4 | 5 |
| 1 | 8 UC 35 | 8 UC 35 | 8 UC 35 | 8 UC 35 | 8 UC 35 | 8 UC 35 | 8 UC 35 |
| 2 | 10 UC 60 | 10 UC 49 | 8 UC 48 | 8 UC 35 | 8 UC 48 | 8 UC 35 | 8 UC 48 |
| 3 | 8 UC 40 | 8 UC 48 | 8 UC 48 | 8 UC 40 | 10 UC 49 | 8 UC 40 | 10 UC 49 |
| 4 | 10 UC 60 | 10 UC 60 | 8 UC 48 | 8 UC 40 | 8 UC 48 | 8 UC 40 | 8 UC 48 |
| 5 | 14 UC 87 | 12 UC 92 | 12 UC 79 | 8 UC 48 | 10 UC 72 | 8 UC 48 | 10 UC 72 |
| 6 | 12 UC 65 | 10 UC 72 | 10 UC 49 | 10 UC 49 | 10 UC 60 | 10 UC 49 | 10 UC 60 |
| 7 | 14 UB 34 | 14 UB 22 | 12 UB 27 | 8 UC 48 | 14 UB 22 | 8 UC 48 | 14 UB 22 |
| 8 | 18 UB 50 | 16 UB 31 | 16 UB 31 | 10 UC 72 | 14 UB 34 | 10 UC 60 | 16 UB 31 |
| 9 | 16 UB 40 | 16 UB 26 | 16 UB 31 | 10 UC 72 | 16 UB 31 | 10 UC 60 | 16 UB 26 |
| 10 | 18 UB 60 | 16 UB 36 | 18 UB 45 | 10 UC 60 | 18 UB 45 | 10 UC 60 | 18 UB 45 |
| 11 | | | | 12 UC 79 | | 10 UC 60 | |
| 12 | | | | 10 UC 72 | | 10 UC 72 | |
| 13 | | | | 12 UB 27 | | 12 UB 25 | |
| 14 | | | | 16 UB 31 | | 14 UB 34 | |
| 15 | | | | 14 UB 30 | | 14 UB 30 | |
| 16 | | | | 16 UB 40 | | 16 UB 40 | |

Anderson[1] designed the same frame using an elastic–plastic design method applied to the sway case and obtained a frame weight of 12·15 ton (see Table 14.2 for details). By this analysis no plastic hinge occurs in a column member below the specified load factor. The uniformly distributed loading (u.d.l.), as shown in Figure 14.3, had to be replaced by an equivalent system of point loads.

Five different designs of the frame have been executed to illustrate various design parameters met with in practice. The first two designs demonstrate the use of different groupings of the member sizes, with no reduction in imposed loading as allowed by the codes. The appropriate grouping of the members imposed on these designs can be found in Figure 14.4 (for the resulting design sizes see Table 14.2). The next two designs represent similar cases, but with due allowance made for the reduction in imposed loading, based on a ratio of imposed loading to total loading of 0·3. The member sizes are given in Table 14.2. In the final design, the design parameters have been made

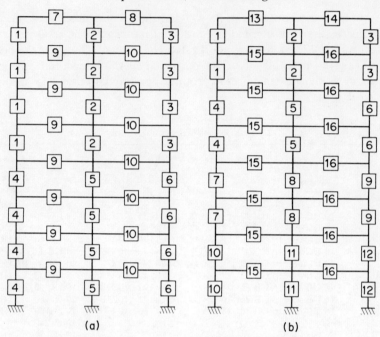

(a)                    (b)

**Figure 14.4**  Group numbering of design frames

identical to the Anderson frame, i.e. an equivalent point-load system acting on the frame (see Table 14.2 for member sizes).

The frame weights in the less restrained designs are lower, and there are marginal decreases when the reduction in loading is taken into account. All the designs obtained by the design system have been checked against an elastic–plastic analysis, and the appropriate load factors for each case given in Table 14.3. Only one of the factors is below that specified, and it is felt

Table 14.3  Summary of designs

| Design case | Elastic | Anderson | 1 | 2 | 3 | 4 | 5 |
|---|---|---|---|---|---|---|---|
| Grouping (Figure 14.4) | (a) | (a) | (a) | (b) | (a) | (b) | (a) |
| Grade of steel | 43 | 43 | 43 | 43 | 43 | 43 | 43 |
| Type of loading | U.D.L. | Point | U.D.L. | U.D.L. | U.D.L. | U.D.L. | Point |
| Load reduction | √ | ? | × | × | √ | √ | √ |
| Load factor: | | | | | | | |
| Vertical loading only | | 1·75 | 1·74 | 1·86 | 1·80 | 1·81 | 1·92 |
| Wind from left | | 1·44 | 1·41 | 1·50 | 1·47 | 1·51 | 1·44 |
| Wind from right | | 1·41 | 1·42 | 1·49 | 1·44 | 1·45 | 1·44 |
| Frame weights (ton) | 15·12 | 12·15 | 12·48 | 12·16 | 12·35 | 11·76 | 12·01 |

that this marginal difference would be acceptable in engineering practice. Comparing the computing time, the Anderson design took 0·53 min as opposed to the 0.31 min taken by the design system, both being executed on the Atlas computer at Manchester.

## 14.9 General assessment of design system

The various procedures discussed have been assembled into a system which automatically analyses and designs multistorey plane frames and can, if required, control sway deflections at working load to within any practical limit. The system is designed to be engineer-orientated, particularly at the computer–engineer interface. The input data has been reduced to a minimum, and the output information is in a form that can be readily understood. Such a system would become too involved if *all* the engineer's requirements had to be included. However, a number of options have been incorporated in the program by which the engineer can simulate a mathematical model with which he will be able to design economical, but safe, structures.

The convergence of the design process is rapid, as it requires no more than three design cycles for each frame, or part of a frame, for any loading case. The program has been shown to be a viable and economical proposition in terms of cost and efficiency. The saving in weight from designs obtained by more conventional methods will more than offset the cost of processing the frame by the design system. Comparisons with other design methods have proved to be favourable.[2] The design system should be viewed as being complementary to a designer, not replacing him.

The natural extension of this design system is to include the design and costing of all connexions. It should be possible to optimize the cost of a joint, provided that a detailed breakdown of the relative cost figures of each fabrication operation involved is known. Such an investigation could lead to standardization for this type of joint in multistorey construction. This fact, in itself, should have the effect of reducing the overall cost of joints.

## References

1. D. Anderson, 'Investigations into the design of plane structural frames', *Ph.D. Thesis*, University of Manchester, August 1969.
2. L. J. Morris, 'Automatic optimum design of multi-storey plane frames, *Ph.D. Thesis*, University of Manchester, May 1971.
3. A. R. Toakley, 'Optimum design using available sections', *J. Struct. Div. ASCE*, **94**, No. ST5, 1219–1241 (1968).
4. R. H. Wood, 'Stability of tall buildings', *Proc. Inst. Civil Engrs.*, **11**, September (1968).
5. M. R. Horne, *Plastic Design of Columns*, British Constructional Steelwork Association Publication No. 23, 1964.

282     *Optimum Structural Design*

7. K. I. Majid and D. Anderson, 'Elastic–plastic design of sway frames by computer', *Proc. Inst. Civil Engrs.*, **41**, December (1968).
8. K. I. Majid, 'The evaluation of the failure loads of plane frames, *Proc. Roy. Soc. A*, **306** (1968).
9. British Constructional Steelwork Association, *The Steel Designer's Manual*, Crosby Lockwood & Sons, 1955.
10. Y. Nakamura and W. A. Litle, 'Plastic design method of multi-storey planar frames with deflexion constraints', MIT Report R68-12, March 1968.
11. B. O. Allwood, H. Heaton and K. Nelson, *Steel Frames for Multi-Storey Buildings*, British Constructional Steelwork Association Publication 16, 1961.

# Chapter 15

# Reinforced-concrete Design

*D. Bond*

## 15.1 Introduction

When concrete is used structurally it is reinforced with steel bars or prestressed wires. During optimization, to search for minimum cost, not only can the geometric configuration and concrete cross-sectional dimensions be adjusted, but the quantities of the relatively expensive steel reinforcement must be varied throughout each structural member to achieve economy. By changing the cement content, the strength and cost of concrete can be varied to suit the circumstances. Voids can be cast in concrete to lighten it, and special lightweight aggregates can be used. These materials specifications can be design variables. When concrete is cast on site, the cost of formwork and its supporting temporary structure is considerable. Formwork construction can make repeated use of the same materials if a project is planned accordingly. This, the alternative of using precast units or other materials, and unknown site conditions also influence practical optimizing procedures.

Since the development of reinforced concrete at the end of the last century, designers have selected concrete dimensions using practical rules and the stress-ratio method described in Chapter 3, with proportions of steel at the most highly stressed sections that have been shown by experience to give reasonable economy.

During the past decade, since computers were first applied to the structural analysis of concrete frameworks, several parallel developments have taken place. Programs which use the finite-element method to analyse more complex structures with given dimensions have been made available to designers. References 1 and 2 describe examples of these. At the same time programs that adjust the cross-sectional dimensions of concrete frames were developed. They normally use the stress-ratio method. Emphasis has justifiably been directed to simplifying input data using problem-oriented languages, and Reference 3 is a good example of this. Using the argument that concrete dimensions can be chosen by experience, and in any event they must frequently conform with an architectural modular coordination scheme, some organizations have developed simple, practical programs that analyse structures having been given the concrete dimensions, and produce details of economic steel reinforcement (Reference 4, for example). Other practical programs have been written which design particular structural elements according to the codes of practice of their countries of origin[5-8].

## 15.2 Structural analysis

Structural optimization depends on the accuracy of structural analysis, and this must be discussed. Concrete is weak in tension, and cracks develop before reinforcing bars can be fully stressed. It is not elastic and does not obey Hooke's law. It creeps under working loads and shrinks as it hardens. Ideally, a reinforced-concrete structure should be analysed repeatedly as cracks develop and the material deforms with decreasing linearity as loads increase. This, however, requires too much computing time to be used for optimization at present. The success of the plastic design of steel structures influenced the direction of reinforced-concrete research, but owing to difficulties that are discussed elsewhere[9], and because serviceability under working loads is important, modern codes of practice in general recommend an elastic structural analysis of concrete structures, which should conform with three criteria:

(1) Under working loads there should be an adequate factor against collapse at any section, in the spirit of lower-bound limit analysis.
(2) No excessive cracking should occur in service.
(3) Short-term deflections and those due to creep must be limited.

(A vibration criterion can be applied if necessary.)

In general, reinforced-concrete structural members are joined integrally and such structures are statically indeterminate. Accepting that an elastic analysis will be used for optimization, the influence of reinforcement and cracking on the nominal stiffness of structural members must be considered. It is of interest to compare gross moments of inertia (no tensile cracks have developed) with those of transformed sections (tensile cracks have reached the neutral plane). It will be seen that the variation of stiffness with the degree of cracking can be greater than that due to practical variations of reinforcement areas. To take into account the reduction of stiffness as stress levels increase requires repeated analysis or its equivalent. This demands too much computing time. In addition, if steel stiffness is included fully, steel areas must become independent design variables. This will increase the number of these variables considerably and the computing time will be excessive. For these reasons it is concluded that, for optimization studies of reinforced-concrete structures in the immediate future, steel areas should be assumed to be dependent variables, the immediate variations of which do not influence the moments and forces in structural members. Codes of practice permit this. Stiffnesses can be computed from nominal steel areas and degrees of cracking, which can be adjusted occasionally during the design procedure.

*Reinforced-concrete Design* 285

## 15.3  Choice of numerical search procedure

As demonstrated in Chapter 2, if a homogeneous elastic structure is optimized and its minimum weight is sought, the objective function can be linear and constraints are in general non-linear. A simple plane frame is used to illustrate this [Figures 15.1(a) and 15.2]. Figure 15.1(b) shows a similar

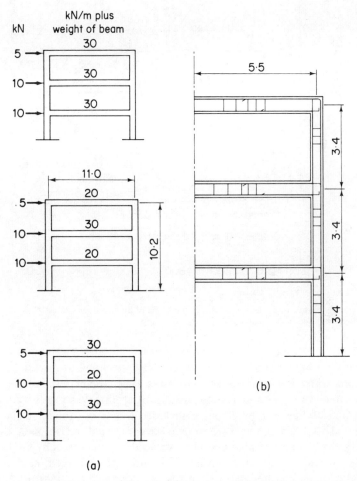

**Figure 15.1**   Plane frames : (a) homogeneous material and (b) reinforced concrete ; all dimensions are in metres and all frames are 305 mm thick

frame that is constructed of reinforced concrete and is subjected to the same loading. It has reinforced areas that are included as dependent design variables and are computed using the load-factor method[10] and an elastic frame analysis. The objective is minimum cost. The design space (Figure 15.3)

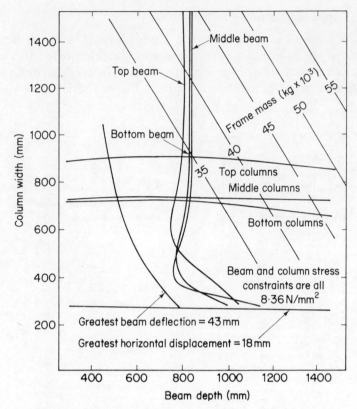

**Figure 15.2** Design space: plane frame of homogeneous material

shows that the total cost of the frame, which is a non-linear function of the unit rates for labour and material for concrete, steel and formwork[11], has a minimum which is away from all constraints. When the material of the frame was assumed to be a homogeneous material, permissible stresses were main constraints. When reinforcement is included this is no longer so, because additional steel can be provided to help the concrete to resist loads. Consequently bending and axial stresses are replaced as constraints by limitations on the amount of steel that can be accommodated in a section. Excessive steel also increases the cost and forces the design into a more economic region of the design space. Situations could arise (a more exacting deflection criterion, or closer limits of steel accommodation, for example) where the design having minimum cost has violated a constraint. In many cases, however, it can occur in the feasible region away from constraints. For these reasons it is considered that unconstrained minimization with a penalty function that can be applied in the region of active constraints, as described

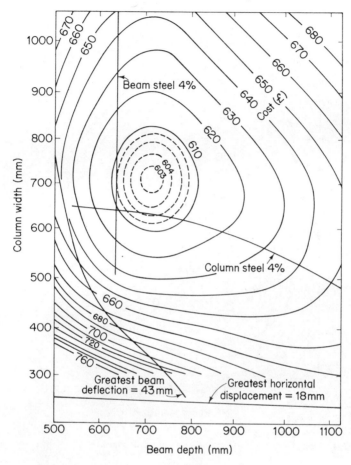

**Figure 15.3** Design space: plane frame of reinforced concrete

in Chapters 5 and 9, is suitable. A McCormick interior penalty function [equation (5.33)] and Powell's direct search method are used for the work that is described in this chapter.

Approximate methods can be used in the early stages of the search, and, for many problems, they may themselves be sufficient. Forces and moments can be held invariant, while cross-sectional dimensions and steel areas are adjusted, using structural analyses to amend their values when necessary. An example is shown in Figure 15.4. A structural analysis is performed at an arbitrary starting point $A_1$ or $A_2$ with $\gamma$ (the weighting factor of the penalty function) equal to zero. When the search arrives at $B_1$ or $B_2$, an analysis directs the search to C. A further analysis will bring the search near the optimum of Figure 15.3. If excessive steel intensities are encountered, the

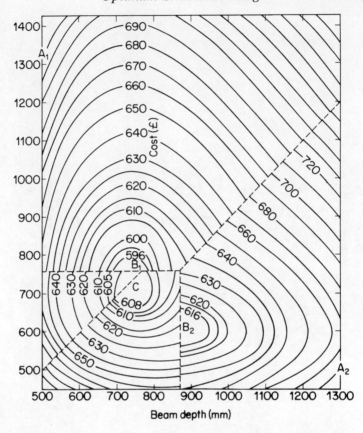

**Figure 15.4**  Design space: plane frame of reinforced concrete with forces and moments invariant

search can continue with $\gamma$ greater than zero. Excessive displacements can be avoided by adjusting dimensions according to simple rules which depend on displacements computed during the most recent structural analysis. If further accuracy is necessary near optima, more frequent structural analyses or their equivalent can be applied.

## 15.4  Slab design

My colleagues and I have limited their investigations to the study of multi-level concrete structures such as buildings and bridges. At each level a slab which will support given loading patterns over a specified area must be provided (except that foundation areas can vary to suit soil pressures). A slab

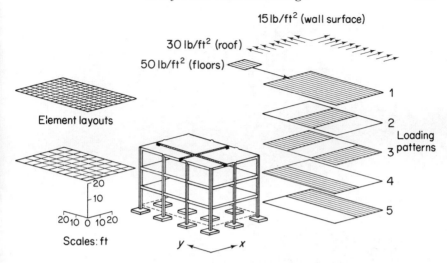

Reproduced from D. Bond, 'A computer program for studying the design of reinforced concrete structures supported on columns', *Proc. Inst. Civil Engrs.*, **43**, 195–216 (1969) by permission of the Council of the Institution of Civil Engineers

**Figure 15.5** Three-storey building: design analysis data

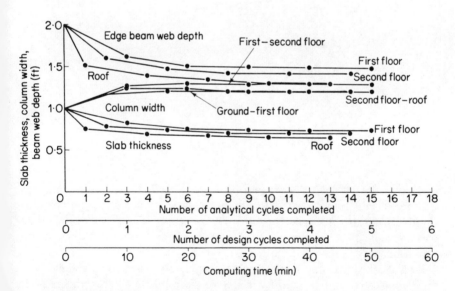

Reproduced from D. Bond, 'A computer program for studying the design of reinforced concrete structures supported on columns', *Proc. Inst. Civil Engrs.*, **43**, 195–216 (1969) by permission of the Council of the Institution of Civil Engineers

**Figure 15.6** Three-storey building: design parameters against design cycles

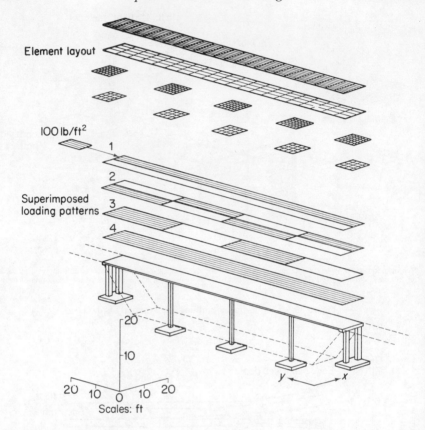

Element layout

100 lb/ft²

Superimposed
loading patterns

1
2
3
4

20
10

20  10  0  10  20
Scales: ft

y ←→ x

Reproduced from D. Bond, 'A computer program for studying the design of reinforced concrete structures supported on columns', *Proc. Inst. Civil Engrs.*, **43**, 195–216 (1969) by permission of the Council of the Institution of Civil Engineers

**Figure 15.7**  Footbridge: design analysis data

can have a flat, stepped, or curved soffit. It can be solid or hollow, and may be reinforced with prestressed wires. Slabs are interconnected by columns at suitable locations and may be stiffened by beams. The problem has thus concentrated on slab optimization, with beam and column design being supplementary to it.

It has been stated that it is difficult to improve on an elastic plate analysis to obtain the minimum quantity of slab reinforcement[12,13]. For this reason, in my main program, slabs are analysed elastically using rectangular finite elements and thin-plate theory[14,15]. Examples of the use of this program in its original form[16] are shown in Figures 15.5 to 15.11. Slabs at each level can be subdivided into regions, each having the same thickness.

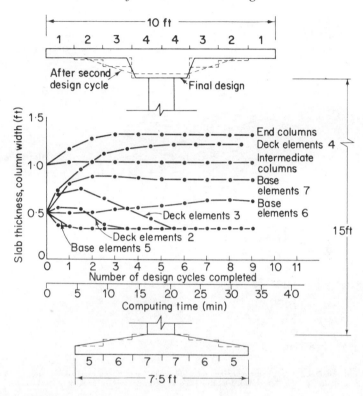

Reproduced from D. Bond, 'A computer program for studying the design of reinforced concrete structures supported on columns', *Proc. Inst. Civil Engrs.*, **43**, 195–216 (1969) by permission of the Council of the Institution of Civil Engineers

**Figure 15.8**  Footbridge: variations—design parameters against design cycles

If all elements at one level have the same thickness, a flat slab will result. This is demonstrated in Figure 15.5. The program allocates a fine layout to the top level and coarse layouts elsewhere. A complete structural analysis is performed. The dimensions of the top level are adjusted according to the stresses and deflections. A fine layout is allocated to the next level and coarse layouts elsewhere, and the procedure is repeated. The dimensions of each level are altered in this manner to complete one design cycle. Further design cycles occur, until no dimension has changed more than a specified amount since the previous design cycle. These dimensions are shown in Figure 15.6. If regions, each having a common slab thickness, are specified longitudinally, a slab having a constant cross-section will develop (Figures 15.7 and 15.8). If these are specified transversely, the slab will have a constant section in this direction (Figures 15.9 to 15.11). If they are defined peripherally,

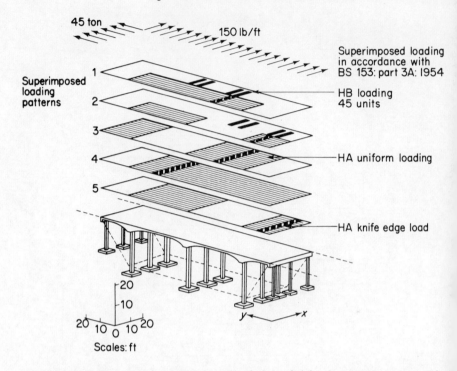

45 ton

150 lb/ft

Superimposed loading in accordance with BS 153: part 3A: 1954

Superimposed loading patterns

1

2

HB loading 45 units

3

4

HA uniform loading

5

HA knife edge load

20
10
20  10  0  10  20

y          x

Scales: ft

Reproduced from D. Bond, 'A computer program for studying the design of reinforced concrete structures supported on columns', *Proc. Inst. Civil Engrs.*, **43**, 195–216 (1969) by permission of the Council of the Institution of Civil Engineers

**Figure 15.9**   Highway bridge: design analysis data

pyramidal bases will result, as shown in Figure 15.8. Any pattern of rectangular elements can be used.

A recent development enables reinforcement zones to be specified within each region, each zone containing the same reinforcement. At each stage during the search procedure, for given slab element thicknesses, the necessary orthogonal reinforcement is computed. Consider bottom steel. Using the method described by Wood[13], orthogonal reinforcement is provided so that its moments of resistance $M_x^*$, $M_y^*$ are adequate to resist the normal moment in any direction caused by one moment triad $M_x$, $M_y$, $M_{xy}$.

Thus

$$M_x^* = M_x + \kappa|M_{xy}| \tag{15.1}$$

$$M_y^* = M_y + \frac{1}{\kappa}|M_{xy}|. \tag{15.2}$$

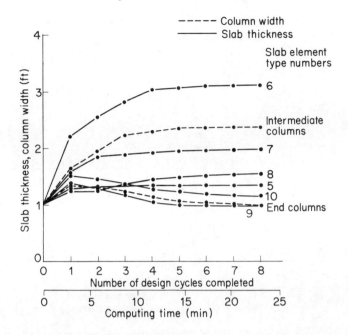

Reproduced from D. Bond, 'A computer program for studying the design of reinforced concrete structures supported on columns'; *Proc. Inst. Civil Engrs.*, **43**, 195–216 (1969) by permission of the Council of the Institution of Civil Engineers

**Figure 15.10**   Highway bridge: design parameters against design cycles

Assuming the same lever arm in both orthogonal directions, Wood shows that $\kappa$ equals unity for the minimum steel to resist one moment triad. This is not necessarily the optimum value of $\kappa$, however, if the steel must resist different moment triads caused by several loading cases and at different locations in the zone. The same method applies to top steel, but with the signs reversed in equations (15.1) and (15.2).

The original program[16] was modified to operate on all moment triads in each zone to optimize the reinforcement using a numerical search as a loop within the main optimizing procedure. The method can be described briefly as follows:

(1) Bottom reinforcement: compute $M_x^*, M_y^*$ for each moment triad separately with $\kappa = 1$.
(2) Put triads in descending order of $M_x^* + M_y^*$.
(3) Form a shortlist by removing triads having both $M_x^*$ and $M_y^*$ less than any pair above.

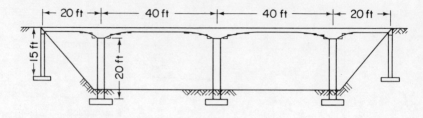

Slab elements types

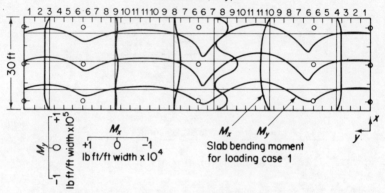

$M_x$
+1   0   -1
lb ft/ft width x $10^4$

$M_x$   $M_y$
Slab bending moment
for loading case 1

Figure 15.11 Highway bridge: plan and final elevation

(4) Give $\kappa$ an initial value of unity. Compute $M^*_{x1}$, $M^*_{y1}$ for the first triad in the shortlist. Compute $M^*_{x2}$, $M^*_{y2}$ for the second triad.

      (a)                or                (b)

If    $M^*_{x2} < M^*_{x1}$                     If    $M^*_{y2} < M^*_{y1}$

then $M^*_{x2} = M^*_{x1}$               then $M^*_{y2} = M^*_{y1}$

and $M^*_{y2} = M_{y2}$                and $M^*_{x2} = M_{x2}$

$$+ \left| \frac{M^2_{xy2}}{M^*_{x1} - M_{x2}} \right|.$$

$$+ \left| \frac{M^2_{xy2}}{M^*_{y1} - M_{y2}} \right|.$$

(5) Repeat search for an initial value of $\kappa$ which gives a minimum of $M^*_{xf} + M^*_{yf}$, being the final values after one run through the shortlist in this manner.

(6) Repeat for the top reinforcement.

(7) Compute the steel areas.

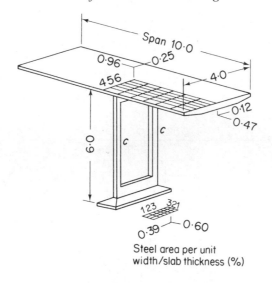

Steel area per unit
width/slab thickness (%)

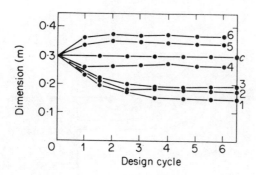

**Figure 15.12** Footbridge: slab thicknesses in final design cycles; all dimensions are in metres

When the reinforcement in each zone has been computed, applying ultimate-load analysis to each section to ensure an adequate factor against collapse, and allowing for minimum steel distribution and overlaps, the cost of each slab region is found. While the slab thicknesses are design variables and reinforcement is computed thus, approximate moments are used. Amended stiffnesses are then computed, including steel areas and allowing for the degree of cracking. These are used during the next design cycle[16] to bring values of moments and displacements up to data. Examples of this are shown in Figures 15.12 and 15.13.

The detailing of reinforcement is beyond the scope of this work, but some consideration of this problem is unavoidable, since realistic constructional costs, which are needed during the search for an optimum, depend on it. It

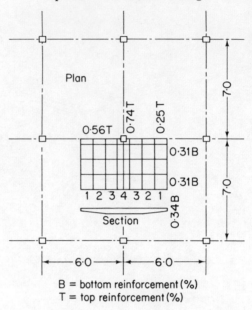

B = bottom reinforcement (%)
T = top reinforcement (%)

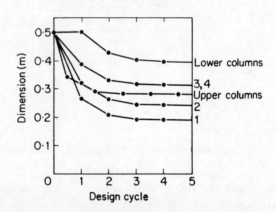

**Figure 15.13** Six-storey building: slab thicknesses in final design cycles; all dimensions are in metres

has been found that bar accommodation is particularly important at beam–column junctions. An example in Figure 15.14 shows a section through a three-dimensional design space for optimizing reinforced-concrete T-beam and column construction, using finite elements and thin-plate theory for slab bending, and for one combination of unit costs. Slab thickness, beam rib depth and rib width are design variables. Reinforcement areas are dependent variables.

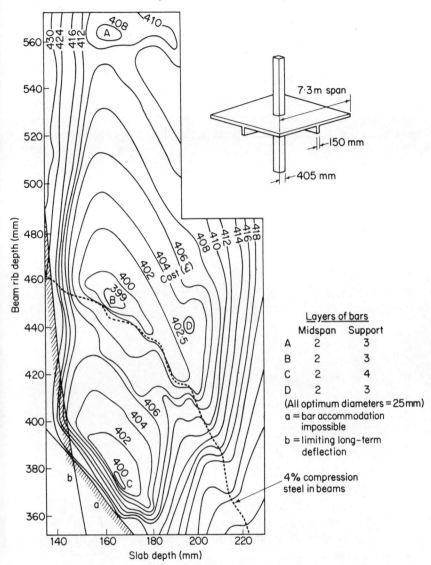

Reproduced from R. Skelton, 'Analytical methods for the optimum design of reinforced concrete slab and beam structures', *Ph.D. Thesis*, Queen's University, Belfast, 1972

**Figure 15.14** Reinforced concrete 'T' beam and column construction: section through design space for rib width = 150 mm

Total costs of the beams and slab in the part structure that is illustrated are represented by the contours. The best standard bar diameter and number of layers in beams is sought for each location in the design space. It can be seen that this causes local optima (for a given beam width) where bar combinations are more efficient. For this width the global optimum is near the 4 per cent limit of compression steel[10]. There is not much difference between costs in these areas, and, when the columns and foundations are included as design variables, it appears that these local optima are eliminated.

## 15.5 Materials specifications

The influence of materials specifications as design variables is being examined. An example of a structural member in isolation is given in Figure 15.15, but, to be meaningful, such examination should be applied to complete

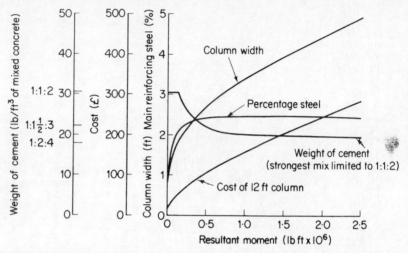

Reproduced from D. Bond, 'A computer program for studying the design of reinforced concrete structures supported on columns', *Proc. Inst. Civil Engrs.*, **43**, 195–216 (1969) by permission of the Council of the Institution of Civil Engineers

**Figure 15.15** Economic column sizes, concrete mixes and amount of mild-steel reinforcement for a square column subjected to a 100,000 lb axial load and a resultant moment acting in a vertical plane through a diagonal

structures. To take into account the interaction of structural members at all levels, and to limit computing costs, the main program for this work uses planes of symmetry and automatically combines the results of symmetric and antisymmetric analyses.

An important variation of materials specification relates to partial prestressing. Figure 15.16 shows concrete slabs and approximate spans in current use. It can be seen that prestressed concrete is more suited to longer spans. This is because stronger steel and concrete are used, and deflections can be reduced because there is less cracking[17]. On the other hand, long lengths of uniform wires or cables are used because they need relatively expensive end anchorages. Unprestressed steel bars can be used with advantage to supplement them where high bending moments occur.

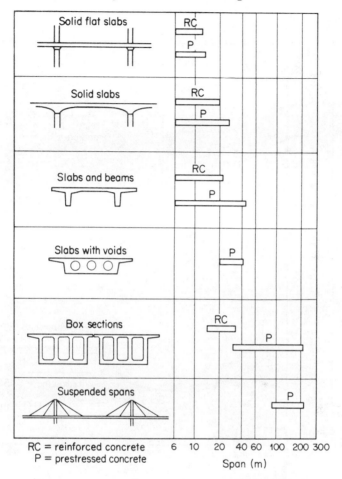

**Figure 15.16** Concrete slabs: approximate ranges of spans

Because, for a given situation, many combinations of wires and steel bars can provide an acceptable design (which may not be the cheapest), it is necessary to include the characteristics of the prestressed wires as independent design variables. The reinforcing bars can still be dependent variables in order to obtain a practical optimum. Research is still in progress, but one simple example can be described.

Figure 15.17 shows a flat slab that is continuous over a large number of supports. It is subjected to ten combinations of HA loading[18]. For a given span, the independent design variables are the slab thickness, concrete strength, area of wires (as a percentage of the total cross-sectional area), and their eccentricity, assuming that their disposition is parabolic along each

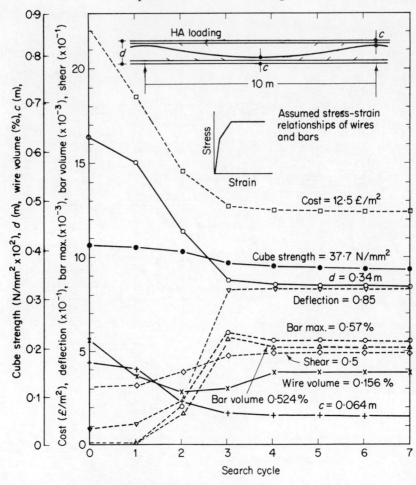

Figure 15.17 Partially prestressed slab: costs and design parameters against search cycles

**Figure 15.17** content (labels):

Cost = 12·5 £/m²

Cube strength = 37·7 N/mm²

d = 0·34 m

Deflection = 0·85

Bar max. = 0·57%

Shear = 0·5

Wire volume = 0·156 %

Bar volume 0·524%    c = 0·064 m

Cost = total cost of slab per unit area
Deflection = (greatest live load deflection)/(permissible value)
Bar max. = (greatest area of main bars top or bottom)/(total area of slab)
Bar volume = (volume of main bars)/(total slab volume)
Shear = (greatest shear stress)/(permissible value)
Wire volume = (volume of wires)/(total slab volume)
c = distance from centre of wires to edge of slab
d = slab thickness

span. Moments and forces are computed by a simple elastic analysis, and load and stress safety factors in the draft British unified code are used to ensure adequate safety against collapse. The stress–strain characteristics of bars and wires in this code are used for the analysis of sections. The shearing stress in the concrete (including punching resistance), a limitation on the

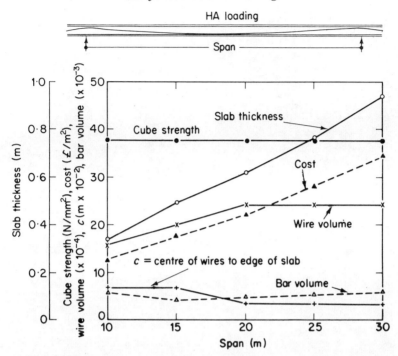

**Figure 15.18** Partially prestressed slab: costs and design parameters against span

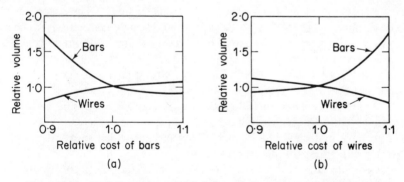

**Figure 15.19** Partially prestressed slab: influence of varying unit costs (15 m span)

accommodation of bars in any section, and the greatest midspan deflection are behaviour functions. The deflection is computed, allowing for steel stiffness and cracking.

An example of a search is shown in Figure 15.17. Generalized unit costs are applied[11], and it is assumed that the installed cost of wires is 4·16 times

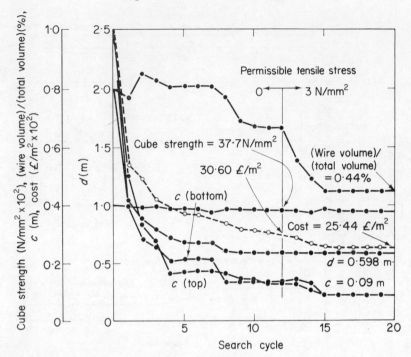

**Figure 15.20** Partially prestressed slab with local damage stresses active: costs and design parameters against search cycle (20 m span)

the cost of reinforcing bars for the same weight of steel[17]. The cost of formwork depends on the weight to be supported during construction. All costs can vary, and these values are used only as an example. Minimum concrete strength becomes an active side constraint. It is limited to this value by codes of practice, particularly to be sufficiently strong to resist end-anchorage forces and bearing stresses.

It can be seen in Figure 15.18 that prestressing wires become more predominant in the designs which have least cost as the span increases. For given unit costs of materials, the intensity of steel bars which supplements prestressing wires remains almost constant.

A sensitivity analysis showed that the optimization procedure attempts to compensate for any increase of a unit cost. For example, if the unit cost of bars is increased, fewer bars and more wires are used in the computed design [Figure 15.19(a)]. Conversely, if the unit cost of wires is greater, the reverse occurs [Figure 15.19(b)].

Unprestressed bars are only stressed effectively when strains are sufficiently high, and these cause small cracks in the surrounding concrete. The extent to which these cracks influence the durability of prestressed concrete is still

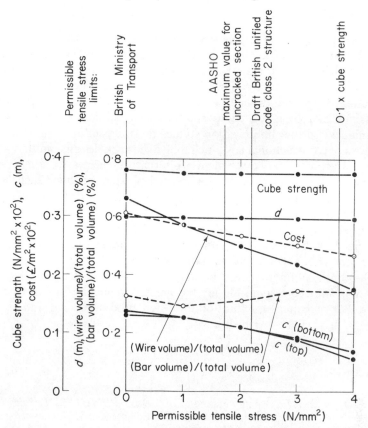

**Figure 15.21** Influence of local-damage permissible tensile stress on partially prestressed slab (20 m span)

under investigation. Cracks at working loads are usually limited by codes of practice by stipulating equivalent permissible concrete tensile stresses. These, together with permissible compressive stresses under working loads, can become local-damage constraints. An example of a search which includes them is shown in Figure 15.20. The search was initially terminated with a zero permissible tensile stress. The influence of a 3 N/mm² permissible tensile stress at working loads is demonstrated by the subsequent search. The influence of permissible values of concrete tensile stress on the design is shown in Figure 15.21.

Using these methods, it will be of interest to see whether there can be a continuous transition from solid slabs with reinforcing bars, through partial prestressing, to fully prestressed box sections (Figure 15.16), or whether the procedure must be discontinuous, and each type of design should be examined separately for a given span and loading.

**15.6   Closure**

As in any branch of the construction industry, a search, from a range of possible alternatives, for the most economic structure that will perform its function with adequate safety and durability requires more than the simple procedures that have been described in this chapter. These have assumed a structural shape and depend on general unit costs of labour and materials that have not been related to specific site conditions or constructional techniques. Further optimization studies must depend on a more accurate analysis of constructional cost, including formwork design, and they should include the preliminary assessment of alternative schemes, the use of precast units applying engineering economics and possibly methods such as geometric programming[19], which depend on simplified structural relationships.

Much research is still being directed to obtain a more complete knowledge of the structural behaviour of reinforced and prestressed concrete. This, combined with probability studies of load applications and materials variations, should contribute to further efficiency.

The continual improvement of management and construction techniques will no doubt be responsible for the greatest reduction of constructional costs, but, if better structural-optimization procedures can add to these savings, they will be worthwhile.

**References**

1. R. D. Logcher, J. J. Connor Jr. and M. F. Nelson, *ICES STRUDL II Engineering User's Manual*, Vols. 1 and 2, Dept. of Civil Engineering, Massachusetts Institute of Technology, Cambridge, Mass., 1968.
2. Genesys Center, *Genesys Subsystems Booklet*, University of Technology, Loughborough, Leics., Feb. 1971.
3. A. A. Palejs, 'Command language interface for automatic design of structures in three dimensions', American Concrete Institute, Fall Convention, St. Louis, Nov. 1970, and *AMECO Design in Concrete*, Ameco Computing Systems Company, Babylon, New York, 1970.
4. P. E. Mast, 'Computer program for the analysis and design of flat plates and continuous concrete frames', Publication SR017.01D, Portland Cement Association, Skokie, Ill., 1968.
5. J. P. Kohli, 'Optimum design of concrete spread footing by computer', *J. American Concrete Institute*, **65**, 384–389 (1968).
6. C. L. Freyermuth, 'Computer program for analysis and design of simple-span precast-prestressed highway or railway bridges', Publication SR033.10E, Portland Cement Association, Skokie, Ill., 1968.
7. A. T. Derecho, 'Computer program for the analysis and design of concrete wall-beam frames', Program description, special report, Portland Cement Association, Skokie, Ill., 1969.
8. G. G. Aperghis, 'Slab bridge design and drawing—an automated process', *Proc. Inst. Civil Engrs.*, **46**, 55–75 (1970).

9. G. Winter, 'Whither inelastic concrete design?', *Proc.*, *International Symposium on Flexural Mechanics of Reinforced Concrete, Miami, Florida, 1964 (Nov.)*, American Concrete Institute Publication SP-12, pp. 581–589.
10. 'The structural use of reinforced concrete in buildings', British Standard Code of Practice CP114, British Standards Institution, London, amended 1965.
11. 'Measured rates', *Civil Engineering and Public Works Review*, **65**, 1037–1039 (1970).
12. R. H. Wood, *Plastic and Elastic Design of Slabs and Plates*, Thames and Hudson, London, 1961.
13. R. H. Wood, 'The reinforcement of slabs in accordance with a predetermined field of moments', *Concrete*, **2**, 69–76 (1968).
14. O. C. Zienkiewicz, *The Finite Element Method in Engineering Science*, McGraw-Hill, London, 1971.
15. S. P. Timoshenko and S. Woinowski-Krieger, *Theory of Plates and Shells*, 2nd ed., McGraw-Hill, New York, 1959.
16. D. Bond, 'A computer program for studying the design of reinforced concrete structures supported on columns', *Proc. Inst. Civil Engrs.*, **43**, 195–216 (1969).
17. T. Y. Lin, *Design of Prestressed Concrete Structures*, Wiley, New York, 1963.
18. 'Specification for steel girder bridges, Part 3A, Loads,' British Standard 153: Part 3A: 1954, British Standards Institution, London, amended 1961.
19. A. B. Templeman, 'Structural design for minimum cost using the method of geometric programming', *Proc. Inst. Civil Engrs.*, **46**, 459–472 (1970).

# Chapter 16

# Structural-steel Design

J. Wright

## 16.1 Introduction

The 1950s represented the period of initial, intensive development of computerized structural analysis. During the 1960s the emphasis swung away from analysis towards automatic design. Developments in automatic structural design included the linear-programming, feasible-direction, penalty-function and dynamic-programming methods discussed already in Chapters 6–10. Iterative design procedures (Chapter 3 and Reference 1) also advanced, and more recently there has been progress in geometric programming[2], integer programming (Chapters 11 and 12) and in optimality-criteria-based procedures (Chapter 4). Despite this progress, it is still a long way from the time when one of the staples of modern civil engineering, structural-steel design, can be left entirely in the hands of a computer.

While the work on automatic design will continue, I believe that the coming decades are likely to be dominated by interactive design, graphics and sophisticated systems (such as Genesys[3]), all of which will enable the practising engineer to use analysis and design programs with a minimum of effort. Structural-steel design should be affected significantly. A review of programs concerned with the design of steel structures, in which I have personally been involved, and encompassing the aspects cited above, is presented in this chapter. A detailed review of developments elsewhere is not feasible within the confines of this chapter.

Section 16.2 contains descriptions of individual programs. Some of them were started during my stay at the University of Newcastle, and have been extended and improved by the Structural Computation Unit at the University of Warwick. The modifications which I consider necessary to update these programs are discussed in Section 16.3. The ultimate goal towards which the Structural Computation Unit is working is a complete system for structural steelwork. The philosophy behind the system, and the data-handling package on which it is based, are described in Section 16.4.

## 16.2 Existing programs

### 16.2.1 General

The frame and loading data required as input to all the following programs have been standardized. Input can be in either metric or imperial units or, indeed, any units at the discretion of the user; it is even possible to mix units at input. The user, however, must specify the units that he requires as output (either imperial or metric) and each has a standard format.

The loading considered at present can be either at a joint or between panel points. Point loads at joints can either be horizontal or vertical. Moment loading at joints is also possible. Loading of members can be either point loads, uniformly distributed loads, or moments. Both point loads and uniformly distributed loads can act horizontally, vertically or normal to a member. Another permissible location and direction in which the uniformly distributed load can act is along the length of a member but vertically downwards—a self-weight type of load. The elastic analysis program (which is described in the next section) requires different input for steel and concrete frameworks; it is thus necessary to input the material of construction.

### 16.2.2  Elastic analysis program

The method of analysis is the stiffness-matrix technique described by Livesley[4]. As far as I am concerned this technique is still the most efficient way of calculating an elastic analysis of a structure. However, certain sophistications are incorporated and include using the upper triangular portion of the symmetric stiffness matrix, a compact storage technique that eliminates the necessity for storing zero values, and a self-segmenting technique that is used when the size of the stiffness matrix is larger than the store available in the computer.

The input includes some basic parameters that allow the program to check the data input by the user, the details of restraints (which include all the foundations), details of the members in the frame, the geometry of the frame and all the loading details. The output from the program is the three deflections of each joint in the frame, and the moment, thrust and shear at each end of every member.

This program covers the analysis of any type of plane frame, including a fully rigid frame, a pin-jointed frame and a rigid frame with pin joints included. The restraints include elastic restraints and applied fixed deflections. Members are grouped together when they have the same section properties and a facility exists for including the self weight of the member if required. The program can recognize, from the input, if the self weight of a particular group of members is to be calculated and considered, and the program will include this additional load in every load case.

Various subprograms can be used with the basic analysis program to enlarge its capability. In practice, these subprograms are automatically included when the user requests the facilities at input. The facility of selecting steel or concrete, metric or imperial units at output, and the varying forms of input data are examples of the additional facilities.

A user can, if he so wishes, produce a line-printer output of the frame geometry. This feature is useful to the designer who wishes to check the validity of the input data presented. He can request a line-printer output of the position and magnitude of the maximum bending moment in any member. The user can also request and obtain a line-printer output of the bending moment, shear force and axial load diagrams for any group of members for each loading condition. At present it is only possible to call for this facility for any particular group of members under every load condition; it is not possible to discriminate between members and between load conditions.

The final feature of the additional facilities for the elastic analysis program is the ability to envelop all the loading conditions and output the position and magnitude of the maximum positive and negative bending moments, shear forces and axial loads. Figure 16.2 gives the envelope diagrams for bending moments in all the members for the frame of Figure 16.1 under the alternative loads shown.

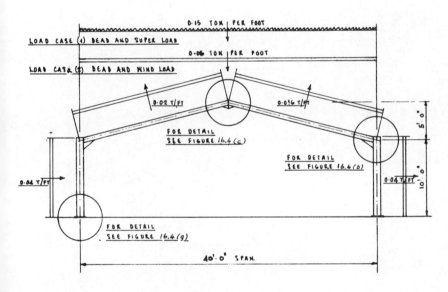

**Figure 16.1**

### 16.2.3 Automatic elastic design of steel frames

This program, which outputs a design in terms of manufactured sections, has been in operation for several years and has been used successfully by practising engineers. It has been described in detail by myself[1] and is outlined only briefly in this section.

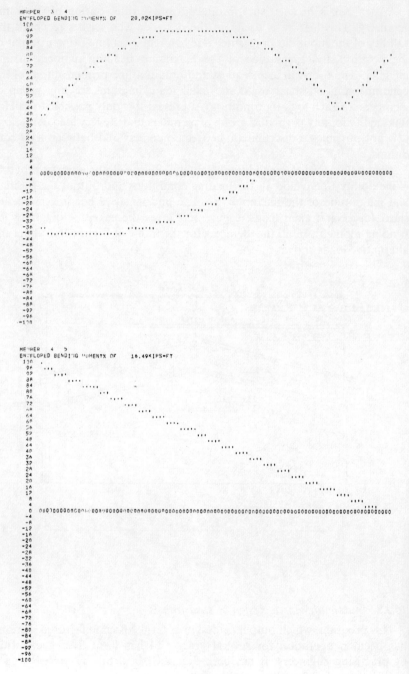

**Figure 16.2**

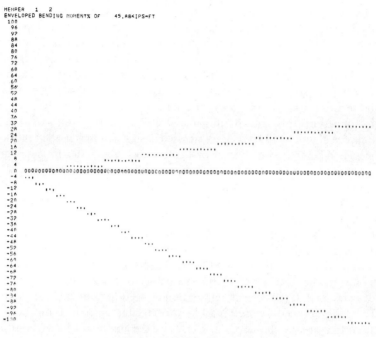

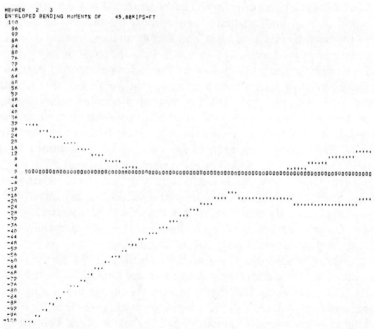

The input for this program is similar to that required for the elastic analysis program already described, with the addition of the effective lengths for every member (which can be grouped if several are similar), a directory of sections arranged in an order of preference, a grouping of members which are to be of the same section, and a design parameter called Simin. A user chooses a value for Simin between 0 and 1. The choice of zero gives an analysis with the appropriate *stress index* for each member, defined as the sum of the ratios of actual stress and permissible stress in bending and axial loading. The choice of unity will give a fully stressed design. Simin is used in the design process and compared with the worst calculated stress index for every group of members. Appropriate action is taken based on this comparison.

A standard directory of sections, arranged in order of weight, is available for use, but a user can prepare his own directory and its order is at his discretion. A facility exists for him to delete or add additional members. The standard directory includes the 105 standard universal beams and columns normally detailed in handbooks. Any of these sections can be omitted and other sections added. Several users have added other directories that are private and therefore not available to other users. These special directories include tubes, rectangular hollow sections and 'autofab' sections.

The design procedure used is an iterative technique, and commences with sections that are the choice of the user, or, in the absence of such a choice, the smallest section in the directory. The program takes each working-load condition in turn and calculates and outputs the deflections and forces. It determines the maximum axial load and maximum bending moment for each member of the frames, and then calculates the stress index.

The program repeats this procedure for each working-load condition in the frame, printing out, for each working-load condition, the deflections, forces, the maximum axial load, the maximum bending moment and the maximum stress index in each member. After completing the final loading condition, it summarizes the condition of stress by selecting the highest value of stress index from each of the working-load conditions for each member, and outputs this with the appropriate working-load case. It then considers each group of members and selects the worst stress index for that group. The program continues by comparing this 'critical' stress index with the value of Simin already mentioned and decides if it should search for another section. The program selects the optimum section that will carry the maximum loads already determined and repeats this procedure for every group. It substitutes any modified section for all the members of the particular group of members in the frame, making the necessary alterations to the stiffness matrix, and modifies the loading vectors for the change in self weight.

The analysis and critical-stress-index determination cycle is repeated until the program is unable to modify any group of members. This constitutes a final design which is then output by the computer.

```
SUMMARY OF CRITICAL LOADINGS FOR EACH MEMBER

MEMBER   MOMENT     AXIAL    STRESS INDEX   CASE   GROUP
 1-     -168.54     3.04        0.90         1       1
 4-      168.54     3.04        0.90         1       1
 2-      168.54     2.12        0.98         1       2
 3-     -168.54     2.12        0.98         1       2


GROUP  1--CRITICAL MEMBER    4/ 5   STRESS INDEX IS   0.90
GROUP  2--CRITICAL MEMBER    2/ 3   STRESS INDEX IS   0.98

         DESIGN TERMINATED                        SUMMARY OF SELECTED SECTIONS :-
GROUP  1 SELECTED SECTION IS  10 X  5.75 X   21.0  LB/FT   STEEL WEIGHT IS   420 LB
GROUP  2 SELECTED SECTION IS   8 X  5.25 X   20.0  LB/FT   STEEL WEIGHT IS   767 LB


TOTAL STEEL WEIGHT IS
 0.530 TONS
```

Figure 16.3

Figure 16.1 shows a frame which has been designed by this program. Figure 16.3 shows a summary of the design cycle and the computer output of the final design.

### 16.2.4   Plastic analysis and design

This program uses the technique of steepest descent described by Livesley[5]. It works in three different modes:

(1) A load factor analysis under any given load condition.
(2) A minimum-weight design for a single-load condition.
(3) A minimum-weight design for a multiload condition.

The input required in addition to the data already specified in the analysis program (see Section 16.2.2) is the fully plastic moment capacity of each group of members for the first mode (load-factor analysis). For the second and third modes (design modes), the relevant load factor must be included in each working load condition. This program has been described fully elsewhere[6] and will not be detailed here.

### 16.2.5   Elastic stability check

A program is available to check the elastic stability of a plane frame. It is used in combination with the plastic design program to assess the reduction in elastic stability with the formation of plastic hinges under an incremental load.

The program gives the load factor against collapse, taken as a factor of all loading or of a particular type of load (e.g. dead-load static factor against collapse for live load). Any number of degrees of elastic instability freedom may be taken into account. For example, torsion may be included or excluded from the analysis.

The approach used in the calculations is a modified form of the method proposed by Livesley[7]. A distinction is drawn between the buckling of individual members and the elastic instability of the frame as a whole. The solution technique involves a repeated analysis of the stability of the frame at a sequence of load factors. The stability is assessed by the evaluation of the change of signature of the stiffness matrix as it is modified by the stability functions. The signature of the stiffness matrix gives an indication of the mode of collapse of the frame, provided that the existence of local buckling is correctly represented in the stiffness matrix (e.g. pin-ended members).

Using the rapid equation-solving techniques developed for the elastic analysis, the solution is relatively quick, requiring, typically, six to ten linear analyses.

*16.2.6    Connexion design*

A program has been written to design joints in a steel frame. This program is a pilot version and only covers the design of the types of joints shown in Figure 16.4. This work has already been published[8] and will not be described in detail here. The input to this program is not sophisticated, but is complicated by having to rely on other programs for details of member sizes and joint loading. Under multiloading conditions, envelope joint loading is required.

It is not intended to use this program as an individual program, because of the complications of input, but only to use it as a suite of subprograms in the system, and consequently it will require no individual input. All information will come from the analysis and design programs.

It is necessary to supply this program with information on materials to be used, and this data is separate from the main program to allow easy updating and modifications. The information includes all details of, and permissible stresses in, bolts, welds, plates, brackets, etc. The user can, at input, specify a particular bolt size he wishes to use, and he must specify the particular type of joint he requires.

As this is a pilot program, there are several limitations to its scope. These include restricting fabrication to shop welding and site bolting only; high-strength friction-grip bolts are used in an arbitrary manner when the shear to be resisted exceeds 20 ton; the bolt sizes are restricted to $\frac{3}{4}$ in, $\frac{7}{8}$ in and 1 in; the program is restricted to imperial units.

The available connexion designs are shown in Figure 16.4. Figures 16.4(a) and (b) show beam/column joints; Figure 16.4(c) shows an apex joint. Figures 16.4(d), (e) and (f) show column splices, and Figures 16.4(g), (h) and (i) show column bases. Figures 16.4(b), (c) and (g) are the computer design for the knee, apex and column base, respectively, for the frame shown in Figure 16.1.

*16.2.7    Price estimation*

Again, this pilot program is described in References, and it has been written in conjunction with the connexion design program to produce an overall price for the frame. This program cannot be used on its own as it is dependent on the member-design program (Sections 16.2.3 and 16.2.4) and the connexion-design program (Section 16.2.6) for fabrication details on which to base a price. When this program is incorporated into the system it will however, be possible to produce a price estimation for members only in a given frame (omitting the pricing of connexions) by inputting the frame and member details only.

To use this facility, it is necessary for the designer to build up a personal costing matrix based on his own costing procedure. A typical costing matrix

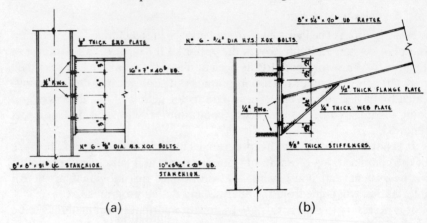

(a)                    (b)

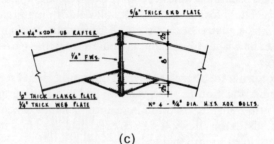

(c)

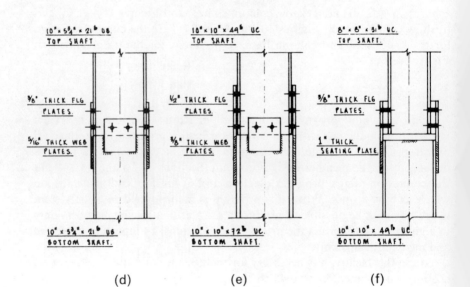

(d)            (e)            (f)

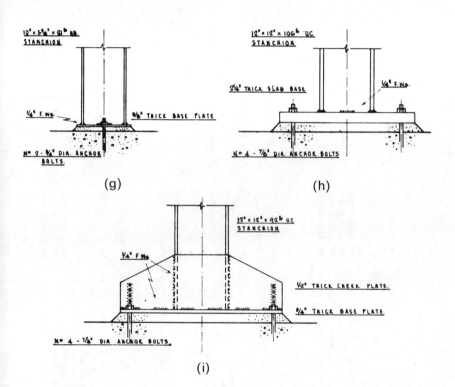

(g)                              (h)

(i)

**Figure 16.4**

CASE NO 1
============

40 FT SPAN PORTAL FRAME

MEMBER 1
========

NO 1 - 10 X 5.75 X 21 UB X 10.667 FT LG   WT 0.0999 TON    & 5.593

END 1 (SB)
----------

NO 1 - 12.000 X 0.375 PLATE X 12.000   IN LG   WT 0.0068 TON   & 0.402
NO 2 - 0.750IN DIA MS HD BOLT X 12.0   IN LG   WT 0.0016 TON   & 0.300
NO 2 - 0.750 IN DIA FLAT WASHERS        IN LG   WT 0.0001 TON   & 0.002
NO 2 - 4.000 X 0.375 PLATE X 4.000    IN LG   WT 0.0015 TON   & 0.091
NO 1 - RUN 0.2500 FW X 37.0

END 2 (NO)

NO CONNECTION DETAIL

MEMBER 1
NO 2 FRAMES AS LISTED
============================

                     0.2198 TON    0.2198 TON   & 12.773   & 12.773   & 1.572
               ----------    ----------   ----------

MEMBER 2
========

NO 1 - 8 X 5.25 X 20 UB X 20.500 FT LG   WT 0.1829 TON    & 10.308

END 1 (KCF)
-----------

NO 1 - 5.875 X 0.750 PLATE X 16.750   IN LG   WT 0.0093 TON   3 0.532
NO 4 - FITTED STIFFENERS OUT OF
NO 2 - 4.000 X 0.375 PLATE X 7.000    IN LG   WT 0.0027 TON   & 0.159
NO 1 - RUN 0.2500 FW X 80.0          IN LG
TOTAL STIFFENER WELD LENGTH
NO 1 - RUN 0.2500 FW X 45.0         IN LG
NO 6 - 0.750 IN DIA HY XOX BOLT X 1.75 IN LG   WT 0.0012 TON   & 0.528
NO 6 - 0.750 IN DIABEVEL WASHERS       IN LG   WT 0.0002 TON   3 0.008
NO 1 - 7.000 X 0.250 PLATE X 16.000   IN LG   WT 0.0035 TON   & 0.228
NO 1 - RUN 0.2500 FW X 13.0        IN LG
NO 1 - 5.375 X 0.500 PLATE X 16.000   IN LG   WT 0.0054 TON   & 0.315
NO 1 - RUN 0.2500 FW X 16.0       IN LG

```
END 2 (ACF)
===========
NO  1 -  5.875 X 0.625 PLATE X  13.750  IN LG   WT 0.0064 TON   @  0.367
NO  1 -  RUN 0.2500 FW X 40.0           IN LG   WT 0.0009 TON   @  0.353
NO  4 -  0.750 IN DIA HY XDX BOLT X 2.25 IN LG  WT 0.0002 TON   @  0.003
NO  4 -  0.750 IN DIA FLAT WASHERS      IN LG   WT 0.0015 TON   @  0.101
NO  1 -  6.000 X 0.250 PLATE X   8.000  IN LG
NO  1 -  RUN 0.2500 FW X  8.0           IN LG
NO  1 -  5.375 X 0.500 PLATE X   8.000  IN LG   WT 0.0027 TON   @  0.159
NO  1 -  RUN 0.2500 FW X  8.0           IN LG
NO  1 -  RUN 0.5000 FW X  5.0           IN LG


MEMBER 2
NO  2 FRAMES AS LISTED
======================
                                    0.4342 TON    @  26.123    @  26.123    @ 14,234

TOTAL STEEL WEIGHT                  0.6539 TON

TOTAL MATERIAL COST                                            @  38.897

TOTAL FABRICATION COST                                         @  15.807    @  15.807
OVERHEAD ON FABRICATION                                        @  39.516
STEEL PRICE UNPAINTED EX SHOP                                  @  94.220
```

Figure 16.5

is not included in this paper, as it is, essentially, a matrix of numbers that would be meaningless unless accompanied by a long explanation.

The final estimated price, including the pricing of connexions, is based on all material costs and fabrication charges. It does not include erection charges, transport charges and painting.

The frame of Figure 16.1 is used to illustrate the pricing output of the program. The elastic design of member size has been described in Section 16.2.3 with the computer output shown in Figure 16.3. The connexions were designed as described in Section 16.2.6. The price estimation program gave a complete listing and price for this frame as shown by the computer output in Figure 16.5.

### 16.2.8  Automatic drawing and detailing

Although the Structural Computation Unit at the University of Warwick has not yet been able to do any work on automatic drawing and detailing, computer work on this subject is possible. To my knowledge, the best example of this type of work is carried out by Construction Modulaires, based in Paris. This firm specializes in a building system consisting of a predetermined kit of parts based upon the Clasp system.

The program developed by this firm actually produces all the drawings necessary for the fabrication of the buildings designed in accordance with the system. These drawings include the foundation drawings, layout drawings, all floor plans, elevations, cross-sections and column details. The same program also gives line-printer listing of all the parts required by the particular structure, updates and keeps a record of stock control, and gives individual prices for each member and the price for the completed structures.

Because this system is based on a standard module, no details are required for the beams. These are designated by numbers and are limited in range. The columns, however, are designed individually, and details of every column with all connexions are necessary. Examples of drawings by Construction Modulaires are included in Figures 16.6 and 16.7.

### 16.3  Proposed extensions to existing programs

The existing programs have been developed over a number of years, and their extensive use has shown up a number of shortcomings. Some of the modifications which would be desirable, but have not yet been implemented, are listed in Section 16.3.1.

### 16.3.1  Automatic elastic design of steel frames

(a) The program works only with one type of steel, i.e. the designer must choose between grades 43, 50 or 55, and the program will use that steel only.

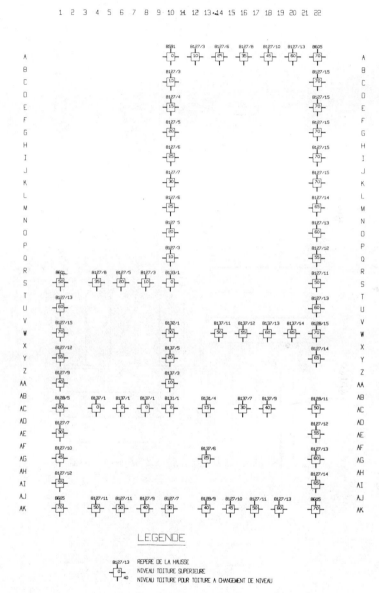

**Figure 16.6**

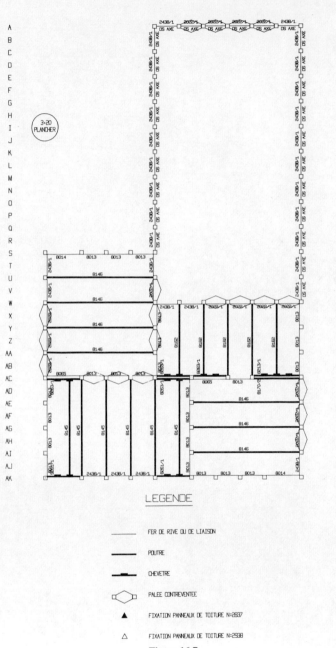

LEGENDE

| | |
|---|---|
| ———— | FER DE RIVE OU DE LIAISON |
| ▬▬▬ | POUTRE |
| ▬■▬ | CHEVETRE |
| ◇ | PALEE CONTREVENTEE |
| ▲ | FIXATION PANNEAUX DE TOITURE N=2837 |
| △ | FIXATION PANNEAUX DE TOITURE N=2938 |

**Figure 16.7**

A proposed amendment is to work in all three steels simultaneously, choosing the most economical material for any particular group of members.

(b) A severe limitation is the use of one directory. Only one directory is permissible for use with any one problem and, using this list, the user has three possible choices: he can use two different types of section in the directory and specify at input that the selected sections must be the best of one type of section, or the best of the other type of section, or the best of the combined directory. In the standard directory, he has the choice of a universal beam, a universal column, or the best of both. It should be possible to extend this facility to use several directories, including angles, double angles, universal beams, universal columns, channel sections, 'autofab' sections, castellated sections, etc. The designer would then have the opportunity of selecting the types of sections that he would be prepared to use from the whole chosen range and allow the program to select the most economical.

(c) There is no provision for the 100 mm eccentricity criterion for a pin-ended beam as laid down in Clause 34a of BS 449[9]. This would require a relatively easy modification.

(d) The loading conditions supplied must be given as working load conditions, i.e. each load condition specified at input must represent a complete working load on a structure that is to be considered by the program. It should be possible to feed in basic load conditions and instruct the program which combinations are required. The program should have the choice of whether to solve basic load conditions, making up the working-conditions solutions after obtaining the basic load solutions, or alternatively working out the working-load conditions initially and obtaining solutions for these. The final solution required by the computer will be the solution of the working-load conditions; it may be more economical, however, to solve for the basic load conditions and add the solutions than to use the present technique.

(e) The program considers only bending about the major axis. Consequently the effective length specified by the user at input is the effective length about the minor axis. The program does not consider the slenderness ratio about the major axis. There are often cases when a user would wish to consider an effective length about both axes. This facility needs to be added.

(f) The program considers only in-plane loading. It is proposed to add joint loading in the plane perpendicular to the frame, i.e. loads coming from minor-axis beams.

(g) The program does not give any consideration to the depth of a section. Facilities should be added to limit depths of beams (when considering headroom) and to consider practical restrictions on column sizes. There are two categories to consider here: when a column has a crane and roof shaft and there must be a certain gap between them at their junction to allow for a crane rail, and, in the case of a stanchion joint, when the program must make sure the top stanchion is equal to or less in depth than the lower stanchion.

(h) As the program allows bending about the major axis only, modifications are necessary to allow members to be placed so that bending in the plane of the frame occurs about their minor axes.

### 16.3.2 Plastic design

The output from modes (2) and (3) of this program (see Section 16.2.4) is in terms of the fully plastic moment requirements for each group of members. The program does not select the section. Work is presently proceeding on a method of assessing the axial load in each member of the structure at collapse. This value will be used in conjunction with the calculated fully plastic moment to produce a minimum plastic modulus requirement for each group of members. The section will then be chosen from the same directory of sections used in the elastic design program.

### 16.3.3 Connexion design

It will be obvious that this program will never be really complete. Particular users will suggest different types of joints because of their particular fabrication techniques, and these will be added. The possible range of joints is almost limitless, but additional joint designs could be added easily at the request of individual users. The pilot version of this program has been written using advice from the structural-steelwork industry on 'typical' joints. Fabrication techniques, which are at present limited to shop welding and site bolting, can be extended, if required, to include shop riveting and site welding.

## 16.4 The structural-steelwork system

### 16.4.1 The purpose of the system

Sections 16.2 and 16.3 describe some programs that are applicable to Structural Steelwork that have been developed at the Universities of Warwick and Newcastle. Each program has its own method of reading input data, storing the data within the program (termed 'data structure') and outputting results.

The traditional view of computing is that, if a program exists, it performs some specific calculation. Input is supplied to the program and the program will produce output. The emphasis is on the *supply* of data (which the computer will reject if it is not in exactly the form specified) and the *production* of results that are a mathematically exact answer to the problem posed. It *may* not be a sound practical solution.

In executing a design, an engineer will use a selection of available programs —the choice of programs and the order in which they are to be used will be at his discretion. He must usually prepare different data for each program.

A more logical approach to the design process would be for the engineer to start with some, possibly vague, ideas of the design, which he inputs to the computer. This information, however, minimal, would be stored as a numerical model within the standard-data-structure model (which may also contain behaviour data such as permissible joint deflections). It would then be transformed by applying analysis or design procedures, or by explicit instructions from the engineer. The emphasis would then be on the numerical model that evolves within the computer until the engineer is satisfied that he has attained a satisfactory design. This approach implies that the engineer is able to interrogate the computer to ascertain the evolution of the design.

This approach has a number of advantages. As the frame is always stored as a standard data structure, the input need only be prepared once. The application routines are then called to operate directly on the data structure, and not on input data. The engineer will only obtain output when it is required—either at the completion of the design or intermediate results at his discretion.

The application programs need not contain any input or output software, thereby avoiding duplication of preparation of input data. A large proportion of the existing programs described previously is concerned with the interpretation of input and output data. Furthermore, since the input and output software need only be developed once, it can be extended to include input via a cathode-ray tube or teletypewriter and output via an incremental plotter.

With the facility for applying several application programs to one problem, there are two modes of use. The problem can be run in batch mode: for example, an engineer may ask for a plastic design, followed by an elastic analysis for deflections and a stability check. The engineer would have to submit the necessary data cards, including commands for the order and choice of application programs, and will then receive the output he requests. Alternatively, the problem may be run online from a teletypewriter terminal or graphical display, when the designer will control the program implementation according to any monitored results.

### 16.4.2 The existing system

The basic design of the system described in the previous section is complete, and the first application programs have been implemented.

The lowest (i.e. base) level within the system is the data-handling package Warden (Warwick Data Engineering) (Larcombe, 1972). The basic element of storage in Warden is the *bead*, and all other subprograms within the structural steelwork system used beads for storage. The Warden routines supply beads as required, handle transfer of beads to the backing store, etc., so that the writer of subprograms for the system need never be concerned with data manipulation such as core-store shortage, data transfers, space allocation, etc.

The data structure in use at present is that developed for use with a cathode-ray-tube (c.r.t.) display and light pen; its size can be altered as the design is modified. Also the data structure can be moved to and from the backing store automatically.

Input to the data structure at present is via cards and via the interactive graphics package. The only applications routine that has been fitted into the system, at the time of writing, is a version of the elastic analysis program described in Section 16.2.2.

### 16.4.3   Operation of the system

An example of the use of the system is provided by considering the operation of the interactive graphics package. Input data of the frame geometry is supplied by any of three different techniques:

(1) A card input is prepared in a manner similar to that already described for the other computer application programs.
(2) A standard frame, the geometry of which is automatically generated by the computer, is selected and displayed on a visual-display unit (v.d.u.); this standard frame is amended to suit the requirements of the user.
(3) The frame is sketched using a light pen. The geometry of the frame can be studied on the v.d.u. and amended until a satisfactory frame is obtained.

Having obtained the basic frame, the user may then add the restraints and pins at the ends of members, and size the members using a directory. Any number of working-load conditions are then input using the light pen or card input, and the frame can be analysed. The user can then study the behaviour of the frame by examining the axial-load, shear-force and bending-moment diagrams. He can also view the deflected-form diagram and the stress-index diagram.

The user has the facility to change any of the basic information, including dimensions, member sizes, geometry and loads, at any time during the study of any frame. He can also change restraints and foundation types, add or remove pins, and add or remove members. Reanalysis is carried out at the request of the user, and the new behaviour diagrams are inspected on the screen.

At the end of a run, the user can plot any of the diagrams on a digital plotter so that he can take away a 'hard copy'. He may also file the job on backing store so that he can return to continue work on it later.

I believe that this is the most sophisticated program of its type available and is the tool of the future for the designer.

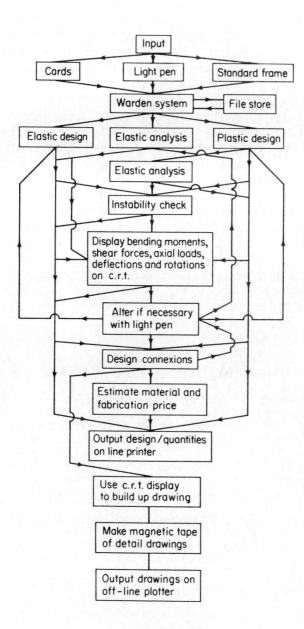

**Figure 16.8**

### 16.4.4    Outline of the complete system

The applications programs discussed in Section 16.2 will be rewritten and incorporated into the system. After the modifications to the existing programs discussed in Section 16.3, the task of incorporating the programs into the system will be a routine task of changing the storage to use the Warden system and adding the appropriate commands to the batch and interactive control routines.

Figure 16.8 shows, in flow-diagram form, all the complete operations envisaged in the system. Work has commenced on the development of this system, and the Warden storage system is working. The graphics programs discussed previously have been incorporated into the system using the Warden storage techniques.

The completed system will enable the designer to monitor completely the whole of the design process. The 'state of the art' is a long way from the time when the designer can input the bare details of a building and have as output, the finished design, complete details and a schedule of prices to be used for a quotation. It is hoped that this system will serve as an intermediate state in which the designer can interrupt the design process at any stage and make necessary amendments, with the program only proceeding at the command of the designer.

### References

1. J. Wright, 'Automatic design of plane steel frameworks', in *Proc. of the IV IKM, Weimar, DDR.*, Vol. 1, 1967, pp. 302–310.
2. A. B. Templeman, 'Structural design for minimum cost using the method of geometric programming', *Proc. Inst. Civil Engrs.*, **46**, 459–472 (1970).
3. D. G. Alcock and B. H. Shearing, 'GENESYS—an attempt to rationalise the use of computers in structural engineering', *The Structural Engineer*, **48**, No. 4, 143–152 (1970).
4. R. K. Livesley, 'Analysis of rigid frames by an electronic digital computer', *Engineering*, 176 (1953).
5. R. K. Livesley, 'The automatic design of structural frames', *Quart. J. Mech. and Applied Math.*, **IX**, Pt. 3 (1956).
6. J. Wright and J. P. Baty, 'Plastic analysis and design of rigid plane frameworks subjected to multi-load conditions', in *Proc. of Int. Symp. on Use of Computers in Structural Engrg., University of Newcastle upon Tyne*, Paper 6.1, 1966.
7. R. K. Livesley, *Matrix Methods of Structural Analysis*, Pergamon Press, 1964, Chap. 10.
8. J. Wright and T. M. Day, 'Automatic design and price estimation for steel frames', *Civil Engineering and Public Works Review*, **66**, 1091–1100 (1971).
9. 'Specification for the use of structural steel in building', British Standard 449: 1959, British Standards Institution, London.

*Chapter 17*

# The Use of Interactive Graphics in Engineering Design

*F. A. Leckie, G. A. Butlin and M. J. Platts*

## 17.1 Introduction

The purpose of the previous chapters has been to discuss the theory and application of methods for obtaining, in some sense, optimum solutions to defined structural problems. The general procedure is to employ the computer to obtain those conditions that define an optimum of some objective function, subject to certain constraints. However, since the mathematics is so specialized, it is unlikely that the expertise required to choose the appropriate procedure will be in the library of knowledge of even the most aware designer. In addition, any solution offered by optimization procedures should satisfy the condition that a true global optimum has been obtained, since a local optimum may yield parameters well removed from those corresponding to the global optimum. Consequently, the traditional method of obtaining the overall 'feel' of a problem by carrying out a number of case studies will certainly continue as an important design procedure, and it is in this connexion that the use of interactive graphics may prove to be of practical value.

As a result of the recent dramatic advances in analytic and computing procedures of structural analysis, there are now very few structural problems that cannot be solved. In spite of this happy state, however, it can still be very difficult for the designer to gain from the computer the insight required when assessing his total design. Clearly, this insight can be strengthened if the output of the computer can be presented in a form that allows easy interpretation of the analysis, if it is easy to make changes in magnitude and topology, and if those changes can be quickly analysed and their significance interpreted.

The initial difficulties that arise with the production of masses of numerical output have been largely circumvented by the introduction of graph plotters and storage display screens. Even with these facilities, however, the deficiencies of batch processing can be very real. A simple example of this deficiency is illustrated in Figure 17.1, which shows some of the geometric parameters associated with the design of the flange of a pressure vessel. The analysis can be readily performed, but the study of the best proportions of flange size, bolt and seal position can be a task of several weeks duration, if it is made within the limits imposed by normal batch processing. It is clear,

*Optimum Structural Design*

in this example, that dynamic interaction with the computer by means of efficient forms of input and output would help the designer enormously, and it is not difficult to think of other situations when this would be the case.

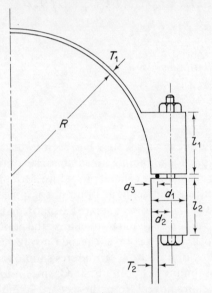

Reproduced by permission from G. A. Butlin and F. A. Leckie, 'Response time in the application of interactive graphics in structural analysis', in *Proceedings of the Symposium of the International Union of Theoretical and Applied Mechanics, Liège, 1970,* University of Liège, 1971, pp. 509–519

**Figure 17.1**   Hemisphere–cylinder flange intersection

To achieve this level of interaction, however, is a rather modest ambition, compared to that required to meet the needs of the designer who does not wish to be constrained by a fixed topology and who wishes to investigate the effect of changes in both topology and component character. The thoughts and reactions of the imaginative designer must vary beyond description, but we can be certain that the facility of greatest use will provide an easy means of making changes and interrogating the corresponding results at any relevant point in the structure.

To achieve such flexibility, it is necessary to design computer systems so that changes in both topology and dimensions can be readily made. In addition, it must be simple to interrogate the appropriate parameters at any chosen point in the structure so that the effect of changes can be quickly appreciated. The design of the system to perform these functions depends on a number of interrelated features. First of all, the relationship of the parameters describing the structure must be studied so that a data structure that provides the design flexibility required can be formulated. Since the computer hardware and operating system place strong restraints on the design of such a system, considerable computer, as well as structural, expertise is required to ensure a successful outcome.

Another problem arises in complex problems, when the time involved in obtaining a solution can be substantial and the advantage of the interactive design is lost if the designer has to spend lengthy periods of idleness waiting for a new set of results. Experience in our own laboratory suggests that a few seconds is the maximum time to avoid irritation. It would appear that two methods can be used to ease this problem. One method is to use a simple, coarse model of the structure which can be readily analysed, and the other is to arrange to use some of the well known approximate methods that involve a minimum of calculation and yield designs that are somewhat conservative. Having gained an overall picture of the problem, the designer can then proceed to refine his design in a series of small discrete steps, each step making only small demands on the computer.

In this chapter an account is given of some of the approaches which have been developed in a research programme into these topics. The first topic is concerned with the use of finite-element procedures to obtain limit loads of plate structures. In this method stress and displacement fields are used to obtain lower- and upper-bound values, respectively, of the limit load. It is found that interactive graphics provides a useful facility for obtaining initial values to insert into standard optimization routines. The second topic is similar in flavour, and is concerned with the location of the yield lines to obtain the collapse load using the yield-line theory of slabs. The graphical procedure is used to obtain the optimum load, and extended further to obtain the optimum position for the location of a hole within a slab.

The third topic is concerned with the layout of high-density housing in which factors such as the amount of daylight, sunlight and privacy provide strong constraints on high-density solutions that are aesthetically pleasing. This is a problem in which a large number of factors interact on each other in an increasingly complex manner as the number of dwellings on the site increases. In this study the computer is used in the unusual and interesting role of producing suitable solutions to the problem, and it is shown how the designer can make best use of the various alternative solutions offered by the computer.

The next topic is concerned with the design of prestressed slab bridges, using a coarse finite-element mesh. It is shown how the easy manipulation of design and the form of the graphical output combine to result in improved design.

The problems of achieving rapid interactions in complex structural situations is covered in the next sections. The study is concerned with the design of an interactive finite-element system that is capable of assembly and modification by addition, deletion and parameter change. Interaction time is kept down by calculating the effect of discrete changes.

There has been considerable and widespread effort put into studying the means of representing the geometry of bodies and surfaces by finite patches. There is a close relationship between this work and recent advances in finite-element research. This relationship is briefly discussed, along with the possible role to be played by interactive graphics.

## 17.2   Finite-element limit-load analysis[1]

The limit-load concept plays an important role in design for structures made of material that can suffer both elastic and plastic strains. Although powerful limit-load calculations can often be completed very easily, occasions do arise when it is necessary, because of structural complexity, to use finite-element methods. Following the usual procedures, the structure is divided into a number of elements, and it is assumed that the stress or strain fields have a simple form within each of the elements. When the stress-fields method is used, the stress fields must satisfy equilibrium and lie within the yield condition for the material. The free parameters of the stress field are chosen to maximize the load subject to the constraint imposed by the yield condition, i.e. the procedure is one of non-linear optimization subject to constraints. The optimum obtained is a lower bound of the true limit load. The finite-element method using strain fields consists of selecting the free parameters to minimize the limit load. This again is a problem in non-linear optimization, but this time without constraints, and the solution this time gives an upper bound to the correct value. Consequently, by using the stress-field and the strain-field methods, it is possible to bracket the true value of the limit or collapse load.

A major difficulty with optimization procedures is that poor convergence can arise when the starting conditions are far removed from the optimum conditions. In this example it is shown that the knowledge and experience of the engineer can play a large part in obtaining a rapid convergence to a satisfactory solution. With the help of the graphical display, which provides instantaneous access to the computed results in an easily comprehensible form, and his knowledge of structural behaviour, the engineer can arrive at a reasonable solution. During this process, the engineer himself will be doing the optimization by changing various parameters on the screen, and the computer will be merely calculating and presenting the bounds for the load factor corresponding to the stress or velocity fields presented to it. With an initial solution thus arrived at, an optimization routine can be entered, and the results continuously displayed for the assessment of the engineer. At any stage of this process he can step in to make any adjustments necessary to avoid a local optimum.

As an example for interactive limit analysis, the limit load for a square cantilever (Figure 17.2) is determined. Lower and upper bounds are found

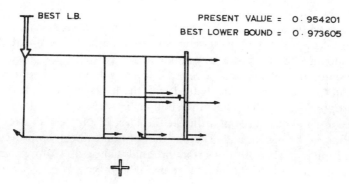

**Figure 17.2**   Display of stress fields

using the von Mises yield condition.

For the interactive analysis, the free parameters are displayed as pointed arrows at relevant positions (Figures 17.2 and 17.3). These are varied by changing the length of the arrows on the screen by using the light pen. Options (chosen by using the buttons provided) are incorporated in the program. The user is able:

(a) to vary parameters without performing bound calculations,
(b) to calculate and display the bound for a selected set of parameters,
(c) to vary the parameters with continuous calculation and display of the bound,
(d) to enter into an optimization subroutine (Rosenbrock[2]) with a selected set of values for the parameters and display the results, or
(e) to obtain a printout or a plot of the current results.

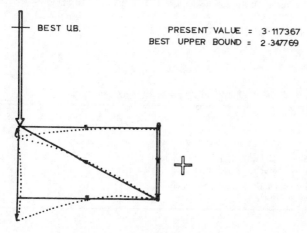

**Figure 17.3**   Display of displacement fields

Throughout the analysis, the bound corresponding to the current set of parameters on the screen and the best bound found so far are displayed. The latter and the associated parameter set are continuously updated and stored, and may be called on at any moment. The current value of the bound is represented by a broad arrow (referred to hereafter as the bound arrow) resting on the cantilever at the loaded end, the length of the bound arrow being proportional to the value of the bound. The best value so far found is indicated by a short horizontal line at the level of the bound arrow corresponding to the best bound. Thus, as the solution is being improved, this line segment is pushed up (lower-bound analysis) or down (upper-bound analysis) by the end of the bound arrow. This is easier than comparing displayed numerical values.

The particular example considered clearly shows the importance of knowing the physical nature of the variables concerned in limit analysis, and the usefulness of the graphical display in utilizing this knowledge. Table 17.1 shows a considerable improvement in the results obtained by interactive analysis over the results from direct optimization. The latter solution probably has reached a local maximum which is much below the global value. The dependency of the attainable stationary point on the

Table 17.1  Lower-bound analysis results

|  | | Bound for $2WL/M_0$ | Time (min) |
|---|---|---|---|
| Interactive analysis | Maximization on the screen, starting from zero field | 0·874 | 6·5 |
| | Using Rosenbrock's method, starting lower bound = 0·874 | 0·947 | 2·5 |
| Direct maximization using Rosenbrock's method, starting from zero field | | 0·742 | 2·6 |
| | | 0·759 | 17·1 |

Table 17.2  Upper-bound analysis results

|  | | Bound for $2WL/M_0$ | Time (min) |
|---|---|---|---|
| Interactive analysis | Minimization on the screen | 1·979 | 10·0 |
| | Using Rosenbrock's method, starting upper bound = 1·979 | 1·531 | 1·3 |
| Direct optimization using Rosenbrock's method, starting from zero field | | 109·277 | 0·8 |
| | | 109·073 | 3·8 |

initial starting point is clearly shown by the results in Table 17.2. The direct optimization starting from the origin gave a result which was highly unacceptable, whereas very considerable improvement was obtained by using a more realistic starting solution. Although, in terms of computer time, the direct optimization seems to overtake the interactive analysis when both methods are used with the same initial field, the use of a time-sharing system will considerably cut down the central-processor time used in the interactive analysis. Some display users have quoted a figure of 10 per cent as the central-processor time used in their graphical display work.

In the present analysis, optimization on the screen was done more or less as a sectional search (in the upper-bound analysis two parameters at a time could be varied continuously), and means have to be found to vary several parameters at a time in a specified way determined by previous experience. With the availability of three-dimensional modelling facilities, the user can make better choices of the parameters at his disposal. Thus the man and the machine can work as a team to arrive at a solution better than that obtained by either one alone.

## 17.3 Yield-line analysis

This program has been developed to determine the collapse loads of slabs and the corresponding yield-line pattern. A typical example is shown in Figure 17.4, in which a slab built in on two adjacent sides and simply supported on the other two is depicted. Other problems can be readily set up by selecting the appropriate options on the screen and adjusting the size by a graphical digital device.

The yield-line pattern can be adjusted by moving the nodal points by means of the tracking cross; the yield load is calculated automatically and the magnitude of the ratio $M/p$ is indicated by the arrow on the left hand side of Figure 17.4. In this ratio $M$ is the yield moment of the slab and $p$ is the distributed load. Hence the yield line pattern is adjusted until its location corresponds to a maximum value of $M/p$. The setting up of the problem and the determination of the yield-line pattern takes a maximum of 7 min, of which about 15 per cent is central-processor-unit time.

An interesting problem in optimization arises when the question is posed of how to place a hole of given dimension to minimize the slab thickness. This involves an optimization within an optimization. First the yield line pattern for each hole location must be determined, then the hole is moved and the procedure repeated until the minimum slab thickness is found. The importance of easy manipulation can be appreciated for such a problem. A new hole location for the slab of Figure 17.4 is shown in Figure 17.5. The reduction in the ratio of $M/p$ from 7175 to 6833 indicates that the hole location in Figure 17.5 is preferable to that shown in Figure 17.4.

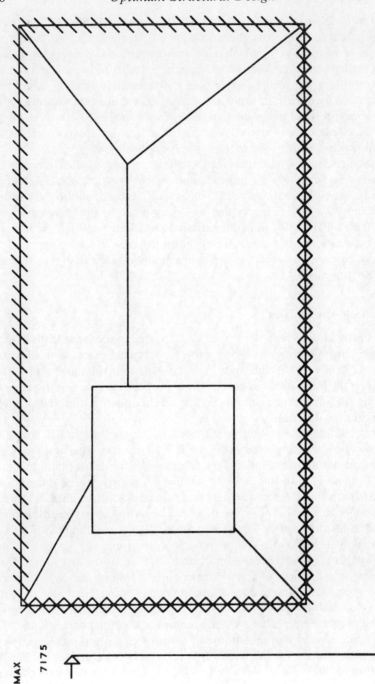

MAX

7175

**Figure 17.4** Yield lines for first hole location

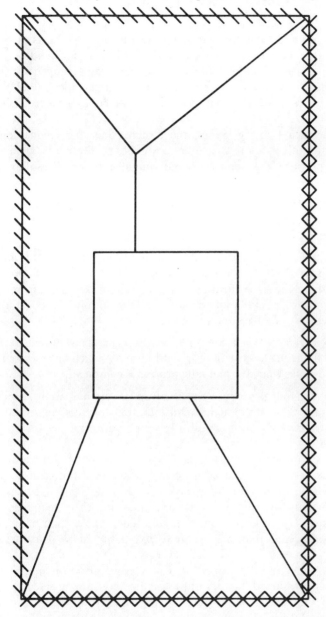

MAX
6833

**Figure 17.5**   Yield lines for second hole location

338 *Optimum Structural Design*

## 17.4 Housing-layout design[3]

The number of large-scale housing developments in the world is rapidly increasing, and designers frequently have to prepare layouts for hundreds, or even thousands, of nearly identical dwellings. This can result in repetitive and often uninteresting schemes which, in spite of their high technical standards, compare badly in aesthetics and social standing with other longer-established developments.

When preparing a housing layout, the designer has many constraints acting on him to limit the otherwise infinite number of alternative arrangements he could propose. These constraints fall roughly into the three following categories:

*Physical constraints*  These include the contours and boundary restrictions of the site, its shape, the types of dwellings, the block heights and density it is proposed to use.

*Social constraints*  These ensure adequate living standards in the final scheme. In Britain these constraints are applied by the Local Planning Authorities, and include a specification of a minimum of sky visible from the living room window, a minimum of sunlight and a minimum distance between the windows of adjacent buildings.

*Subjective constraints*  They are not clearly defined, and reflect the designer's own judgement, sense of order and taste by which he assesses his own work and decides when the planning process is complete.

In the past the layout procedure has involved the manual application of all these constraints, but the practical considerations of satisfying physical and social constraints are such as to push all else, temporarily at least, from the designer's mind.

However, if the process of satisfying physical and social constraints is mechanized, the designer is relieved from the tedious and time-consuming work and is left with the interesting task of exercising his judgement in assessing the machine-produced layouts and finalizing them manually. It also allows him to try a much greater range of alternatives than he might otherwise consider, and thus helps him to avoid the more unattractive layouts.

The Basic Architectural Investigation and Design program number 1 (Baid 1) has been developed to carry out these time-consuming calculations encountered in the design of medium- to high-density housing layouts. In Baid 1, the architect first supplies the computer with data defining the physical constraints and the various minimum requirements of the social constraints. The social-constraint checking program is then run several times in batch mode (because each run takes about 30 min on an ICL 4130) to generate some possible layouts randomly and store them on the backing

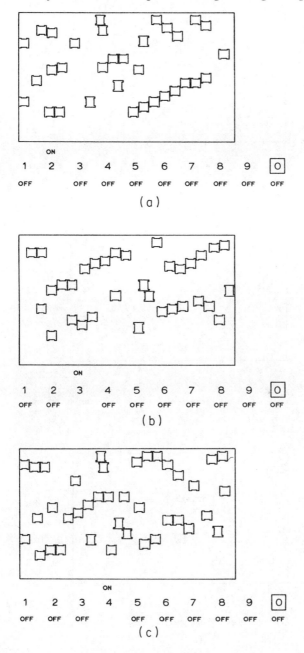

**Figure 17.6** (a) First layout suggested by batch computer run, with (b) second layout and (c) third layout

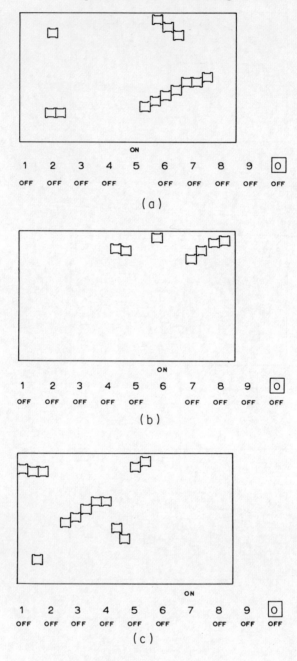

**Figure 17.7** (a) House sites selected, by light pen on a c.r.t. screen, from suggested layout, and selected sites from (b) second suggested layout and (c) third suggested layout

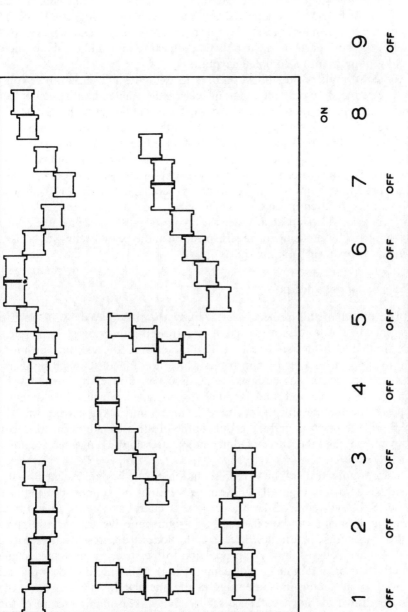

**Figure 17.8**   Final layout derived from selections of Figures 17.7(a)–(c)

store of the computer. This stored data is then accessed by an interactive graphics program that permits the viewing, editing, modification and amalgamation of these layouts by means of the light pen and function buttons. These facilities enable the designer to use his own judgement and his sense of order and aesthetics to analyse the layouts 'suggested' by the computer. He then modifies the layouts manually with the light pen as he sees fit, freezes those parts of the layouts that appeal to him, and combines these local groupings into a complete layout.

As an example of the utilization of Baid 1, Figure 17.6 shows three computer-produced layouts for a site in Newcastle, England, and Figure 17.7 shows sections of these layouts selected by the architect to be of interest. Figure 17.8 shows the amalgamation of those three selections along with some additions and minor adjustments. (Each of these figures represents accurately the display on a c.r.t.) The last stage is to feed this complete layout back, via the disk backing store, as an input to the first batch program. On choosing the appropriate option, the computer proceeds to check for the social constraints in batch mode once again. Should the layout be unsatisfactory, the process may be repeated. When the layout of pads is accepted, the design is completed by choosing the precise position and type of dwelling to be used within each pad site (Figure 17.8).

## 17.5　Slab-bridge design[4]

The object of this project was to study the role played by interactive graphics in the process of designing prestressed-slab bridges. The project was limited to the design of rectangular bridges, since the awkward positioning of supports can lead to problems which are difficult enough by themselves. The system was designed to analyse slab bridges by means of the finite-elements method, and facilities are included which allow different types of bridge loading to be placed at any point in the carriageway. In addition, the position and the profile of the prestressed cables can be easily adjusted so that the effect of different prestressing arrangements can be quickly evaluated.

The axonometric graphical representation of one bridge designed by the system is shown in Figure 17.9. It can be seen that the support bearings are close to the centreline of the bridge, so that lateral as well as longitudinal bending moments can be expected to be significant. Deflections are represented by vectors at the nodes. Stresses at the nodes are represented by crosses, with the arm lengths drawn in proportion to the stress magnitude (a single line for tension, a double line for compression). This simplified drawing has proved unsatisfactory in some ways, as the fine detail is not shown on the plotter drawings, but it still successfully gives a reasonable picture of the action of a slab, and in some cases is beautifully expressive (Figure 17.10).

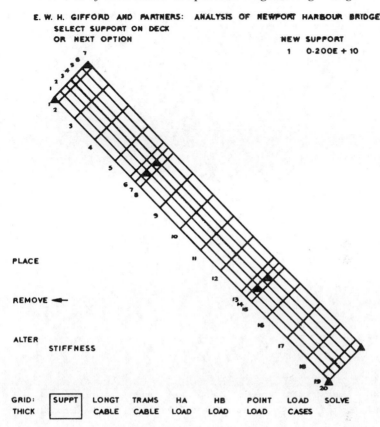

E. W. H. GIFFORD AND PARTNERS:   ANALYSIS OF NEWPORT HARBOUR BRIDGE
SELECT SUPPORT ON DECK
OR NEXT OPTION                                        NEW SUPPORT
                                                      1    0·200E + 10

**Figure 17.9**  Axonometric view of bridge

The interactive facility has also proved very successful in allowing loads to be positioned and moved visually on the deck to seek the worst loading condition (a judgement which is difficult without pictorial aid). The ability to control the program means that alteration and investigation can be done at will.

The treatment of prestressing cables was the most successful part of the program. It starts with the consideration that only two items of information are relevant—the vertical position of the cable in the slab (relevant for practical details) and the curvature of the cable (relevant because it produces the vertical loads on the slab and also because it causes cable friction). To allow considerable exploration, cable profiles are defined by a series of cubics (the cable is, in fact, defined as a series of beam finite elements), and the vertical position at each node is set by the engineer (Figure 17.11). The cable automatically assumes a minimum-curvature line through these points, the

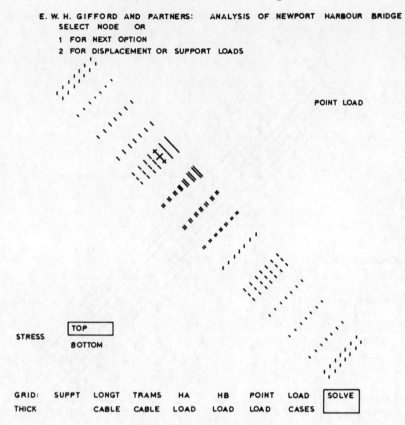

POINT LOAD

STRESS     TOP

BOTTOM

GRID:   SUPPT   LONGT   TRAMS   HA     HB     POINT   LOAD   SOLVE
THICK           CABLE   CABLE   LOAD   LOAD   LOAD    CASES

**Figure 17.10**   Stresses in bridge

curvature along the cable being part of the displayed picture. In this way practical detail and desired curvature are manipulated directly. The classical method of designing prestressing-cable profiles in terms of concordant profiles and parasitic profiles has always been a very difficult method to grasp, and many people must be very wary of using prestressed concrete because of it. Computer graphics reduces it to the level of a straightforward structural concept, so that the engineers involved now find it easy to design prestressed concrete. Investigating the different cable profiles on the bridge is extremely simple.

The availability of the interactive graphics allows the bridge design to approach a more satisfactory and significantly simpler solution than that obtained by conventional means. The ability to experiment with the prestressing cables results in a two-dimensional cable topography that greatly reduces the peaky behaviour of bending moments associated with constant longitudinal prestressing profiles.

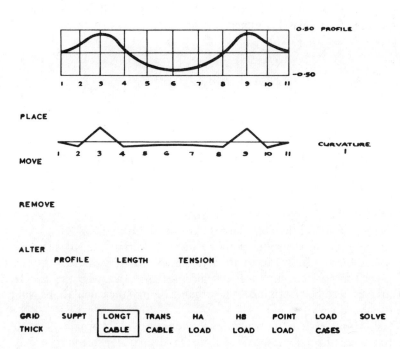

**Figure 17.11**   Graphical representation of prestressing cables

## 17.6   Finite-element interactive system[5]

In this section a description is given of a finite-element system designed to meet the needs of easy communication and speed of interaction. The problem of change and interrogation at an arbitrary location within a structure has been dealt with by careful design of the data structure, and the problem of the interaction time has been dealt with by using a standard piece of structural theory which bypasses the need for repeated inversion of the whole stiffness matrix.

The structural analysis follows a standard finite-element procedure of forming the stiffness matrix of the structure from the stiffness matrices of each of the elements, together with an appropriate connexion matrix. After setting the displacement boundary conditions, the assembled stiffness matrix is inverted to give the flexibility matrix. The values of the nodal displacements, and hence the stresses, are then readily determined following the standard procedure.

When a modification in geometry or element property is made, it is not necessary to repeat this procedure. If, for example, a structure is modified by increasing the stiffness of some of the elements by $\delta K$, it is not difficult to show that, if the flexibility matrix for the unmodified structure is

$$\begin{bmatrix} \Delta_0 \\ \Delta_1 \end{bmatrix} = \begin{bmatrix} f_{00} & f_{01} \\ f_{10} & f_{11} \end{bmatrix} \begin{bmatrix} P_0 \\ P_1 \end{bmatrix}$$

where the subscript 1 denotes parameters of elements whose stiffness is changed, then the flexibility matrix for the modified structure is given by

$$\begin{bmatrix} \Delta_0 \\ \Delta_1 \end{bmatrix} = \begin{bmatrix} f_{00} - f_{01}\,\delta K C^{-1} f_{10} & f_{01}(I - \delta K C^{-1} f_{11}) \\ C^{-1} f_{10} & C^{-1} f_{11} \end{bmatrix} \begin{bmatrix} P_0 \\ P_1 \end{bmatrix}$$

where $C = I + f_{11}\,\delta K$.

The formation of the new matrix involves an inversion of a matrix which is the same size as the stiffness matrix $\delta K$. In addition, it is not necessary to complete the flexibility matrix if only some of the stresses are to be viewed. In this circumstance only those displacements necessary for the calculation of stresses need be determined. The organization of the calculations will now be discussed.

The whole system of programs (amounting to approximately 10,000 Fortran statements) is controlled from the display console and is made up of four sections, each one using relocatable data structures stored on disk.

*Section 1* provides a powerful set of facilities for the generation and modification of data representing a triangular-finite-element idealization of a plan structure.

*Section 2* reads the data structure stored on disk by Section 1, relocates it in a free storage area in the core, and provides facilities for setting displacement boundary conditions. It evaluates element stiffness matrices, assembles the sparse structural stiffness matrix and re-stores all the data on disk.

*Section 3* inverts the stiffness matrix and files the flexibility matrix on disk.

*Section 4* provides the means for continuously modifying the structure while maintaining an almost immediate response of structural behaviour on the screen. As in the other two interactive sections of the system (Sections 1 and 2), a menu-organizer program presents the user with a sequence of menus of options on the screen.

On first entering Section 4, the user is able to select which nodes are to be moved, which element stiffnesses are to be changed, which stresses are to be

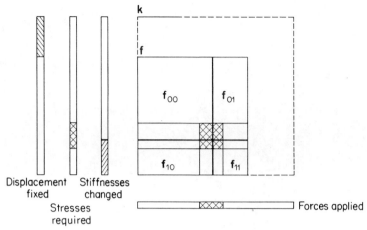

Reproduced by permission from G. A. Butlin and F. A. Leckie, 'Response time in the application of interactive graphics in structural analysis', in *Proceedings of the Symposium of the International Union of Theoretical and Applied Mechanics, Liège, 1970*, University of Liège, 1971, pp. 509–519

**Figure 17.12** The two major menus of the recalculation program

viewed and which forces are to be applied. These items can be reselected until the user is satisfied with his choice (see Figure 17.12). As the user makes his choice as to which stresses are to be viewed, a mask (see Figure 7.13) is generated to indicate which displacement parameters will be required. This mask is then used later to determine the particular 'slice' of rows of the flexibility matrix that will be required for the recalculation routine. Similarly, as the user makes his choice of the nodes to be moved or the element stiffnesses to be changed, so a mask is set up to define the partitioning of the flexibility matrix. As the user makes his choice of applied forces, so a mask is set up to

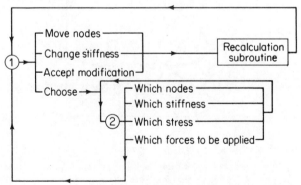

Reproduced by permission from G. A. Butlin and F. A. Leckie, 'Response time in the application of interactive graphics in structural analysis', in *Proceedings of the Symposium of the International Union of Theoretical and Applied Mechanics, Liège, 1970*, University of Liège, 1971, pp. 509–519

**Figure 17.13** Definition of masks

BUT 1 TO CHANGE SCALE: 2-4 TO CHANGE STRESS: 5-6 DIRECTION: 7 EXIT.

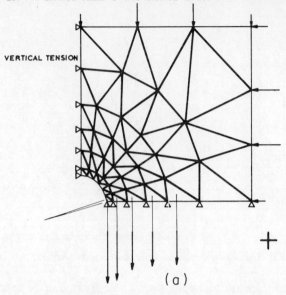

(a)

BUT 1 TO CHANGE SCALE: 2-4 TO CHANGE STRESS 5-6 DIRECTION: 7 EXIT.
WARNING: NOT ALL STRESS AND DISPLACEMENTS HAVE BEEN UPDATED
FOR LAST MODIFICATION

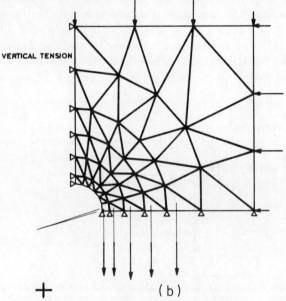

(b)

**Figure 17.14** Display of element mesh on screen showing (a) initial stresses and (b) after a series of modifications to thickness of two elements pointed to by diverging lines

define the slice of columns of the flexibility matrix that will be required for the recalculation routine. At the same time as forming these masks, further data structures are formed. One has been designed to direct the numerical changes in stiffness matrices to locations convenient for rapid processing with the flexibility matrix, and another designed to direct the back-substitution matrices to locations convenient for rapid recalculation of the stresses.

The operation of continuous modification and assessment of stresses (illustrated in Figure 17.14) then proceeds with the pressing of function buttons and identification with the light pen. A thickness can be changed, a button depressed to initiate the recalculation routines, stresses viewed and assessed, the thickness changed again and the cycle performed again. Or, with the depression of an appropriate button, the user can return to the menu, inviting the reselection of the items to be changed and the items to be viewed. This cycle of recalculation typically takes about 5 s, whereas the time for stiffness calculation, matrix-assembly inversion and back substitution takes about 7 min. The problem of interaction time has thus been solved by using a standard piece of structural theory that bypasses the need for repeated inversion of the whole stiffness matrix. The more stresses selected for viewing, the more changes in stiffnesses or the more applied forces, the longer the recalculation takes. On the other hand, the fewer stresses selected for viewing, the fewer changes in stiffness or the fewer applied forces, the faster will be the recalculation.

## 17.7    Geometric representation of bodies and finite elements

Considerable interest has been shown in the geometric representation of bodies on a display screen. It is obvious that two-dimensional drawings can be reproduced on the screen and, while these can be of considerable advantage, because of the ease with which parameters can be adjusted, such drawings suffer the same drawbacks as conventional engineering drawings in that it is not always easy to visualize the three-dimensional object that they represent. It has been shown, however, that the display screen can enjoy considerable advantages because of its ability to project different views very rapidly, so that a 'dynamic' effect can be achieved which can help in visualizing an object.

There are a variety of methods of representing surfaces, but the technique that appears to be most common is to represent the surface by means of a number of patches, the form of the object being described by a finite number of parameters at the nodes of each patch. The patches are joined together by applying continuity conditions, such as displacement and slope, along the boundaries. Connected in this way, the patches form a continuous surface. The accuracy of the representation will depend, of course, on the number of patches used and the number of parameters required for patch description.

*Optimum Structural Design*

Various forms of shape functions have been suggested[6], and it is not surprising that many of these shape functions are well known to those working with finite elements. The work on surface representation has now reached a stage at which it is possible to mould shapes by observing the effects of parameter changes that are inserted by appropriate manipulation from the screen, or by using a teletypewriter input.

This patch technique has been applied in a variety of applications such as ship, aeroplane and automobile design, and it has been shown to be possible to manipulate the various parameters that define the surface to provide, in some sense, the desired result. At the same time that these efforts on surface description were taking place, developments in the field of finite elements were directed towards the structural analysis of complex shapes. In addition to the problems of the division of the body into suitable finite elements, there was the additional problem of determining suitable displacement fields within the elements. In addition to satisfying the usual compatibility conditions, displacement fields had to be designed so that body movement of the elements could be achieved without causing strain. An ingenious solution to this problem has been offered by Zienkiewicz *et al.*[7] in the form of iso-parametric elements for solid bodies, and by Argyris and Scharpf[8] for the solution of shell-type structures. The approach used by each of these groups is similar in that the functions used to define the shapes of the finite elements are identical to those used to describe the displacements within the finite element. This feature means that the 'body movement without strain' requirement can be fulfilled, and such elements are proving to be very successful, as is evidenced by the rapidly growing literature in that field.

The intellectual gap between the problem in computer graphics of surface description and that of structural analysis by finite elements is now bridged, but, to our knowledge, this existing technique remains to be implemented. It can only be a matter of time, however, before surface and body description on a display screen can be analysed for structural performance.

## References

1. M. P. Ranaweera and F. A. Leckie, 'Interactive use of the lightpen in finite element limit load analysis', in *Proceedings of the Symposium on Computer Graphics, Brunel University, 1970.*
2. H. H. Rosenbrock, 'An automatic method for finding the greatest or least value of a function', *Computer Journal*, **3**, 175–184 (1960).
3. B. Auger, G. A. Butlin and R. J. Hubbold, 'Computer graphics workshop report 2. A high density housing layout program', Leicester University Engineering Department, Report No. 71-9, March 1971.
4. R. A. Britton and M. J. Platts, 'Computer graphics workshop report 1. A bridge design program', Leicester University Engineering Department, Report No. 71-8, March 1971.

5. G. A. Butlin and F. A. Leckie, 'Response time in the application of interactive graphics in structural analysis', *Proceedings of the Symposium of the International Union of Theoretical and Applied Mechanics, Liège, 1970*, University of Liège, 1971, pp. 509–519.

6. S. A. Coons, 'Surfaces for computer-aided design of space forms', Project MAC, MIT, MAC-TR-41, June 1967.

7. O. C. Zienkiewicz, *The Finite Element Method in Engineering Science*, McGraw-Hill, 1972, Chaps. 8 and 9.

8. J. H. Argyris and D. W. Scharpf, 'The Sheba family of shell elements for the matrix displacement method', *The Aeronautical Journal of the Royal Aeronautical Society*, Technical note, 873–883 (1968).

# Author Index

Numbers in italics refer to the lists of references at the end of each chapter.

# Subject Index

Aircraft wing, 44, 47

Bleich–Melan theorem, 101
Branch-and-bound algorithm, 204, 219
Bridge, foot, 291
  highway, 292, 342

Conjugate-direction method, 53
Conjugate gradient, 55
Constraints, 7
Constraints, equality and inequality types,
  11, 52
Convexity, 12
Cost function, 9
Cutting-plane method, 24, 67, 121

Dams, 111
  arch, 124
Decomposition method, 68
Design space, 7, 9
Design variable, 7
DFP method, 57
Discrete-variable methods, 201
DSC method, 53
Duality, 90, 233
  of limit theorem, 234, 268
Dynamic programming, 179

Fail-safe design, 255
Feasible-design region (exterior region),
  14
Feasible-direction method, 56, 66, 127
Finite elements, 41, 109, 228, 290, 332
Force derivative, 114
Frame structures, 85
  concrete, 285
  multistorey, 267
Fully stressed design, 2, 19

Gradient-projection method, 128
Gradients, 10, 12, 109

Hill-climbing method, 130

Infeasible-design region (interior region),
  14
Instability, 274, 314
Integer programming, 233, 273
Interactive graphics, 329

Koiler theorems, 101
Kuhn–Tucker condition for optimality,
  29, 53

Lagrange multipliers, 63, 196
Layout theory, 1, 223
Least-squares methods, 61
Limit design (see plastic limit design),
  223, 332
Limited-move methods, 123
Linear programming, 3, 24, 63, 79
  sequential, 109, 140

Mathematical programming, 2, 51
Merit function, 9
Minima, global and local, 16, 52
von Mises criterion, 41, 333

Network problems, 180
Newton method, 56, 61

Objective function, 7
Optimality criteria, 2, 33
Optimality principle, 187

Penalty-function method, 68, 143
  exterior and interior, 69, 144

357

358 *Subject Index*

Plastic-limit theorems, lower and upper
 bounds, 80, 95, 101, 225
Plastic load factor, 88
Probability of failure, 243

Quadratic programming, 62

Radiotelescope, 212
Random variables, 243
Reduced-gradient method, 65, 67
Reinforced concrete, design, 283
 beams, 257
 slabs, 288, 342
Reliability, 241
Restricted-step methods, 58

Scaling of variables, 156
Search, 155
Sensitivity, 109, 209
 vector, 121
Sequential linear programming, 109
Shape optimization, 109, 133, 223

Ship, bulkhead, 160
 tanker, 173
Simplex method, 53, 81
 revised, 62
Steepest descent method, 53
Stochastic programming, 260
Stress-ratio method, 21
Structural-steel design, 267, 307
Suboptimisation, 210

Transmission tower, 42, 193
Trusses, 9, 183, 212
Turbine disk, 113, 133

Unconstrained optimization, 53, 151
Unidirectional scores, 153

Weakest-link method, 246, 253

Yield-line analysis, 335

Zero–One programming, 223